Nonequilibrium Atmospheric Pressure Plasma Jets

Nonequilibrium Atmospheric Pressure Plasma Jets

Fundamentals, Diagnostics, and Medical Applications

XinPei Lu

Stephan Reuter

Mounir Laroussi

DaWei Liu

CRC Press
Taylor & Francis Group
Boca Raton London New York

CRC Press is an imprint of the
Taylor & Francis Group, an **informa** business

CRC Press
Taylor & Francis Group
6000 Broken Sound Parkway NW, Suite 300
Boca Raton, FL 33487-2742

First issued in paperback 2021

ISBN-13: 978-0-367-77988-7 (pbk)
ISBN-13: 978-1-4987-4363-1 (hbk)

Library of Congress Cataloging-in-Publication Data

Names: Lu, XinPei, author. | Reuter, Stephan, 1974- author. | Laroussi, M. (Mounir), 1955- author. | Liu, DaWei, author.
Title: Nonequilibrium Atmospheric Pressure Plasma Jets : fundamentals, diagnostics, and medical applications / XinPei Lu, Stephan Reuter, Mounir Laroussi, DaWei Liu.
Description: Boca Raton : CRC Press, Taylor & Francis Group, [2019] | Includes bibliographical references.
Identifiers: LCCN 2018043420 | ISBN 9781498743631 (hardback : alk. paper)
Subjects: LCSH: Plasma jets. | Cancer--Treatment--Research.
Classification: LCC TP159.P57 L8 2019 | DDC 530.4/4--dc23
LC record available at https://lccn.loc.gov/2018043420

Visit the Taylor & Francis Web site at
http://www.taylorandfrancis.com

and the CRC Press Web site at
http://www.crcpress.com

Nonequilibrium Atmospheric Pressure Plasma Jets
Fundamentals, Diagnostics, and Medical Applications

XinPei Lu

Stephan Reuter

Mounir Laroussi

DaWei Liu

CRC Press
Taylor & Francis Group
Boca Raton London New York

CRC Press is an imprint of the
Taylor & Francis Group, an **Informa** business

CRC Press
Taylor & Francis Group
6000 Broken Sound Parkway NW, Suite 300
Boca Raton, FL 33487-2742

First issued in paperback 2021

ISBN-13: 978-0-367-77988-7 (pbk)
ISBN-13: 978-1-4987-4363-1 (hbk)

Library of Congress Cataloging-in-Publication Data

Names: Lu, XinPei, author. | Reuter, Stephan, 1974- author. | Laroussi, M. (Mounir), 1955- author. | Liu, DaWei, author.
Title: Nonequilibrium Atmospheric Pressure Plasma Jets : fundamentals, diagnostics, and medical applications / XinPei Lu, Stephan Reuter, Mounir Laroussi, DaWei Liu.
Description: Boca Raton : CRC Press, Taylor & Francis Group, [2019] | Includes bibliographical references.
Identifiers: LCCN 2018043420 | ISBN 9781498743631 (hardback : alk. paper)
Subjects: LCSH: Plasma jets. | Cancer--Treatment--Research.
Classification: LCC TP159.P57 L8 2019 | DDC 530.4/4--dc23
LC record available at https://lccn.loc.gov/2018043420

Visit the Taylor & Francis Web site at
http://www.taylorandfrancis.com

and the CRC Press Web site at
http://www.crcpress.com

*Mounir Laroussi dedicates this book to
his wife, Nicole Mache Laroussi.*

Contents

Contents

Contents

Contents

Foreword

Nowadays nonthermal or cold atmospheric plasma (CAP) can have tremendous applications in biomedical technology. CAP can potentially offer a minimally invasive surgery that allows specific cell targeting without influencing the whole tissue. It is well established that CAP triggers various effects, including cell detachment, without affecting cell viability, controllable cell death, modified cell migration, etc. As such, plasma treatment offers advantages that were never considered in most advanced laser surgery. Due to the myriad of potential applications for cold plasma in biomedical science, it is critical to understand the mechanisms regulating plasma generation and propagation.

The variety of different effects of plasma can be explained by the plasma's complex chemical composition and variations in the way that CAP is generated. CAP can be described as a cocktail containing a variety of reactive oxygen species (ROS), reactive nitrogen species (RNS), charged particles, an electric field, UV photons, etc. Overall, CAP formation can be classified into direct and indirect methods. *Direct plasmas* employ the target (such as living tissue or organs) as one of the electrodes, and thus the target directly participates in the active discharge plasma processes. Some current may flow through the living tissue in the form of small conduction current, displacement current, or both. On the other hand, *indirect plasmas* are produced between two electrodes and are then transported to the area of application entrained in a gas flow.

There is a great variety of different configurations of indirect plasma sources depending on the size, type of gas, and applied power. The main focus of this book is on nonequilibrium atmospheric pressure plasma jets (N-APPJ), which can be classified as indirect plasma sources. These jets are fascinating devices that have rich plasma physics and chemistry. To this end, this book offers a comprehensive description of multiple experimental and theoretical approaches, allowing better understanding of the mechanisms of operation of N-APPJs.

This book on nonequilibrium atmospheric pressure plasma jets will be useful not only as a reference book, but also as a primary or supplemental textbook for graduate students.

Michael Keidar
George Washington University

Preface

Nonequilibrium atmospheric pressure plasma jets (N-APPJs) are unique sources of nonequilibrium plasmas because they provide controllable and repeatable low-temperature plasmas in ambient conditions and are not confined by electrodes. Today, these plasma jets are routinely used in material processing and biomedical applications. These include surface modification, etching of silicon, ashing of photoresists, thin film deposition, etc. The medical applications include sterilization/decontamination, wound healing, and the killing of cancer cells.

In the last decade, various advanced diagnostics techniques ranging from fast photography to spectroscopy have been employed to characterize N-APPJs and to elucidate their ignition and propagation mechanisms. Early investigations showed that these plasma jets are made of a series of plasma packets/bullets traveling at high velocities. Experimental and modeling works concluded that these plasma bullets are in fact fast ionization waves guided within a well-defined gas channel, hence the name "guided ionization waves." At first, photoionization was proposed as the mechanism whereby the propagation of the plasma bullets is enabled, but further investigations showed that in addition to photoionization, the electric field at the head of the ionization front plays an important role. The strength of this electric field can reach up to 30 kV/cm.

The motivation behind this book is to thoroughly cover the physics associated with N-APPJs within one manuscript. The book presents detailed information on the ignition, propagation, and structure of the plasma bullets that make up the plasma jets. Diagnostic techniques that have been employed to fully characterize N-APPJs are presented in great detail. These include optical emission spectroscopy, absorption spectroscopy, laser-based spectroscopy (LIF, TALIF, etc.), and other techniques such as molecular beam mass spectrometry and electron paramagnetic resonance spectroscopy. Finally, applications of N-APPJs in biomedicine are covered, specifically their use for cancer treatment. The book intends to close the gap between two recent, excellent multi-author monographs on the topic of plasma medicine. While one focuses on applied plasma medicine, the other gives an overview of plasma medical research. The present book engages with a detailed view on the physics and the diagnostics of plasma jets used in plasma medical research and relates these to future perspectives in plasma medicine.

This book is aimed at a wide readership that includes plasma physicists/researchers, engineers who use plasma technology in their industrial processes, graduate students who conduct research in nonequilibrium plasma science, and healthcare professionals and researchers who seek to discover new technologies that can be adopted in the development of novel therapies.

Acknowledgments

Dr. XinPei Lu is grateful to a number of graduate students who have worked on plasma jets. Their work makes this book possible. Dr. XinPei Lu also would like to thank the National Natural Science Foundation of China (Grant No: 10875048, 51077063, 51277087, 51477066, 51625701), ChangJiang Scholar Program, and IFSA Collaborative Innovation Center, Shanghai Jiao Tong University for their financial support.

Dr. Stephan Reuter would especially like to thank his colleagues from Princeton University in the Applied Physics Group of Prof. R. Miles for valuable and inspiring discussions and to acknowledge funding of his research stay at Princeton University by the Alexander von Humboldt Foundation, during which part of the manuscript was written. Dr. Reuter thanks the members of his research group for their highly valued work at the Centre for Innovation Competence *plasmatis,* funded by the German Federal Ministry of Education and Research, at the INP Greifswald. For their good collaboration, Dr. Reuter wants to thank—representative of the many colleagues at the INP Greifswald—Prof. K.-D. Weltmann and Prof. Th. von Woedtke. Finally, much scientific progress on plasma liquid interaction was rendered possible through the European COST Action TD1208 "Electrical Discharges with Liquids for Future Applications." The open dialogue and tight network initiated by the COST action is greatly appreciated.

Dr. Mounir Laroussi's research was partly supported by the Directorate of Physics and Electronics of the US Air Force Office of Scientific Research (AFOSR). This directorate has sponsored research on nonequilibrium atmospheric pressure plasma for more than three decades and has played a key role in the flourishing of this field in the United States. AFOSR was also the first government agency in the world that supported and promoted the biomedical applications of nonthermal atmospheric pressure plasma and therefore was instrumental in the early success of plasma medicine. Dr. Laroussi is also thankful to the IEEE-NPSS for supporting plasma medicine very early on and for helping in the dissemination of the early research results in its journals and conferences. Finally, Dr. Laroussi is grateful to all his students and Post-Doctoral Associates who contributed to his research. He enjoyed working with them, teaching them, and learning from them.

Authors

XinPei Lu received a PhD degree in electrical engineering from the Huazhong University of Science and Technology, Wuhan, Hubei province, China. Then he worked at Old Dominion University, Norfolk, Virginia. In 2007, he joined the Huazhong University of Science and Technology, where he is now a chair professor in the School of Electrical and Electronic Engineering. He serves as a guest editor for IEEE Transactions on Plasma Science, Europe Physic Letter, and High Voltage. He is a board member of IEEE Transactions on Radiation and Plasma Medical Science, Plasma Research Express, Plasma Science and Technology, and High Voltage. He served as session chair at many conferences, including several times at the International Conference on Plasma Science. He has written one book and several book chapters. He has been asked to give plenary talks at many international conferences, including the IEEE International Conference on Plasma Science. His research interests include low-temperature plasma sources and their biomedical applications, modeling of low-temperature plasmas, and plasma diagnostics. He is the author or coauthor of about 150 peer-reviewed journal articles, with a citation of about 6000 and an H-index of 36, and he holds 12 patents in these areas. His work has been reported several times by scientific medium, including Science, Nature, the American Institute of Physics, and the Institute of Physics in the UK.

Stephan Reuter recently finished a research stay at Princeton University, Princeton, New Jersey, as a Feodor-Lynen fellow funded by the Alexander von Humboldt Foundation, where he performed ultrafast laser spectroscopy on plasmas. He received his PhD at University of Duisburg-Essen, North Rhine-Westphalia, Germany, in 2007. After a research fellow appointment at the Centre for Plasma Physics at Queen's University Belfast, Belfast, Northern Ireland, he started his own research group at the Centre for Innovation Competence plasmatis, INP Greifswald in 2010, focusing on cold atmospheric pressure plasmas interacting with biologically relevant liquids for plasma medicine. After a visiting professorship at Lublin Technical University, he is currently visiting Professor at Université Paris-Saclay, Department of Electrical, Optical and bio Engineering, Lublin, Poland. He has more than 80 peer-reviewed journal articles and has given more than 45 invited talks at conferences, workshops, and seminaries. He is co-initiator of the Young Professional Workshop Series on Plasma Medicine. In 2018, he wrote a comprehensive review on the kINPen atmospheric pressure plasma jet, the only cold-plasma jet certified as a medical device to date. His interests are nonequilibrium plasma diagnostics and fundamentals of plasma liquid interaction. His main topics of applied research include plasma material synthesis and plasma medicine.

Mounir Laroussi from Old Dominion University, Norfolk, Virginia, is one of the pioneers who discovered the plasma bullet phenomena in atmospheric pressure plasma jets and one of the pioneers in the field of plasma medicine. He obtained his PhD in electrical engineering from the University of Tennessee, Knoxville, Tennessee. After a post-doc and a few years as a research assistant professor at the University of Tennessee, Dr. Laroussi joined the Old Dominion University Applied Research Center in 1998 as a research faculty member, where he established the Applied Plasma Technology Laboratory. He now holds a full professor position at the Electrical and Computer Engineering Department and is the director of the Plasma Engineering & Medicine Institute. He has written 1 book and several book chapters, more than 10 reviews and, more than 100 refereed journal papers, and he is the holder of several patents. His reviews in IEEE Transactions on Plasma Science and several other leading journals that span several fields of research are highly cited. Dr. Laroussi is a fellow of IEEE and is the recipient of the 2012 IEEE Merit Award, the highest technical award given by the plasma science society, for his pioneering work in the field of non-equilibrium atmospheric pressure plasmas and their biomedical applications.

DaWei Liu received a PhD degree in electrical engineering from Loughborough University, United Kingdom, in 2009. In 2009, he joined the Huazhong University of Science and Technology, Wuhan, Hubei province, China, where he is now a professor in the School of Electrical and Electronic Engineering. He is the author of several book chapters (edited books), and author or coauthor of more than 40 journal papers. His H-index is 19 achieved over approx. He was asked to organize several domestic conferences on atmospheric pressure plasma science and their application in China. His research interests include 2D modeling and optical diagnostics of atmospheric pressure plasma and the development of plasma sources and their biomedical and environmental applications. His findings on the interactions between plasma and biological targets in μm scale and on the production mechanisms of the main reactive species of plasma jets help with the understanding of plasma medicine.

PART I

Basic Concept of Nonequilibrium Atmospheric Pressure Plasma Jets (N-APPJs)

Introduction

1.1 Atmospheric Pressure Plasma—Equilibrium and Nonequilibrium Plasma

1.1.1 Definition of Plasma

Plasma is often referred to as the fourth state of matter, the others being solid, liquid, and gas. Plasma is the most abundant form of ordinary matter in the universe. Most plasma is in the rarefied intergalactic regions, particularly the intracluster medium, and in stars, including the sun. Plasmas make up more than 99% of visible matter in the universe. They consist of positive ions, electrons or negative ions, and neutral particles. When a solid (the first state of matter) obtains energy from an outer source, such as heating, the particles in it get sufficient energy to loosen their structure and thus melt to form a liquid (the second state of matter). After obtaining sufficient energy, the particles in a liquid escape from it and vaporize to gas (the third state of matter). This leads to phase transitions, which occur at a constant temperature for a given pressure. The amount of energy required for the phase transition is called the latent heat. Subsequently, when a significant amount of energy is applied to the gas through mechanisms such as an electric discharge, the electrons that escape from atoms or molecules not only allow ions to move more freely but also produce more electrons and ions via collisions after accelerating rapidly in an electric field. Eventually, the higher number of electrons and ions change the electrical property of the gas, which thus becomes ionized gas or plasma. However, this transition from a gas to a plasma is not a phase transition in the thermodynamic sense because it occurs gradually with increasing temperature.

Nonequilibrium Atmospheric Pressure Plasma Jets (N-APPJs)

From a scientific point of view, not all media containing charged particles can be classified as plasma. It must satisfy certain conditions, or criteria, to be classified as plasma. The first criterion is macroscopic neutrality, which is related to Debye length (λ_D) [1–3].

$$\lambda_D = \left(\frac{\epsilon_0 k T_e}{n_e e^2} \right)^{1/2},$$

(1.1)

where ϵ_0 is the permittivity of free space, k is Boltzmann's constant (1.38×10^{-23} J/K), T_e is the electron temperature, n_e is electron density, and e is electron charge. Assuming L is a characteristic dimension of plasma, the first criterion to be a plasma is

$$L \gg \lambda_D.$$

(1.2)

Within the distance of λ_D, the condition of macroscopic electrical neutrality may not be satisfied. Beyond the distance, macroscopic neutrality is maintained.

In addition, the number of electrons inside a Debye sphere be very large. A second criterion for the definition of a plasma is therefore

$$n_e \lambda_D^3 \gg 1.$$

(1.3)

This means that the average distance between electrons, which is roughly given by $n_e^{-1/3}$, must be very small compared to λ_D. The quantity defined by

$$g = 1/(n_e \lambda_D^3)$$

(1.4)

is known as the plasma parameter and the condition $g \ll 1$ is called the plasma approximation.

The third criterion is time domain. It is required that the electron-neutral collision frequency (v_{en}) be smaller than the electron plasma frequency v_{pe}, i.e.,

$$v_{pe} \gg v_{en},$$

(1.5)

where the electron plasma frequency

$$v_{pe} = \frac{1}{2\pi} \left(\frac{n_e e^2}{m_e \epsilon_0} \right)^{1/2}$$

(1.6)

and m_e is electron mass. If **Equation (1.5)** is true, it means the electrons will be able to behave in an independent way; otherwise, it will be forced by collisions to be in equilibrium with the neutrals, and thus the medium can be treated as a neutral gas. Equation (1.5) can also be alternatively written as

$$\omega \tau \gg 1,$$

(1.7)

where $\tau = 1/v_{en}$, which represents the average time an electron travels between collisions with neutrals, and ω stands for the angular frequency of typical plasma oscillations. When **Equation (1.7)** is true, it means that the average time between electron-neutral collisions is large compared to the characteristic time during which the physical plasma parameters are changing.

1.1.2 Equilibrium and Nonequilibrium Atmospheric Pressure Plasma

Plasma can be generated in a laboratory under different gas pressures. However, a vacuum system is needed for plasma to be generated at low pressures. To avoid the expensive and complicated vacuum system, plasmas generated at atmospheric pressure have been explored in the past several decades. At a pressure of 1 atm, the neutral particle density is about $2.4 \times 10^{25}/m^3$. The electron-neutral collision frequency $v_{en} = n_0(2kT_e/m_e)^{1/2}\sigma_0$ is on the order of $10^{11-12}/s$, where n_0, T_e, m_e, and σ_0 are, respectively, the neutral particle density, electron temperature, electron mass, and electron-neutral collision cross section, and k is the Boltzmann constant [4]. When electric discharge is used to generate plasma, if the electric field is high enough, an electron obtains more energy from the electric field than it loses through collision with the neutral particles, so it can accelerate under the electric field and result in an avalanche that finally induces the breakdown of the gas. Because of the high electron-neutral collision frequency, a significant part of the electron energy is transferred to the neutral particles. This is especially true when molecular gas is present in the working gas because the energy levels of the rotational and vibrational states of the molecular gas are much lower than the ionization potential of the molecule. Under such a condition, the neutral particle temperature T_n is significantly increased and can be close to the electron temperature T_e. When such a condition is satisfied, the plasma is classified as equilibrium plasma. The typical equilibrium plasma made in a lab is the arc discharge driven by DC power. Both the neutral particle temperature T_n and the electron temperature T_e of an arc plasma are typically several thousand Kelvin. Arc plasmas are used widely in welding, cutting, and waster material treatment. However, a gas temperature of several thousand Kelvin is too high for many other applications. For example, the application of plasma medicine, one of the fastest-growing fields in atmospheric pressure plasma, requires the gas temperature T_n of the plasma to stay close to room temperature. Such plasma with a gas temperature T_n much lower than the electron temperature T_e is classified as nonequilibrium plasma.

Several techniques have been reported to generate nonequilibrium plasma at atmospheric pressure. In general, these techniques can be classified into two categories, i.e., by controlling the discharge through either temporal or spatial methods. One of the most commonly used temporal-controlling methods is the dielectric barrier discharge (DBD), which limits the discharge current lasting time by covering one or both electrode(s) with dielectric; thus, the discharge current duration is limited, the actual discharge becomes pulsed, and the total energy delivery to the plasma is significantly reduced [5–15]. The typical electron temperature of a DBD is a few electron volts (eV), and the gas temperature of a DBD is normally slightly above or close to room temperature. The typical peak electron density of a DBD is on the order of 10^{11} cm^{-3}.

Another widely used temporal-controlling technique is using nanosecond pulsed applied voltage to drive the discharge [16–20]. Nanosecond pulsed applied voltage only generates plasma when the voltage is applied. Thus, the total energy is also limited. Besides, nanosecond pulsed discharge has several other advantages. Because of the fast rising time of the applied voltage, the electric field applied to the discharge gap can achieve significantly high overvoltage, which can result in a significantly high electron temperature and high peak discharge current. The peak electron temperature and electron density of an atmospheric pressure plasma driven by nanosecond pulsed applied voltages can reach about 10 eV and 10^{13} cm^{-3}. Because of the high electron temperature and high peak current, the reactive species concentration generated by nanosecond pulsed voltage is higher. In addition, studies found that uniform atmospheric pressure can be generated at atmospheric pressure even when air is used as the working gas. Finally, various plasma parameters, such as reactive species concentration and gas temperature, can be precisely adjusted by controlling the pulse rising time, pulse frequency, pulse width, and amplitude of the applied voltage.

Regarding spatial-controlling methods, one well-studied technique is to confine the plasma in small dielectric cavities, which limits the discharge to a small volume with dimensions in submillimeter ranges [21–30]. Because of the large surface-to-volume ratio, the conduction cooling helps keep the gas temperature in the discharge relatively low and so keep its nonequilibrium property. In addition, diffusion to the walls effectively dampens small fluctuations in the plasma that would otherwise lead to filamentation.

1.2 Atmospheric Pressure Plasma Jets—Equilibrium and Nonequilibrium Plasma Jets

For traditional discharges, plasma is generated as long as the applied electric field across the discharge gap is high enough to initiate a breakdown. However, at a pressure of 1 atm, the electric field required to initiate the discharge is quite high. For example, when air is used, the required electric field is about 30 kV cm^{-1}. That is why the discharge gaps for most atmospheric-pressure discharges vary in size from mm to several cm.

On the other hand, from the applications point of view, the short discharge gaps significantly limit the size of the objects to be treated if direct treatment (i.e., when the object is placed between the gaps) is desired. If indirect treatment (i.e., when the object is placed next to the gaps and the active radicals of the plasma reach the object by flowing with the gas) is applied, active radicals with short lifetimes and charged particles may already disappear before reaching the sample to be treated. To overcome the shortcomings of traditional nonequilibrium atmospheric pressure plasmas, plasmas generated in an open space rather than in a confined discharge gap are needed. However, when a plasma is launched in an open space where the applied electric field is normally quite low, it is extremely difficult to sustain the existence of the nonequilibrium plasma.

Fortunately, various methods were developed to overcome these challenges and several sources based on different designs were subsequently reported.

APPJs are generated in open space rather than in confined gaps. Thus, they can be used for direct treatment, and there is no limitation on the size of the object to be treated. This is extremely important for many applications.

As described above, atmospheric pressure plasma can be classified into equilibrium and nonequilibrium plasma. Accordingly, atmospheric pressure plasma jets also can be classified into two categories, i.e., equilibrium atmospheric pressure plasma jets (E-APPJs) and nonequilibrium atmospheric pressure plasma jets (N-APPJs). E-APPJs have attained noteworthy industrial significance, including the DC electric arc, which is used for welding, cutting, and so on [31–35]. Basically, E-APPJs transfer electrical energy to gas, normally air, forming equilibrium plasma. The applications of E-APPJs take advantage of the high gas temperature of the plasma.

On the contrary, N-APPJs have a much lower gas temperature. Applications in plasma medicine even require that the gas temperature of the plasma be close to room temperature [36–55]. When N-APPJs are used, the reactive oxygen and nitrogen species (RONS) generated by the plasma are playing the main role in various applications. In some circumstances, ultraviolet (UV) or vacuum ultraviolet (VUV) radiation, charge particles, and the electric field could also play some direct roles. Nevertheless, from an energy point of view, N-APPJs receive electrical energy from a power supply and transfer the significant part of the energy through ionization, excitation, dissociation, and so on, to form RONS rather than heat up the gas temperature. The concentration of the RONS of an N-APPJs can be many orders higher than the gas mixture at the same gas temperature.

1.3 Applications of Nonequilibrium Atmospheric Pressure Plasma Jets (N-APPJs) in Medicine

Research on the biomedical applications of nonequilibrium atmospheric plasmas, a field today known as plasma medicine, started in the mid-1990s under a US AFOSR program seeking to use nonequilibrium atmospheric pressure plasma to decontaminate both biotic (such as tissue, skin) and abiotic media and surfaces [56]. The aim was to use plasma technology to sterilize/decontaminate tools and gear and to disinfect wounds for speedy healing. At the time, only a few university-based US researchers were involved in this endeavor. By the early 2000s, proof of principal scientific work was accomplished, and most of the outstanding issues and challenges were identified. By the middle of the first decade of the 2000s, many of the major research groups active in plasma medicine today entered the field and gave it a new impetus. A global research community formed and the field took off exponentially. In the meantime, several centers focused on plasma medicine were established, including the INP in Germany, PBRC in South Korea, and the Drexel Plasma Institute and the ODU Plasma Engineering and Medicine Institute in the United States.

As mentioned above, N-APPJs generate plasma in an open space rather than in a confined space, which is much more suitable for plasma medicine applications. On the other hand, the advances achieved with N-APPJs in the past decade have greatly pushed forward the research in plasma medicine. In addition, the urgent demand for suitable plasma sources has attracted significant effort to develop, optimize, and understand N-APPJs. The number of publications in both plasma medicine and N-APPJs has grown exponentially in the past decade.

Today, work on plasma-aided wound healing, blood coagulation, dentistry, cancer treatment, and much more is going on in earnest at several laboratories worldwide.

2

Nonequilibrium Atmospheric Pressure Plasma (N-APP)

2.1 Nonequilibrium Atmospheric Pressure Plasma (N-APP) Sources

2.1.1 Corona Discharge

In order to get electrical breakdown in a gas, a high electric field is needed. Corona discharge normally has a pin or a thin wire electrode. With such a configuration, the electric field near the pin or the wire electrode is higher, but it decreases quickly with the increase of the distance from the pin or the wire electrode. Because of this characteristic, the gas breakdown only happens near the pin or the wire electrode. Such a discharge is called a corona. A corona will occur when the strength of the electric field around an electrode is high enough to form a conductive region, but not high enough to cause electrical breakdown or arcing to nearby objects [57–60].

In many high-voltage applications, a corona is an unwanted side effect. For example, corona discharge from high-voltage electric power transmission lines constitutes an economically significant waste of power for utilities. In a high-voltage power supply, the current leakage caused by coronas can constitute an unwanted load, and it can even destroy the power supply. Corona discharges can often be suppressed by improved insulation, the layout of the wires, and making high-voltage electrodes in smooth, rounded shapes. On the other hand, controlled corona discharges are used in a variety of processes, such as air filtration, photocopiers, and ozone generators.

If the electrode is positive, it has a positive corona; if it is negative, it has a negative corona. The physics of positive and negative coronas are strikingly different. This asymmetry is caused by the great difference in mass between electrons and positively charged ions.

With a positive corona, the electrons resulting from ionization are attracted to the curved electrode, and the positive ions are repelled from it. By undergoing inelastic collisions closer and closer to the curved electrode, more molecules are ionized through an electron avalanche.

For a positive corona, secondary electrons are created by ionization caused by the photons emitted from the plasma. The electrons resulting from the ionization of a neutral gas molecule are then electrically attracted back toward the curved electrode and attracted into the plasma, which begins the process of creating further avalanches inside the plasma.

Therefore, a positive corona is divided into two regions, i.e., the inner region around the electrode and the outer region surrounding the inner region. The inner region contains electrons and positive ions; the electrons avalanche in this region, creating many more ion/electron pairs. The outer region consists mainly of slowly migrating positive ions and secondary electrons that are liberated by photons and then accelerated into the inner region.

For a negative corona, the electrons move outward and away from the curved electrode. The dominant process for generating secondary electrons in this case is the photoelectric effect from the surface of the electrode itself. The high-energy photons are from atoms/molecules within the plasma. The work function of the electrons of an electrode is normally considerably lower than the ionization energy of air at standard temperatures and pressures, making it a more liberal source of secondary electrons. Away from the curved electrode, because of the accumulation of positive ions, the electric field is weakened, and thus the ionized neutral gas as a source of ionization is further attenuated. However, under certain conditions, the collision of the positive species with the curved electrode can also cause electron liberation.

A negative corona can be divided into three regions, i.e., the inner region, the intermediate region, and the outer region. In the inner region, high-energy electrons inelastically collide with neutral atoms and cause avalanches. In the intermediate region, electrons combine to form negative ions, but they typically have insufficient energy to cause avalanche ionization. The outer region mainly contains negative ions but fewer free electrons. The inner two regions are known as the corona plasma. The outer region is known as the unipolar region.

2.1.2 Dielectric Barrier Discharge

It was Theodose du Moncel who first discovered in 1853 that a discharge can be induced between two conducting plates separated by two glass plates [61]. To drive the discharge, he used a Ruhmkorff coil, which is an induction coil that allows for the generation of high AC voltages from a low-voltage DC source (see **Figure 2.1**). Du Moncel wrote, "If we maintain a distance of 3–4 mm between the glass plates C, D, E, F (**Figure 2.1**) which are covered by metallic plates, A, B, G, H, we see in the darkness a rain of fire of a beautiful blue color in the gap between the glass plates." He also noticed that a surface discharge was visible between the metal plate and the glass plate it covers.

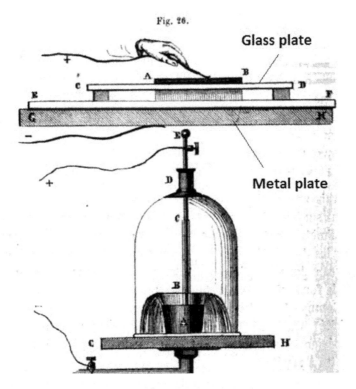

FIGURE 2.1 Th. du Moncel's experiment: Top schematic shows a DBD arrangement with metal electrodes (A, B, G, H), glass plates (C, D, E, F), a 3–4 mm gap between the glass plates where an atmospheric pressure air plasma is generated. (From Du Moncel, T., *Notice sur l'appareil d'Induction Electrique de Ruhmkorff et sur les Experiences que l'on peut faire Avec cet Instrument*, Hachette et Cie publishers, New York, 1855.)

Du Moncel's discovery was followed by the work of Werner von Siemens, who in 1857 reported on the design and application of a dielectric barrier discharge (DBD) that generated ozone [62]. Siemens's DBD apparatus had a cylindrical geometry with tin foils as electrodes and used glass as a dielectric. However, detailed scientific investigations to generate nonequilibrium, low-temperature atmospheric pressure plasmas were conducted much later, in the 1930s, by Von Engel, who tried to generate such plasma by controlling the temperature of the cathode (by water cooling) [63]. These early works were not pursued much further until about five decades later (late 1980s and early 1990s) when the generation of relatively large volume, nonequilibrium, diffuse atmospheric pressure plasma was finally successful by the use of a DBD. These successful experiments were first reported by Kanazawa et al., Massines et al., and Roth et al. [64–66]. These investigators used mostly a planar geometry and sinusoidal voltages in kV at frequencies in the kHz range. The initial performance of the DBD was greatly improved upon when fast rise time voltage pulses with pulse widths in the nanoseconds-microseconds range were employed [67–69]. These pulses preferentially couple the applied energy to the electron population and provide a means to control the electron energy distribution function (EEDF): Controlling the EEDF allows for the enhancement of the plasma chemistry [68–70]. The DBD as a source of low-temperature plasma (LTP) has been used in various plasma-processing applications and was the first device used in the mid-1990s in the groundbreaking research that ultimately resulted in the emergence of the field of plasma medicine [71–74].

Operation of the Dielectric Barrier Discharge (DBD)

One of the most widely used approaches for generating low-temperature atmospheric pressure plasmas is the dielectric barrier discharge (DBD). The original concept of a dielectric barrier discharge was improved upon by several investigators over several decades [64–66,75–83]. DBDs use a dielectric material to cover at least one of the electrodes. The electrodes are driven by voltages of several kV at frequencies in the kHz range. In the past few decades, DBDs have been extensively used in material processing, such as the modification of the hydrophilicity or hydrophobicity of materials, surface modification, their use as flow control actuators, the generation of ozone [76,84,85], and since the mid-1990s in biomedical applications [71–74].

Common dielectric materials used in DBDs include glass, quartz, ceramics, and polymers. The gap distance between electrodes varies considerably, from less than 0.1 mm in plasma displays to several millimeters in ozone generators and up to several centimeters in CO_2 lasers. DBD devices can be made in many different configurations, including planar, using parallel plates separated by a dielectric or a cylinder, or using coaxial plates with a dielectric tube between them. **Figure 2.2** shows schematics of different types of DBD, which include the type generated in a volume (V-DBD) and the type generated on a surface (S-DBD). For V-DBD, the plasma is generated between two electrodes, such as between two parallel plates with one or two dielectrics in between. For S-DBD, the plasma is generated on the surface of a dielectric, and thus the discharges are limited to the surface.

DBDs usually generate plasmas with filamentary structures, resulting in nonuniform material treatment. However, since the late 1980s, development of DBDs that can generate nonfilamentary or diffuse plasma led to interesting results. Several reports were published that showed that under some conditions DBDs can produce diffuse, relatively homogenous plasmas at atmospheric pressure [64–66,77–83].

Under sinusoidal excitation, the electrodes of the DBD are energized by high sinusoidal voltages with amplitudes in the 1–20 kV range at frequencies in the kHz range. The electrode arrangement is generally contained within a vessel to allow for the introduction of a select gaseous mixture. Surface charges accumulate on the dielectric when a discharge is ignited in the gas. These surface charges create an electrical potential, which counteracts the externally applied voltage, resulting in self-limitation of the discharge current. Although generally DBDs produce filamentary plasmas, under some conditions, homogeneous plasmas can also be generated.

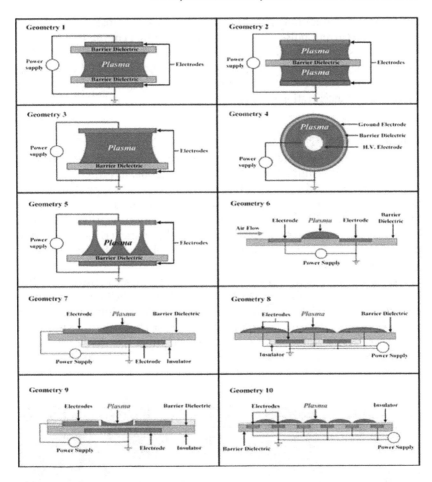

FIGURE 2.2 Schematics of dielectric barrier discharge with different configurations.

The charge accumulation on the surface of the dielectric material covering the electrodes plays a crucial role in maintaining the nonequilibrium nature of the plasma.

The operating conditions that lead to a diffuse plasma were originally suggested by Kanazawa et al. [64] as: (1) Helium must be used as the operating gas; and (2) the frequency of the applied voltage must be in the kHz range. However, other investigators were able to generate diffuse plasmas with other gases and at different frequency ranges [79–83]. Massines et al. proposed the mechanisms which lead to a diffuse plasma: Seed electrons and metastable atoms that are available between current pulses make gas breakdown possible under low electric field conditions. These seed particles allow for either a Townsend-type breakdown or a glow discharge depending on the gas used (nitrogen vs. helium). If helium is the gas used, a density of seed electrons greater than $10^6 \, \text{cm}^{-3}$ was found to be sufficient to keep the plasma ignited under low field conditions [78]. The seed electrons are those left over from the previous current pulse and/or those produced by Penning ionization. If nitrogen is used, it is the metastables that play the dominant role in keeping the discharge ignited between pulses. In this case, the surface of the dielectric, which can be a source of quencher species, plays an important role in setting the available concentration of metastable atoms.

Pulsed Excitation

The electron energy distribution in nonequilibrium discharges plays the most important role in defining the chemistry of the plasma. It is through electron impact excitation and ionization that the charged particles, excited species, and radicals are produced. To achieve an increase of ionization and to extend the electron energy distribution to higher values, short high-voltage pulses have been used. Pulses with widths less than the characteristic time of the onset of the glow-to-arc transition help keep the plasma stable and maintain its nonequilibrium nature.

Mildren et al. [67] and Laroussi et al. [69,70] investigated pulsed dielectric barrier discharges (P-DBD) in xenon and in helium (or helium/air mixture), respectively. The P-DBD could generate large volume, diffuse plasma in atmospheric pressure when a He, He/O$_2$, He/air, or He/N$_2$ mixture was used (see **Figure 2.3**). Laroussi, Lu et al. [69] reported that two discharges occur for every applied voltage pulse. The first discharge (or primary discharge) was

FIGURE 2.3 Photograph of a diffuse P-DBD discharge. Gas: helium; Pulse width: 500 ns; Rep rate: 5 kHz.

ignited at the rising edge of the applied voltage pulse while the second discharge (or secondary discharge) was self-ignited during the falling edge of the applied voltage pulse.

2.1.3 Resistive Barrier Discharge

To extend the operating frequency range of barrier discharges, a few methods were proposed. Okazaki used a dielectric wire mesh electrode in a DBD to generate a glow discharge at a frequency of 50 Hz [79]. Laroussi and co-workers proposed what came to be known as the resistive barrier discharge (RBD) [86]. The RBD can be operated with DC or AC (60 Hz) power supplies. This discharge is based on the dielectric barrier (DBD) configuration, but instead of a dielectric material, a high resistivity (a few MΩ.cm) sheet is used to cover one of the electrodes. The high resistivity sheet plays the role of a distributed resistive ballast, which inhibits the discharge from localizing and the current from reaching a high value and, therefore, prevents arcing. It was found that if helium was used as the ambient gas between the electrodes and if the gap distance was below 5 cm, spatially diffuse plasma could be maintained for time durations of several tens of minutes. However, if air was added to the helium (>1%), the discharge formed filaments, which randomly appeared within a background of more diffuse plasma. This occurred even when the gap distance was small [86].

The RBD can be operated under DC or AC excitation modes. Under both modes, the discharge current exhibits a series of pulses, suggesting that like the DBD, the RBD is also a self-pulsed discharge. The current pulses are a few microseconds wide and occur at a repetition rate of a few tens of kHz. The pulsing of the discharge current can be explained by the combined resistive and capacitive nature of the device. When the gas breaks down and a current of sufficient magnitude flows, the equivalent capacitance of the electrodes becomes charged to the point where most of the applied voltage starts appearing across the resistive layer of the electrodes. The voltage across the gas then becomes too small to maintain a discharge and the plasma extinguishes. At this point, the equivalent capacitor discharges itself through the resistive layer, hence lowering the voltage across the resistive layer and increasing the voltage across the gas until a new breakdown occurs [87].

When helium was used as the ambient gas between the electrodes and if the gap distance was not too large (5 cm and below), a spatially diffuse discharge could be maintained for time durations of several tens of minutes as shown in **Figure 2.4**. However, if 1% of the air was mixed with helium, the discharge formed filaments that randomly appeared within its volume. This occurred even for small gaps.

The RBD offers a practical method for generating relatively large volumes of low-temperature plasma for material processing and biomedical applications [88]. The homogeneity of the plasma is enhanced when helium is the main component of the ambient gas mixture between the electrodes, while an admixture of air renders the discharge filamentary. If only air is used, plasma can still be initiated for small gaps (millimeters). However, in this case, the structure of the plasma is spatially nonuniform.

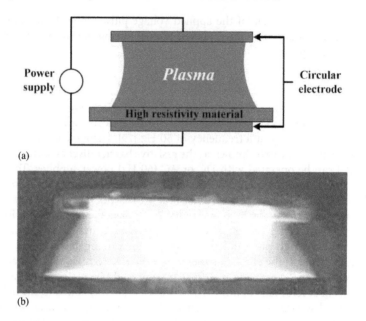

(a)

(b)

FIGURE 2.4 (a) Configuration of the RBD. (b) Appearance of the discharge with helium as the ambient gas. The gap is 4.5 cm wide, and the applied voltage is 13 kV at 60 Hz.

2.1.4 Resistor-Capacitor Barrier Discharge

One more way to control the discharge current and the power delivery to the plasma during each discharge pulse is resistor-capacitor barrier discharge [89,90]. As shown in **Figure 2.5**, the plasma is driven by a pulsed DC power supply. In this setup, the capacitor controls the total power that can be delivered to the plasma during each discharge pulse and the resistor limits the charging current. As long as the capacitance is small enough and the resistance of the resistor is not too big, the discharge frequency is determined by the frequency of the pulse frequency of the applied voltage.

The current and voltage waveforms of the discharge are shown in **Figure 2.6**, where V_a is the applied voltage, I_{tot} is the total current (with gas flow: plasma on), and I_{no} is the displacement current (without gas flow: plasma off). It should be mentioned that the voltage waveform remains the same whether the plasma is on or off. This figure clearly shows that the actual discharge current, i.e., the difference between I_{tot} and I_{no}, has a peak value of about 10 mA. The displacement current waveform behaves as that of a typical resistor-capacitor (RC) charge and discharge circuit. The voltage on the needle V_{needle} has a peak of about 6 kV. According to the current and voltage waveforms of the discharge, the power deposited into the plasma can be estimated to be less than 0.1 W for an applied voltage of 8 kV, a pulse width of 500 ns, and a pulse frequency of 10 kHz.

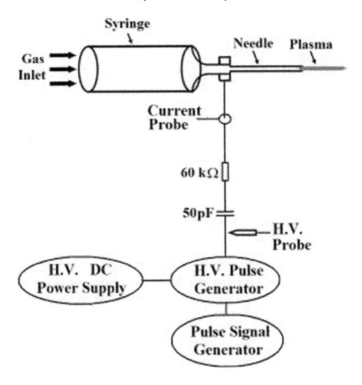

FIGURE 2.5 Schematic of a resistor-capacitor barrier discharge.

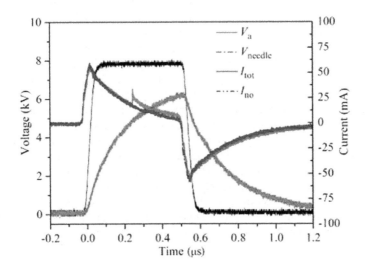

FIGURE 2.6 Current and voltage waveforms of the discharge. Applied voltage: V_a; voltage on the needle: V_{needle}; total current: I_{tot} (plasma on); and displacement current: I_{no} (plasma off).

2.2 Nonequilibrium Atmospheric Pressure Plasma Jets (N-APPJs)

For nonequilibrium atmospheric pressure plasma (N-APP), the electron temperature is far higher than the temperature of the heavy particles. Due to the high collision frequency between electrons and heavy particles, the electrons lose their energy in a short period. If molecular gas is present, the electrons could quickly transfer their energy to molecular rotational and vibrational states because the energy levels of the rotational and vibrational states of the molecules can be much lower than that of the electrons' excitation and ionization levels. This makes it a difficult task to obtain N-APP with high electron energy. Thus, the ionization efficiency in such a case is low. Furthermore, when an electronegative gas, such as O_2, is present, the electrons could be absorbed by the gas on a time scale of tens of nanoseconds, or even shorter, which makes it even harder to obtain N-APP with electronegative gases.

Nevertheless, for traditional discharges, plasma is generated as long as the applied electric field across the discharge gap is high enough to initiate a breakdown. However, at a pressure of 1 atm, the electric field required to initiate the discharge is quite high. For example, when air is used, the required electric field is about 30 kV/cm. That is why the discharge gaps for most atmospheric-pressure discharges vary in size from mm to several cm. On the other hand, from the applications point of view, the short discharge gaps significantly limit the size of the objects to be treated if direct treatment (i.e., when the object is placed between the gaps) is desired. If indirect treatment (i.e., when the object is placed next to the gaps and the active radicals of the plasma reach the object by flowing with the gas) is applied, active radicals with short lifetimes and charged particles may already disappear before reaching the sample to be treated. To overcome the shortcomings of traditional N-APP, nonequilibrium plasmas generated in an open space rather than in a confined discharge gap, i.e., N-APPJs, are needed. However, when a plasma is launched in an open space where the applied electric field is normally quite low, it is extremely difficult to sustain the existence of the plasma.

Briefly, as pointed out above, there are two facts that make it a big challenge to generate N-APPJs: one is the high electron heavy particle collision frequency, and the other is the low applied electric field. Fortunately, various methods were developed to overcome these challenges and several sources based on different designs, driven by different types of power supplies, were subsequently reported.

N-APPJs are generated in open space rather than in confined gaps. Thus, they can be used for direct treatment, and there is no limitation on the size of the object to be treated. This is extremely important for applications such as plasma medicine. In the next section, first, the development of N-APPJs will be briefly reviewed. Second, one of the most interesting phenomena of N-APPJs, i.e., the dynamic behavior of N-APPJs, widely called the plasma bullet, is discussed.

2.2.1 N-APPJ Sources

Various types of atmospheric pressure nonequilibrium plasma jets (APNP-Js) with different configurations have been reported, where most of the jets are working with noble gas mixed with a few percent of reactive gases, such as O_2. Plasma jets operating with noble gases can be classified into four categories, i.e., dielectric-free electrode (DFE) jets, dielectric-barrier-discharge (DBD) jets, DBD-like jets, and single electrode (SE) jets, as shown in **Figure 2.7**. On the other hand, noble gases are more costly and not convenient for some applications. To overcome such shortages, N_2 or even air has been used as the working gas. **Figure 2.8** show several plasma jet devices developed worldwide using noble gases, N_2, or air as the working gas.

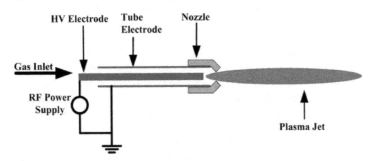

FIGURE 2.7 Schematic of a dielectric-free electrode (DFE) jet. (From Lu, X. et al., *Plasma Sources Sci. Technol.*, 21, 034005, 2012.)

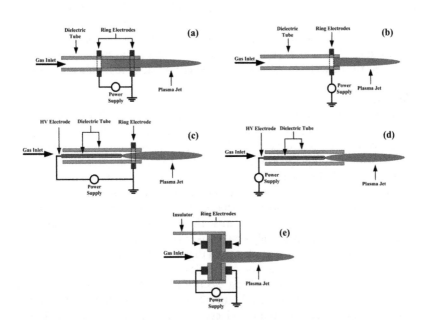

FIGURE 2.8 Schematics showing various configurations of DBD-based plasma jets: (a) Using two external ring electrodes; (b) Using a single ring electrode; (c) Using an axial pin electrode and an external ring electrode; (d) Using a single axial pin electrode; (e) Using two annular ring electrodes on centrally perforated dielectric disks.

Dielectric-Free Electrode (DFE) Jets

One of the early APNP-Js, developed by Hicks' group, was a DFE jet, as shown in **Figure 2.7** [55,91]. The jet is driven by a radio frequency power source at 13.56 MHz. It consists of an inner electrode, which is coupled to the power source, and a grounded outer electrode. A mixture of He with reactive gases is fed into the annular space between the two electrodes. Cooling water is needed to keep the jet from overheating, and the gas temperature of the plasma jet varies from 50°C to 300°C, depending on the RF power. The window of stable operation without apparent arcing of the device is: He flow rates greater than 25 L/min, O_2 concentrations of up to 3.0% by volume, CF4 concentrations of up to 4.0% by volume, and RF power between 50 and 500 W.

Several notable characteristics of the DFE jet are, first, arcing is unavoidable when stable operation conditions are not met. Second, compared to DBD and DBD-like jets that will be discussed below, the power delivered to the plasma for the DFE jet is much higher. Third, due to the high power delivered, the gas temperature of the plasma is quite high and out of the acceptable range for biomedical applications. Fourth, for this DFE jet, which is driven by an RF power supply, the peak voltage is only a few hundred volts, so the electric field within the discharge gap is relatively low and its direction is radial (perpendicular to the gas flow direction). The electric field in the plasma plume region is even lower, especially along the plasma plume propagation direction (gas flow direction). Finally, since the electric field along the plasma plume propagation direction is very low, the generation of this plasma plume is probably gas-flow driven rather than electrically driven.

On the other hand, because relatively high power can be delivered to the plasma and the gas temperature is relatively high, the plasma is very reactive. This kind of plasma jet is suitable for applications such as material treatment as long as the material to be treated is not very sensitive to high temperatures.

DBD Jets

For DBD jets, as shown in **Figure 2.8a–e**, there are many different configurations. As shown in **Figure 2.8a**, which was first reported by Teschke et al. [35], the jet consists of a dielectric tube with two metal ring electrodes on the outer side of the tube. When a working gas (He, Ar) is flowing through the dielectric tube and the kHz high voltage power supply is turned on, a cold plasma jet is generated in the surrounding air. The plasma jet only consumes several watts of power. The gas temperature of the plasma is close to room temperature. The gas flow velocity is less than 20 m/s. The plasma jet, which looks homogeneous to the naked eye, is actually a bullet-like plasma volume with a propagation speed of more than 10 km/s. It's believed that the applied electric field plays an important role in the propagation of the plasma bullet.

Figure 2.8b eliminates one ring electrode [92]. So, the discharge inside the dielectric tube is weakened. **Figure 2.8c** replaces the high-voltage (HV) ring electrode with a centered pin electrode, which is covered by a dielectric tube with one end closed [93]. With this configuration, the electric field along the plasma plume is enhanced. Walsh and Kong's studies show that the high electric field along the plasma plume is favorable for generating long plasma plumes and more active plasma chemistry. **Figure 2.8d** further removes the ground ring electrode of **Figure 2.8c** [94]. So, the discharge inside the tube is also weakened. On the other hand, a stronger discharge inside the discharge tube (in the case of **Figure 2.8a** and **c**) helps the generation of more reactive species. With the gas flow, the reactive species with relatively long lifetimes may also play an important role in various applications. The configuration of **Figure 2.8e** is different from the previous four DBD jet devices. The two ring electrodes are attached to the surface of two centrally perforated dielectric disks. The holes in the center of the disks are about 3 mm in diameter. The distance between the two dielectric disks is about 5 mm. With this device, a plasma plume up to several centimeters in length can be obtained [36].

All the DBD jet devices discussed above can be operated either by kHz AC power or by pulsed DC power. The length of the plasma jet can easily reach several cm or even longer than 10 cm as reported by Lu et al. [93]. This capability makes the operation of these plasma jets easy and practical. There are several other advantages of the DBD jets. First, due to the low power density delivered to the plasma, the gas temperature of the plasma remains close to room temperature. Second, because of the use of the dielectric, there is no risk of arcing whether the object to be treated is placed far away or close to the nozzle. These two characteristics are very important for applications such as plasma medicine, where safety is a strict requirement.

DBD-Like Jets

All the plasma jet devices shown in **Figure 2.9** are named DBD-like jets. This is based on the following facts. When the plasma plume is not in contact with any object, the discharge is more or less like a DBD. However, when the plasma plume is in contact with an electrical conductor (a non-dielectric material), especially a ground conductor, the discharge is actually running between the high voltage electrode and the object to be treated (ground conductor). In such circumstances, it does not operate as a DBD anymore. The devices shown in **Figure 2.9** can be driven by kHz AC power, by RF power, or by pulsed DC power.

Figure 2.9b replaces the solid HV electrode in **Figure 2.9a** with a hollow electrode [95,96]. The benefit of this kind of configuration is that two different gases can be mixed in the device. Normally, gas inlet 2 is used for reactive gas, such as O_2 flow, and gas inlet 1 is for a noble gas. It was found that the plasma plume is much longer with this kind of gas control than that using a premixed

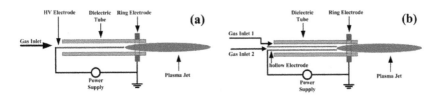

FIGURE 2.9 Schematics of DBD-like plasma jets: (a) Using an axial pin electrode and an external ring electrode; (b) Using a hollow tube electrode, an external ring electrode, and two gas inlets.

gas mixture with the same percentage [95]. The role (and advantage) of the ring electrode in **Figure 2.9a** and **b** is the same as in the case of DBD jets.

When DBD-like plasma jets are used for plasma medicine applications, the object to be treated could be cells or a whole tissue. In this case, these types of jet devices should be used carefully because of the risk of arcing. On the other hand, if they are used for the treatment of conductive materials, since there is no dielectric, more power can be easily delivered to the plasma. So, as long as arcing is carefully avoided, DBD-like jets have their own advantages.

Single Electrode (SE) Jets

The schematics of single electrode (SE) jets are shown in **Figure 2.10a–c**. **Figure 2.10a** and **b** are similar to the DBD-like jets except there is no ring electrode on the outside of the dielectric tube. The dielectric tube only plays the role of guiding the gas flow. These two jets can be driven by DC, kHz AC, RF, or pulsed DC power.

Because of the risk of arcing, the plasma plumes generated by **Figure 2.10a** and **b** are not the best for biomedical applications due to the safety issues [97]. In order to overcome this problem, Lu et al. developed a similar single electrode jet as shown in **Figure 2.10c** [89]. The capacitance C and resistance R are about 50 pF and 60 kΩ respectively. The resistor and capacitor are used for controlling the discharge current and voltage on the hollow electrode (needle). This jet is driven by a pulsed DC power supply with a pulse width of 500 ns, a repetition frequency of 10 kHz, and an amplitude of 8 kV. The advantage of this jet is that the plasma plume or even the hollow electrode can be touched without any danger of harm, making it suitable for plasma medicine applications.

One of the potential applications is in dentistry, such as root canal treatment. Due to the narrow channel geometry of a root canal, which typically has a length of a few centimeters and a diameter of one millimeter or less, the plasma generated by a plasma jet is not efficient enough to deliver reactive agents into the root canal for disinfection. Therefore, to have a better killing efficacy, a plasma needs to be generated inside the root canal. When plasma is

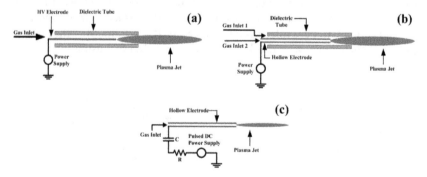

FIGURE 2.10 Schematics of SE plasma jets: (a) Using an axial pin electrodes inside a dielectric tube; (b) Using a hollow tube electrode inside a dielectric tube, and two gas inlets; (c) Using just one hollow tube electrode.

generated inside the root canal, reactive agents, including short-lifetime species such as charge particles, could play some role in the killing of bacteria. By using the device of **Figure 2.10c**, a cold plasma could be generated inside a root.

N₂ Plasma Jets

As pointed out in the introduction, it is difficult to generate atmospheric pressure nonequilibrium nitrogen (N_2) plasma jets. Up to now, only a few N_2 plasma jets have been reported [98–100]. **Figure 2.11** shows the schematic of two N_2 plasma jets. The N_2 plasma jet as shown in **Figure 2.11a** was reported by Hong and Uhm [98]. A 20 kHz AC power supply is connected to two electrodes with a 3 mm thickness and a center hole that is 500 μm in diameter. The two electrodes are separated by a dielectric disk with a center hole that has the same diameter. With this configuration, they are capable of generating a N_2 plasma up to 6.5 cm long. When the N_2 gas flow rate is 6.3 SLM (standard liters per minute), the gas is ejected from the hole at a speed of about 535 m/s. The gas temperature of the plasma plume at 2 cm from the nozzle is below 300 K. **Figure 2.11b** shows a slightly different N_2 plasma jet device, which replaces the inner perforated HV electrode of **Figure 2.11a** with a pin electrode [99]. The inner electrode also can be replaced by a tube, which Hong et al. did [100].

Air Plasma Jets

Due to the presence of the electro-negative oxygen (O_2) in air, it is difficult to sustain an atmospheric pressure nonequilibrium air plasma jet. Nevertheless, several different air plasma jets have been reported [101–105]. Mohamed et al. reported a micro plasma jet device that can operate in various gases, including air [102]. The schematic is shown in **Figure 2.12a**. A discharge channel through an insulator with a thickness of about 0.2–0.5 mm and a diameter of 0.2–0.8 mm separates the anode and cathode, which have a center hole with the same diameter. The ballast resistor is 51 kΩ. When air is flowing through the hole and a DC voltage of a few hundred volts (up to a kV) is applied between the anode and cathode (depending on the thickness of the insulator separating the electrodes), a relatively low temperature air plasma is generated in the surrounding air with a length up to 1 cm long, depending on the gas flow rate and discharge current. However, the gas temperature of the plasma can still be quite high. The gas temperature within the

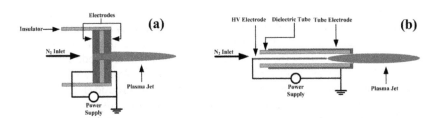

FIGURE 2.11 Schematics of N_2 plasma jets: (a) Using two hollow disk electrodes separated by a dielectric disk; (b) Using an axial pin electrode and external tube electrodes covering a dielectric tube.

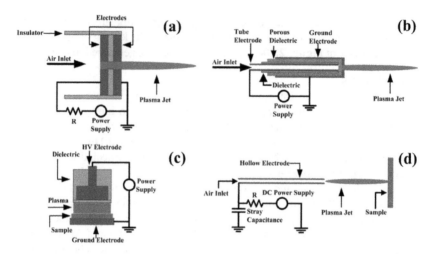

FIGURE 2.12 Schematics of air plasma jets: (a) Using two hollow disk electrodes separated by a dielectric disk; (b) Using a central tube electrode covered by a porous dielectric tube covered by a grounded tube electrode; (c) Floating electrode DBD with a grounded electrode; (d) Using just one hollow tube electrode, DC voltage, and a target electrode.

micro gap is about 1000 K. However, it drops quickly as the gas propagates in the surrounding air. It is about 50°C at 5 mm away from the nozzle for an air flow rate of 200 ml/min and discharge current of 19 mA.

Hong et al. reported another type of air plasma jet device as shown in **Figure 2.12b** [101]. One of the notable characteristics of this device is that a porous alumina dielectric is used to separate the HV stainless-steel (typical injection needle) electrode and the outer ground electrode. The alumina used in this device has approximately 30 vol% porosity and has an average pore diameter of 100 μm. The ground electrode is fabricated from stainless steel and has a centrally perforated hole of 1 mm in diameter through which the plasma jet is ejected to the surrounding ambient air. When a 60 Hz HV power supply is applied and the flow rate of air is at several standard liters per minute (SLM), a APNP-J up to about 2 cm is generated in the surrounding air. During one voltage cycle, there are multiple discharges. The increase of the input power results in more current pulses. The shortcoming of this device is the same as the previous one, i.e., the gas temperature of the plasma is quite high. It is about 60°C at 10 mm away from the nozzle for an air flow rate of 5 SLM. For lower flow rates, the gas temperature is even higher.

Figure 2.12c and **d** are the schematics of two "floating" electrode air plasma jets [104,105]. They are not strictly plasma jets since the plasmas are generated within a gap. However, because the secondary electrode (ground electrode) can be a human body, we still categorize them as plasma jets in this paper. Both jets could generate room-temperature air plasmas. They are completely safe from an electrical perspective and not damaging to animals and human beings.

For **Figure 2.12c**, a kHz AC or a pulsed DC voltage with an amplitude of 10–30 kV is used to drive the device. The discharge ignites when the powered electrode approaches the surface to be treated at a distance (discharge gap) less than about 3 mm, depending on the form, duration, and polarity of the driving voltage. This jet is suitable for large smooth surface treatment.

On the contrary, the jet shown in **Figure 2.12d** is more suitable for localized three-dimensional treatments. This jet is driven by a homemade DC power supply. The output voltage of the power supply can be adjusted up to 20 kV. The output of the power supply is connected to a stainless-steel needle (typical injection needle) electrode through a resistor R of 120 MΩ, which is several orders of magnitude higher than those reported [106]. When a counter electrode, such as a finger, is placed close to the needle, a plasma is generated, as shown in **Figure 2.8**. The plasma is similar to the positive corona discharge. However, this jet can be touched by the human body directly, which is not the case for the traditional corona discharge. The jet has no risk of glow-arc transition. The maximum length of the plasma is about 2 cm. The gas temperature of the plasma is kept at room temperature. It is interesting to point out that the discharge is actually pulsed. It appears periodically with a pulse frequency of tens of kHz, depending on the applied voltage and the distance between the tip of the needle and the object to be treated.

Brief Summary

Although noble gas plasma jets are relatively easy to generate, noble gas plasma jets are not as reactive as air plasma jets. That's the reason why an amount of only a few percent or less of reactive gases is added to the noble gas when the plasma jets are used for various applications. The noble gas serves as the carrier gas to generate the plasma. For biomedical applications, O_2 or H_2O_2 is usually added. For etching applications, CF_4 or O_2 could be used. Regarding N_2 plasma jets, they are also not as reactive as air plasma jets. It is also recommended to add an amount of only a few percent or less of reactive gases to the carrier gas.

2.2.2 Dynamics of N-APPJs

The discrete nature of the structure of plasma jets was first observed by Teschke et al. using an RF-driven plasma jet [35] and by Lu and Laroussi using a pulsed-DC plasma jet (the plasma pencil) [36]. Using fast imaging, these investigators found that the plasma plume, which appeared continuous to the naked eye, was in fact made up of fast-moving plasma structures. Teschke et al. found that the small volume of plasma, or plasma bullet, travels at a velocity of about 1.5×10^4 m/s [35] while Lu and Laroussi measured velocities as high as 1.5×10^5 m/s [36]. Comparatively, the estimated upper limit of the drift velocity of electrons under the external applied electric field is only 1.1×10^4 m/s, and the estimated upper limit of the drift velocity is 2.2×10^2 m/s. Because these speeds are far slower than the measured bullet-like plume velocity, Lu and Laroussi invoked a streamer propagation model based on photoionization, in the manner that was proposed by Dawson and Winn [107] for streamers, to explain the properties of these so-called plasma bullets.

However, there are some notable differences between the streamer-like APNP-J and positive corona discharges that are typically used to study cathode-directed streamers. For example, streamers developed in pulsed positive corona discharges are typically not very repeatable due to the stochastic nature of their initiation. In contrast, the plasma bullet behavior is mostly very repeatable. In addition, experiments revealed ring-shaped profiles of plasma bullet radiation, of the densities of nitrogen ions and metastable He atoms, with maxima shifted from the jet axis. Such a pattern is different from that of a typical streamer in uniform media, which does not have a donut-shape structure. Besides, the propagation of the plasma plume left a dark channel between the plume head and the electrode. It is not clear yet whether the conductivity of the dark channel is high enough to affect the propagation of the plasma plume like in a streamer discharge or not. Furthermore, photoionization plays an important role in positive streamers. However, it is not clear whether photoionization plays a similar role in the propagation of the plasma bullet either. Finally, in a streamer discharge, the discharge behaves differently when the polarity of the voltage is changed from positive to negative. How plasma jets will behave for different polarities also needs investigation. Because of so much difference between streamers and the bullet-like plasma plume, many studies have been carried out to investigate the so-called "plasma bullet" behavior in the past several years [108–143]. Some of these questions are much better understood now. Details about the plasma bullet will be given in Chapter 4 of this book.

References for Part I

1. Y. Raizer, *Gas Discharge Physics*, Springer-Verlag, Moscow, Russia, 1987.
2. F.F. Chen, *Introduction to Plasma Physics and Controlled Fusion, Vol. 1: Plasma Physics*, Plenum Press, New York, 1974.
3. M. Lieberman and A. Lichtenberg, *Principles of Plasma Discharges and Materials Processing*, John Wiley & Sons, New York, 2006.
4. X. Lu and M. Laroussi, Electron density and temperature measurement of an atmospheric pressure plasma by millimeter wave interferometer, *Appl. Phys. Lett.* 92 (2008) 051501.
5. F. Massines, P. Ségur, N. Gherardi, C. Khamphan and A. Ricard, Physics and chemistry in a glow dielectric barrier discharge at atmospheric pressure: Diagnostics and modelling, *Surf. Coat. Technol.* 174 (2003) 8–14.
6. U. Kogelschatz, Dielectric-barrier discharges: Their history, discharge physics, and industrial applications, *Plasma Chem. Plasma Process.* 23 (2003) 1–46.
7. U. Kogelschatz, B. Eliasson and W. Egli, From ozone generators to flat television screens: History and future potential of dielectric-barrier discharges. *Pure Appl. Chem.* 71 (1999) 1819–1828.
8. T. Corke, C. Enloe and S. Wilkinson, Dielectric barrier discharge plasma actuators for flow control, *Annu. Rev. Fluid Mech.* 42 (2010) 505–529.
9. U. Konelschatz, B. Eliasson and W. Egli, Dielectric-barrier discharges. Principle and applications, *J. Phys.* IV FRANCE 7 (1997) 47–68.
10. K. Buss, Die elektrodenlose Entladung nach Messung mit dem Kathodenoszillographen, *Arch. Elektrotech.* 26 (1932) 261–265.
11. K. Honda and Y. Naito, On the nature of silent electric discharge, *J. Phys. Soc. Jpn.* 10 (1955) 1007–1011.
12. B. Eliasson and U. Kogelschatz, UV excimer radiation from dielectric-barrier discharges, *Appl. Phys. B* 46 (1988) 299–303.
13. J.-P. Boeuf and P.C. Litchford, Calculated characteristics of an ac plasma display panel cell, *IEEE Trans. Plasma Sci.* 24 (1996) 95–96.
14. U. Kogelschatz, Advanced ozone generation, in: *Process Technologies for Water Treatment,* S. Stucki (Ed.), Plenum Press, New York, 1988.
15. D. Braun, V. Gibalov and G. Pietsch, Two-dimensional modelling of the dielectric barrier discharge in air, *Plasma Sources Sci. Technol.* 1 (1992) 166–172.
16. S. Liu and M. Neiger, Excitation of dielectric barrier discharges by unipolar submicrosecond square pulses, *J. Phys. D Appl. Phys.* 34 (2001) 1632.
17. H. Ayan, G. Fridman, A. Gutsol, V. Vasilets, A. Fridman and G. Friedman, Nanosecond-pulsed uniform dielectric-barrier discharge, *IEEE Trans. Plasma Sci.* 36 (2008) 504–508.
18. G. Nersisyan and W.G. Graham, Characterization of a dielectric barrier discharge operating in an open reactor with flowing helium, *Plasma Sources Sci. Technol.* 13 (2004) 582.
19. S. Liu and M. Neiger, Electrical modelling of homogeneous dielectric barrier discharges under an arbitrary excitation voltage, *J. Phys. D* 36 (2003) 3144.
20. R. Mildren and R. Carman, Enhanced performance of a dielectric barrier discharge lamp using short-pulsed excitation, *J. Phys. D* 34 (2001) L1.
21. J. Eden and S. Park, Microcavity plasma devices and arrays: A new realm of plasma physics and photonic applications, *Plasma Phys. Control. Fusion* 47 (2005) B83.

22. K.H. Schoenbach, R. Verhappen, T. Tessnow, F.E. Peterkin and W.W. Byszewski, Microhollow cathode discharges, *Appl. Phys. Lett.* 68 (1996) 13.

23. A. Habachi and K.H. Schoenbach, Emission of excimer radiation from direct current, high-pressure hollow cathode discharges, *Appl. Phys. Lett.* 72 (1998) 22.

24. R. Stark and K.H. Schoenbach, Direct current glow discharges in atmospheric air, *Appl. Phys. Lett.* 74 (1999) 3770.

25. J. Frame, D. Wheeler, T. DeTemple and J.G. Eden, Microdischarge devices fabricated in silicon, *Appl. Phys. Lett.* 71 (1997) 1165.

26. S.J. Park, J. Chen, C. Liu and J.G. Eden, Silicon microdischarge devices having inverted pyramidal cathodes: Fabrication and performance of arrays, *Appl. Phys. Lett.* 78 (2001) 419.

27. C.J. Wagner, S.J. Park and J.G. Eden, Excitation of a microdischarge with a reverse-biased pn junction, *Appl. Phys. Lett.* 78 (2001) 709.

28. M. Kushner, Modelling of microdischarge devices: Plasma and gas dynamics, *J. Phys. D App. Phys.* 38 (2005) 1633.

29. C. Qu, P. Tian, A. Semnani and M. Kushner, Properties of arrays of micro-plasmas: Application to control of electromagnetic waves, *Plasma Sources Sci. Technol.* 26 (2017) 105006.

30. K. Schoenbach and K. Becker, 20 years of microplasma research: A status report, *Eur. Phys. J. D* 70 (2016) 29.

31. V. Nemchinsky, Dross formation and heat transfer during plasma arc cutting, *J. Phys. D Appl. Phys.* 30 (1997) 2566.

32. C. Pardo, J. González-Aguilar, A. Rodríguez-Yunta and M. Calderón, Spectroscopic analysis of an air plasma cutting torch, *J. Phys. D Appl. Phys.* 32 (1999) 2181.

33. M. Moisan, G. Sauve, Z. Zakrzewski and J. Hubert, An atmospheric pressure waveguide-fed microwave plasma torch: The TIA design, *Plasma Sources Sci. Technol.* 3 (1994) 584.

34. Z. Duan and J. Heberlein, Arc instabilities in a plasma spray torch, *J. Therm. Spray Technol.* 11 (2002) 44–51.

35. M. Teschke, J. Kedzierski, E.G. Finantu-Dinu, D. Korzec and J. Engemann, High-speed photographs of a dielectric barrier atmospheric pressure plasma jets, *IEEE Trans. Plasma Sci.* 33 (2005) 310–311.

36. X. Lu and M. Laroussi, Dynamics of an atmospheric pressure plasma plume generated by submicrosecond voltage pulses, *J. Appl. Phys.* 100 (2006) 063302.

37. S. Reuter, K. Niemi, V. Schulz-von der Gathen and H.F. Dobele, Generation of atomic oxygen in the effluent of an atmospheric pressure plasma jet, *Plasma Sources Sci. Technol.* 18 (2009) 015006.

38. J. Walsh and M. Kong, Room-temperature atmospheric argon plasma jet sustained with submicrosecond high-voltage pulses, *Appl. Phys. Lett.* 91 (2007) 221502.

39. J. Kolb, A. Mohamed, R. Price, R. Swanson, A. Bowman, R. Chiavarini, M. Stacey and K. Schoenbach, Cold atmospheric pressure air plasma jet for medical applications, *Appl. Phys. Lett.* 92 (2008) 241501.

40. D. Graves, The emerging role of reactive oxygen and nitrogen species in redox biology and some implications for plasma applications to medicine and biology, *J. Phys. D Appl. Phys.* 45, (2012) 263001.

41. N. Barekzi and M. Laroussi, Dose-dependent killing of leukemia cells by low temperature plasma, *J. Phys. D Appl. Phys.* 45 (2012) 422002.

42. S. Reuter, J. Winter, S. Iseni, S. Peters, A. Schmidt-Bleker, M. Dunnbier, J. Schafer, R. Foest and K. Weltmann, Detection of ozone in a MHz argon plasma bullet jet, *Plasma Sources Sci. Technol.* 21 (2012) 034015.

43. Z. Xiong and M.J. Kushner, Surface corona-bar discharges for production of pre-ionizing UV light for pulsed high pressure plasmas, *J. Phys. D Appl. Phys.* 43 (2010) 505204.

44. Z. Xiong and M. Kushner, Atmospheric pressure ionization waves propagating through a flexible high aspect ratio capillary channel and impinging upon a target, *Plasma Sources Sci. Technol.* 21 (2012) 034001.

45. D. Breden, K. Miki and L. Raja, Computational study of cold atmospheric nanosecond pulsed He plasma jet in air, *Appl. Phys. Lett.* 99 (2011) 111501.

46. Z. Xiong, E. Robert, V. Sarron, J. Pouvesle and M.J. Kushner, Dynamics of ionization wave splitting and merging of atmospheric-pressure plasmas in branched dielectric tubes and channels, *J. Phys. D Appl. Phys.* 45 (2012) 275201.

47. G. Naidis, Simulation of streamers propagating along helium jets in ambient air: Polarity-induced effects, *Appl. Phys. Lett.* 98 (2011) 141501.

48. Y. Sakiyama, D. Graves, J. Jarrige and M. Laroussi, Finite elements analysis of ring-shaped emission profile in plasma bullets, *Appl. Phys. Lett.* 96 (2010) 041501.

49. Y. Xian, P. Zhang, X. Lu, X. Pei, S. Wu, Q. Xiong and K. Ostrikov, From short pulses to short breaks: Exotic plasma bullets via residual electron control, *Sci. Reports* 3 (2013) 1599.

50. X. Lu, Z. Jiang, Q. Xiong, Z. Tang, Z. Xiong, J. Hu, X. Hu and Y. Pan, Effect of E-field on the length of a plasma jet, *IEEE Trans. Plasma Sci.* 36 (2008) 988–989.

51. E. Karakas, M.A. Akman and M. Laroussi, The evolution of atmospheric pressure low temperature plasma jets: Jet current measurements, *Plasma Sources Sci. Technol.* 21 (2012) 034016.

52. Z. Xiong, E. Robert, V. Sarron, J.M. Pouvesle and M.J. Kushner, Atmospheric-pressure plasma transfer across dielectric channels and tubes, *J. Phys. D Appl. Phys.* 46 (2013) 155203.

53. S. Wu, Q. Huang, Z. Wang and X. Lu, The effect of nitrogen diffusion from surrounding air on plasma bullet behavior, *IEEE Trans. Plasma Sci.* 39 (2011) 2286–2287.

54. E. Robert, V. Sarron, D. Ries, S. Dozias, M. Vandamme and J.M. Pouvesle, Characterization of pulsed atmospheric-pressure plasma streams (PAPS) generated by a plasma gun, *Plasma Sources Sci. Technol.* 21 (2012) 034017.

55. X. Lu, M. Laroussi and V. Puech, On atmospheric-pressure non-equilibrium plasma jets and plasma bullets, *Plasma Sources Sci. Technol.* 21 (2012) 034005.

56. M. Laroussi, The biomedical applications of plasma: A brief history of the development of a new field of research, *IEEE Trans. Plasma Sci.* 36 (2008) 1612.

57. J. Chang, P. Lawless and T. Yamamoto, Corona discharge processes, *IEEE Trans. Plasma Sci.* 19 (1991) 1152–1166.

58. B. Sun, M. Sato and J. Clements, Optical study of active species produced by a pulsed streamer corona discharge in water, *J. Electrostatics* 39 (1997) 189–202.

59. A. Sharma, B. Locke, P. Arce and W. Finney, A preliminary study of pulsed streamer corona discharge for the degradation of phenol in aqueous solutions, *Hazard. Waste Hazard. Mater.* 10 (2009) 209–219.

60. K. Adamiak and P. Atten, Simulation of corona discharge in point–plane configuration, *J. Electrostatics* 61 (2004) 85–98.

61. T. Du Moncel, *Notice sur l'appareil d'Induction Electrique de Ruhmkorff et sur les Experiences que l'on peut faire Avec cet Instrument*, Hachette et Cie publishers, New York, 1855.

62. W. von Siemens, Poggendorfs, Ozone production in an atmospheric-pressure dielectric barrier discharge, *Ann. Phys. Chem.* 12 (1857) 66.

63. A. Von Engel, R. Seelinger and M. Steenbeck, Glow discharge at high pressure, *Z. Phys.* 85 (1933) 144.

64. S. Kanazawa, M. Kogoma, T. Moriwaki and S. Okazaki, Stable glow at atmospheric pressure, *J. Phys. D Appl. Phys.* 21 (1988) 838–840.

65. F. Massines, C. Mayoux, R. Messaoudi, A. Rabehi and P. Ségur, Experimental study of an atmospheric pressure glow discharge application to polymers surface treatment, in: *Proceedings of the GD-92*, Swansea, UK, Vol. 2, pp. 730–733 (1992).

66. J.R. Roth, M. Laroussi and C. Liu, Experimental generation of a steady-state glow discharge at atmospheric pressure, *Proc. IEEE Int. Conf. Plasma Sci.*, pp. 170–171 (1992).

67. R.P. Mildren and R.J. Carman, Enhanced performance of a dielectric barrier discharge lamp using short-pulsed excitation, *J. Phys. D Appl. Phys.* 34 (2001) 3378.

68. X. Duten, D. Packan, L. Yu, C.O. Laux and C.H. Kruger, DC and pulsed glow discharges in atmospheric pressure air and nitrogen, *IEEE Trans. Plasma Sci.* 30 (2002) 178.

69. M. Laroussi, X. Lu, V. Kolobov and R. Arslanbekov, Power consideration in the pulsed DBD at atmospheric pressure, *J. Appl. Phys.* 6 (2004) 3028–3030.

70. X. Lu and M. Laroussi, Temporal and spatial emission behavior of homogeneous dielectric barrier discharge driven by unipolar sub-microsecond square pulses, *J. Phys. D Appl. Phys.* 39 (2006) 1127–1131.

71. M. Laroussi, Sterilization of contaminated matter with an atmospheric pressure plasma, *IEEE Trans. Plasma Sci.* 24 (1996) 1188–1191.

72. M. Laroussi, Low temperature plasmas for medicine? *IEEE Trans. Plasma Sci.* 37 (2009) 714–725.

73. G. Fridman, G. Friedman, A. Gutsol, A.B. Shekhter, V.N. Vasilets and A. Fridman, Applied plasma medicine, *Plasma Process. Polym.* 5 (2008) 503–533.

74. K.-D. Weltmann, E. Kindel, T. von Woedtke, M. Hähnel, M. Stieber and R. Brandenburg, Atmospheric-pressure plasma sources: Prospective tools for plasma medicine, *Pure Appl. Chem.* 82 (2010) 1223–1237.

75. R. Bartnikas, Note on discharges in helium under AC conditions, *Brit. J. Appl. Phys. (J. Phys. D.) Ser. 2* 1 (1968) 659–661.

76. K.G. Donohoe, *The Development and Characterization of an Atmospheric Pressure Nonequilibrium Plasma Chemical Reactor*, PhD Thesis, California Institute of Technology, Pasadena, CA (1976).

77. T. Yokoyama, M. Kogoma, T. Moriwaki and S. Okazaki, The mechanism of the stabilized glow plasma at atmospheric pressure, *J. Phys. D Appl. Phys.* 23 (1990) 1125–1128.

78. F. Massines, A. Rabehi, P. Decomps, R.B. Gadri, P. Ségur and C. Mayoux, Experimental and theoretical study of a glow discharge at atmospheric pressure controled by a dielectric barrier, *J. Appl. Phys.* 8 (1998) 2950–2957.

79. S. Okazaki, M. Kogoma, M. Uehara and Y. Kimura, Appearance of a stable glow discharge in air, argon, oxygen and nitrogen at atmospheric pressure using a 50 Hz source, *J. Phys. D Appl. Phys.* 26 (1993) 889–892.

80. N. Gherardi, G. Gouda, E. Gat, A. Ricard and F. Massines, Transition from glow silent discharge to micro-discharges in nitrogen gas, *Plasma Sources Sci. Technol.* 9 (2000) 340–346.

81. N. Gheradi and F. Massines, Mechanisms controlling the transition from glow silent discharge to streamer discharge in nitrogen, *IEEE Trans. Plasma Sci.* 29 (2001) 536.

82. J.J. Shi, X.T. Deng, R. Hall, J.D. Punnett and M. Kong, Three modes in a radio frequency atmospheric pressure glow discharge, *J. Appl. Phys.* 94 (2003) 6303.

83. F. Massines, N. Gherardi, N. Naude and P. Segur, Glow and townsend dielectric barrier discharge in various atmosphere, *Plasma Phys. Contr. Fusion* 47 (2005) B557.

84. U. Kogelschatz, Silent discharges for the generation of ultraviolet and vacuum ultraviolet excimer radiation, *Pure Appl. Chem.* 62 (1990) 1667–1674.

85. U. Kogelschatz, B. Eliasson and W. Egli, Dielectric-barrier discharges: Principle and applications, *J. Phys. IV*, 7, C4 (1997) 47–66.

86. M. Laroussi, I. Alexeff, J.P. Richardson, and F.F. Dyer The resistive barrier discharge, *IEEE Trans. Plasma Sci.*, 30, 158–159, (2002).

87. X. Wang, C. Li, M. Lu and Y. Pu, Study on atmospheric pressure glow discharge, *Plasma Sources Sci. Technol.* 12 (2003) 358–361.

88. M. Laroussi, J.P. Richardson and F.C. Dobbs, Effects of non-equilibrium atmospheric pressure plasmas on the heterotrophic pathways of bacteria and on their cell morphology, *Appl. Phys. Lett.* 81 (2002) 772–774.

89. X. Lu, Y. Cao, P. Yang, Q. Xiong, Z. Xiong, Y. Xian and Y. Pan, An RC plasma device for sterilization of root canal of teeth, *IEEE Trans. Plasma Sci.* 37 (2009) 668.

90. X. Lu, Z. Xiong, F. Zhao, Y. Xian, Q. Xiong, W. Gong, C. Zou, Z. Jiang and Y. Pan, A simple atmospheric pressure room-temperature air plasma needle device for biomedical applications, *Appl. Phys. Lett.* 95 (2009) 181501.

91. S. Babayan, J. Jeong, V. Tu, J. Park, G. Selwyn and R. Hicks, Etching materials with an atmospheric-pressure plasma jet, *Plasma Sources Sci. Technol.* 7 (1998) 286.

92. Q. Li, J. Li, W. Zhu, X. Zhu, and Y. Pu, Effects of gas flow rate on the length of atmospheric pressure nonequilibrium plasma jets, *Appl. Phys. Lett.*, 95, 141502 (2009).

93. X. Lu, Z. Jiang, Q. Xiong, Z. Tang, X. Hu and Y. Pan, An 11cm long atmospheric pressure cold plasma plume for applications of plasma medicine, *Appl. Phys. Lett.* 92 (2008) 081502.

94. X. Lu, Z. Jiang, Q. Xiong, Z. Tang and Y. Pan, A single electrode room-temperature plasma jet device for biomedical applications, *Appl. Phys. Lett.* 92 (2008) 151504.

95. V. Leveille and S. Coulombe, Design and preliminary characterization of a miniature pulsed RF APGD torch with downstream injection of the source of reactive species, *Plasma Sources Sci. Technol.* 14 (2005) 467.

96. A. Shashurin, M. Shneider, A. Dogariu, R. Miles and M. Keidar, Temporal behavior of cold atmospheric plasma jet, *Appl. Phys. Lett.*, 94 (2009) 231504.

97. E. Stoffels, I. Kieft and R. Sladek, Superficial treatment of mammalian cells using plasma needle, *J. Phys. D Appl. Phys.* 36 (2003) 2908.

98. Y.C. Hong and H.S. Uhm, Microplasma jet at atmospheric pressure, *Appl. Phys. Lett.* 89 (2006) 221504.

99. T.L. Ni, F. Ding, X.D. Zhu, X.H. Wen and H.Y. Zhou, Cold microplasma plume produced by a compact and flexible generator at atmospheric pressure, *Appl. Phys. Lett.* 92 (2008) 241503.

100. Y.C. Hong, H.S. Uhm and W.J. Yi, Atmospheric pressure nitrogen plasma jet: Observation of striated multilayer discharge patterns, *Appl. Phys. Lett.* 93 (2008) 051504.

101. Y.C. Hong and H.S. Uhm, Air plasma jet with hollow electrodes at atmospheric pressure, *Phys. Plasmas* 14 (2007) 053503.

102. A.H. Mohamed, J.F. Kolb and K.H. Schoenbach, Method and device for creating a micro plasma jet, US Patent 7,572,998 B2 (2009).

103. Y. Hong, W. Kang, Y. Hong, W. Yi and H. Uhm, Atmospheric pressure air-plasma jet evolved from microdischarges: Eradication of E. coli with the jet, *Phys. Plasmas* 16 (2009) 123502.

104. G. Fridman, Blood coagulation and living tissue sterilization by floating-electrode dielectric barrier discharge in air, *Plasma Chem. Plasma Process.* 26 (2006) 425.

105. S. Wu, X. Lu, Z. Xiong and Y. Pan, A touchable pulsed air plasma plume driven by DC power supply, *IEEE Trans. Plasma Sci.* 38 (2010) 3404.

106. Z. Machala, C. Laux and C. Kruger, Transverse dc glow discharges in atmospheric pressure air, *IEEE Trans. Plasma Sci.* 33 (2006) 320.

107. G.A. Dawson and W.P. Winn, A model for streamer propagation, *Z. Phys.* 183 (1965) 159.

108. R. Leiweke and B. Ganguly, Effects of pulsed-excitation applied voltage rising time on argon metastable production efficiency in a high pressure dielectric barrier discharge, *Appl. Phys. Lett.* 90 (2007) 241501.

109. C. Jiang, M.T. Chen and M.A. Gundersen, Polarity-induced asymmetric effects of nanosecond pulsed plasma jets, *J. Phys. D Appl. Phys.* 42 (2009) 232002.

110. Z. Xiong, X. Lu, Y. Xian, Z. Jiang and Y. Pan, On the velocity variation in atmospheric pressure plasma plumes driven by positive and negative pulses, *J. Appl. Phys.* 108 (2010) 103303.

111. J. Oh, J. Walsh and J. Bradley, Plasma bullet current measurements in a free-stream helium capillary jet, *Plasma Sources Sci. Technol.* 21 (2012) 034020.

112. G.B. Sretenovic, I.B. Krstic, V.V. Kovacevic, B.M. Obradovic and M.M. Kuraica, Spectroscopic study of low-frequency helium DBD plasma jet, *IEEE Trans. Plasma Sci.* 36 (2012) 2870–2878.

113. K.N. Ostrikov, M.Y. Yu and N.A. Azarenkov, Nonlinear effects of ionization on surface waves on a plasma–metal interface, *J. Appl. Phys.* 84 (1998) 4176.

114. W.C. Zhu, Q. Li, X.M. Zhu and Y.K. Pu, Characteristics of atmospheric pressure plasma jets emerging into ambient air and helium, *J. Phys. D Appl. Phys.* 42 (2009) 202002.

115. Q. Li, X.M. Zhu, J.T. Li and Y.K Pu, Role of metastable atoms in the propagation of atmospheric pressure dielectric barrier discharge jets, *J. Appl. Phys.* 107 (2010) 043304.

116. H.S. Park, S.J. Kim, H.M. Joh, T.H. Chung, S.H. Bae and S.H. Leem, Optical and electrical characterization of an atmospheric pressure microplasma jet with a capillary electrode, *Phys. Plasmas* 17 (2010) 033502.

117. S. Wu, Z. Wang, Q. Huang, X. Tan, X. Lu and K. Ostrikov, Atmospheric-pressure plasma jets: Effect of gas flow, active species, and snake-like bullet propagation, *Phys. Plasmas* 20 (2013) 023503.

118. J.J. Liu and M.G. Kong, Sub-60°C atmospheric helium–water plasma jets: Modes, electron heating and downstream reaction chemistry, *J. Phys. D Appl. Phys.* 44 (2011) 345203.

119. R.J. Leiweke, B.L. Sands and B.N. Ganguly, Effect of gas mixture on plasma jet discharge morphology, *IEEE Trans. Plasma Sci.* 39 (2011) 2304–2305.

120. S. Hofmann, A. Sobota and P. Bruggeman, Transitions between and control of guided and branching streamers in DC nanosecond pulsed excited plasma jets, *IEEE Trans. Plasma Sci.* 40 (2012) 2888–2899.

121. B.L. Sands, S.K. Huang, J.W. Speltz, M.A. Niekamp and B.N. Ganguly, Role of penning ionization in the enhancement of streamer channel conductivity and Ar(1s5) production in a He–Ar plasma jet, *J. Appl. Phys.* 113 (2013) 153303.

122. G.V. Naidis, Simulation of streamer dynamics in atmospheric pressure plasma jets, in: *19th International Symposium on Plasma Chemistry—ISPC*, Bochum, Germany, July 26–31 (2009), Paper 22.

123. Y. Xian, X. Lu, J. Liu, S. Wu, D. Liu, and Y. Pan, Multiple plasma bullet behavior of an atmospheric-pressure plasma plume driven by a pulsed DC voltage, *Plasma Sources Sci. Technol.* 21 (2012) 034013.

124. M.A. Akman and M. Laroussi, Insights into sustaining a plasma jet: Boundary layer requirement, *IEEE Trans. Plasma Sci.* 41 (2013) 839.

125. M. Laroussi and M.A. Akman, Ignition of a large volume plasma with a plasma jet, *AIP Adv.* 1 (2011) 032138.

126. C. Douat, G. Bauville, M. Fleury, M. Laroussi and V. Puech, Dynamics of colliding microplasma jets, *Plasma Sources Sci. Technol.* 21 (2012) 034010.

127. A. Bourdon, Z. Bonaventura and S. Celestin, Influence of the pre-ionization background and simulation of the optical emission of a streamer discharge in preheated air at atmospheric pressure between two point electrodes, *Plasma Sources Sci. Technol.* 19 (2010) 034012.

128. X. Lu, Q. Xiong, Z. Xiong, J. Hu, F. Zhou, W. Gong, Y. Xian et al., Propagation of an atmospheric pressure plasma plume, *J. Appl. Phys.* 105 (2009) 043304.

129. X. Yan, Z. Xiong, F. Zou, S. Zhao, X. Lu, G.H. Yang, G. He and K. Ostrikov, Plasma-induced death of HepG2 cancer cells: Intracellular effects of reactive species, *Plasma Process. Polym.* 9 (2012) 59.

130. M. Ishaq, M. Evans and K. Ostrikov, Effect of atmospheric gas plasmas on cancer cell signalling, *Int. J. Cancer* 134 (2014) 1517.

131. Z. Xiong, S. Zhao, X. Mao, X. Lu, G. He, G. Yang, M. Chen, M. Ishaq and K. Ostrikov, Selective neuronal differentiation of neural stem cells induced by nanosecond microplasma agitation, *Stem Cell Res.* 12 (2014) 387.

132. M. Keidar, A. Shashurin, O. Volotskova, M.A. Stepp, P. Srinivasan, A. Sandler and B. Trink, Cold atmospheric plasma in cancer therapy, *Phys. Plasmas* 20 (2013) 057101.

133. A. Mai-Prochnow, A.B. Murphy, K.M. McLean, M.G. Kong and K. Ostrikov, Atmospheric-pressure plasmas: Infection control and bacterial responses, *Int. J. Antimicrob. Agents* 43 (2014) 508.

134. W. Yan, Z.J. Han, W.Z. Liu, X.P. Lu, B.T. Phung and K. Ostrikov, Designing atmospheric-pressure plasma sources for surface engineering of nanomaterials, *Plasma Chem. Plasma Proc.* 33 (2013) 479.

135. N. Barekzi and M. Laroussi, Effects of low temperature plasmas on cancer cells, *Plasma Process. Polym.* 10 (2013) 1039.

136. A. Vogelsang, A. Ohl, R. Foest and K.-D. Weltmann, Fluorocarbon plasma polymer deposition by an atmospheric pressure microplasma jet using different precursor molecules: A comparative study, *Plasma Process. Polym.* 10 (2013) 364.

137. S.E. Marshall, A.T.A. Jenkins, S.A. Al-Bataineh, R.D. Short, S.H. Hong, N.T. Thet, T. Naing, J.S. Oh, J.W. Bradley and E.J. Szili, Studying the cytolytic activity of gas plasma with self-signalling phospholipid vesicles dispersed within a gelatin matrix, *J. Phys. D Appl. Phys.* 46 (2013) 185401.

138. N. Jiang, J.L. Yang, F. He and Z. Cao, Interplay of discharge and gas flow in atmospheric pressure plasma jets, *J. Appl. Phys.* 109 (2011) 093305.

139. J.S. Oh, O.T. Olabanji, C. Hale, R. Mariani, K. Kontis and J.W. Bradley, Imaging gas and plasma interactions in the surface-chemical modification of polymers using micro-plasma jets, *J. Phys. D Appl. Phys.* 44 (2011) 155206.

140. E. Robert, V. Sarron, T. Darny, D. Ries, S. Dozias, J. Fontane, L. Joly and J.-M. Pouvesle, Rare gas flow structuration in plasma jet experiments, *Plasma Sources Sci. Technol.* 23 (2014) 012003.

141. Z. Cao, Q. Nie, D.L. Bayliss, J.L. Walsh, C.S. Ren, D.Z. Wang and M.G. Kong, Spatially extended atmospheric plasma arrays, *Plasma Sources Sci. Technol.* 19 (2010) 025003.

142. M. Ghasemi, P. Olszewski, J.W. Bradley and J.L. Walsh, Interaction of multiple plasma plumes in an atmospheric pressure plasma jet array, *J. Phys. D Appl. Phys.* 46 (2013) 052001.

143. N.Y. Babaeva and M.J. Kushner, Interaction of multiple atmospheric-pressure micro-plasma jets in small arrays: He/O_2 into humid air, *Plasma Sources Sci. Technol.* 23 (2014) 015007.

PART II

Physics of Nonequilibrium Atmospheric Pressure Plasma Jets (N-APPJs)

Basic Physical Phenomena and Theoretical Models of N-APPJs

3.1 Overview of Theoretical Descriptions of N-APPJs

Recent experimental studies have increased our understanding of the discharge process within N-APPJs [1,2]. The main characteristic of N-APPJs is that the cold plasma jets produced by nanosecond pulsed DC discharge in a region of working gas are actually propagating plasma bullets that are guided along the axis of the noble gas channel as it flows into stagnant ambient air. The image of the plasma jet with nanosecond exposure time reveals that the plasma bullet is a luminous zone confined to the head of the streamer that travels at the speed of ~10^5 m/s to several centimeters downward of the jet nozzle. The cross-section image of the plasma jet captures the light-emitting zone of the plasma bullet and shows that it is ring-shaped in the radial direction. The laser-induced fluorescence also indicates that most of the excited species of plasma jets are produced in this ring-shaped region. The optical emission spectroscopy and power measurement suggest that there are three distinct operating modes: the chaotic mode, the bullet mode, and the continuous mode, depending on the input power [3]. The main plasma physics and chemistry related to each phenomenon are still unclear and need studying.

Several numerical models have been developed through simulation groups' efforts to understand the main plasma physics and chemistry of plasma jets. Sakiyama et al. developed a one-way, coupled model of neutral gas flow and plasma dynamics to explain the ring-shaped emission pattern that has been observed experimentally in plasma bullets. The local field approximation in one-dimensional, cylindrical coordinates that correspond to a cross-section of a plasma bullet was solved by a fluid model. Time and spatially resolved spectroscopic measurements support the simulation results [4]. Naidis developed a two-dimensional numerical model to study the plasma bullet propagation dynamics in atmospheric pressure helium jets injected into ambient air. There are two types of plasma jets: one with the maxima of electric field and electron density at the jet axis, and another with the maxima of these parameters near the boundary between the jet and the surrounding air. Plasma jets depend on the jet width and the initial radial distribution of electron density [5]. Bouef et al. developed a two-dimensional fluid model to explain the experimental features of plasma jets. They found that a plasma jet is similar to a cathode streamer (ionization wave) guided by a helium jet, and the properties of the helium streamer and the plasma channel behind the streamer head are a function of parameters such as electrode geometry, voltage pulse waveform, and preionization density [6]. Other work by Naidis shows that the radial position where the concentration of electrons and metastable nitrogen molecules reaches a maximum coincides with the position where the air molar fraction is about 1% in the mixing layer. He also found that the ring structure is attributed primarily to electron-impact ionization processes and not Penning reactions [7]. Breden et al. developed a self-consistent, multi-species, multi-temperature plasma model with detailed finite-rate chemistry and photoionization effects. They found that the fluid mechanical mixing layer between the helium jet core and the ambient air is instrumental in guiding the propagation direction of the streamer, and it gives the plasma jet a visibly collimated appearance. The key chemical reaction which drives the streamer propagation is the electron-impact ionization of helium neutral and nitrogen molecules. Photoionization plays a role in enhancing the propagation speed of the streamer, but it is not necessary to sustain the streamer [8].

Because of the increasingly broad applications of APPJs for plasma medicine, from deactivation of bacteria to cancer treatment, recent computational investigations have been conducted into the behavior of APPJs interacting with surfaces. The nonPDPSIM is a two-dimensional simulator which uses Poisson's equation for electric potential and transport equations for charged and neutral species to analyze the interaction between plasma filaments in dielectric barrier streamer discharges and liquid-covered wounds. The sizes of the wounds are of the order of the plasma filaments, and the liquid within the wound, an approximation of blood serum, contains idealized blood platelets. The electrical properties of a wound, such as the deformation of the vacuum electric fields due to the shape, the effective capacitance of the wound, and the discontinuities in electrical permittivity, have significant effects on the spreading of the plasma on the wound surface. This, in turn, affects the penetration of the electric field to cells and tissues immersed in liquid [9]. Further numerical study of the interaction between pulsed DC driving plasma jets and the water layer covering tissue indicates that when the plasma bullet touches the water surface, not only were more dissolved OH_{aq}, H_2O_{2aq}, and O_{3aq} found, but also more charged species were produced in the water layer, with $H_3O^+_{aq}$, $O_3^-{}_{aq}$, and $O_2^-{}_{aq}$ being the dominant terminal species. Simultaneously, N_xO_y species do not accumulate in the volume; therefore, aqueous nitrites, nitrates, peroxynitrite, and HNO_{3aq} have low densities [10]. Another 2D plasma jet model was used to analyze the interaction of negative streamers produced by air dielectric barrier discharge with bacteria biofilm on an apple surface. The ionization near the biofilm facilitates the propagation of negative streamers when the streamer head is 1 mm from the biofilm. The structure of the biofilm results in the nonuniform distribution of ROS and RNS captured by flux and time fluence of these reactive species. The mean-free path of charged species in μm scale allows the plasma to penetrate into the cavity of the biofilm; therefore, although the density of ROS and RNS decreases by 6 ~ 7 orders of magnitude, the diffusion results in the uniform distribution of ROS and RNS inside the cavity during the pulse off period [11].

3.2 Approaches to Modelling the Dynamics and Structure of Guided Streamers

The computational model used to simulate the N-APPJ usually comprises a coupled set of governing equations for charged and neutral species continuity, electron energy transport, and self-consistent electrostatic potential. In addition, a set of reactions between each species of plasma and a model to account for the photoionization process in air are included.

3.2.1 Plasma Governing Equations

Species Continuity

The continuity equations of each species are solved to obtain the number densities of the species,

$$\frac{\partial n_k}{\partial t} + \vec{\nabla} \cdot \vec{\Gamma}_k = \mathring{R}_k, \, k = 1, 2, ..., k_g \, (k \neq k_b) \tag{3.1}$$

for all but one dominant neutral species (assumed to be nitrogen for this case) [8]. The individual species chemistry source terms in $\mathring{R}_k$ are computed using a mass-action kinetics formulation with reaction pathways and rate coefficients that are discussed later.

Although the background species vary spatially (helium, nitrogen, and oxygen) for helium N-APPJ, the relative changes in all three over the duration of the pulse are negligible. The initial concentrations of these three background species can be calculated by Navier-Strokes, continuity and convection-diffusion equations as follows:

$$-\nabla \cdot [\eta(\nabla u + (\nabla u)^T) + \rho(u \cdot \nabla u) + \nabla p = F \tag{3.2}$$

$$\nabla \cdot u = 0 \tag{3.3}$$

$$DV^2 c - u \cdot \nabla c = 0, \tag{3.4}$$

where u is the velocity field, ρ is the density (Helium 0.1664 kg/m³, air 1.293 kg/m³), p is the pressure, η is the dynamic viscosity (Helium 1.94×10^{-5} Pa.s and air 1.82×10^{-5} Pa.s), D is the diffusion coefficient (Helium in air 7.2×10^{-5} m²/s), c is the helium mole fraction mol/m³, and F is the body force field (buoyancy). The resulting plasma is typically weakly ionized (on the order of 1 ppm), and the number densities of the three bulk species remain unchanged over the pulse duration.

Drift–Diffusion

The separate momentum equation for the heavy species and the drift–diffusion approximation is solved to obtain the species number flux term Γ_k. The drift–diffusion approximation is valid if the plasma can be considered highly collisional. The mean-free path of the ions and neutral species (μm) is much smaller than the characteristic length scales of the discharge (millimeters) at atmospheric pressure and room temperature. Therefore, the drift–diffusion approximation is usually used for plasma fluid models at atmospheric pressure [12].

$$\vec{\Gamma_k} = n_k \vec{u_k} = -\mu_k n_k \vec{\nabla}\phi - D_s \vec{\nabla} n_k \tag{3.5}$$

The terms in the drift–diffusion equation are the species mobility, μ_k, the species diffusion coefficient, D_k, and the electric field represented by the negative gradient of the electrostatic potential, $-\vec{\nabla}\phi$.

Electrostatic Potential

Poisson's equation is used to calculate the self-consistent electrostatic field ϕ, and it is computed by,

$$E = -\nabla\phi, \quad \nabla^2\phi = -4\pi\rho, \tag{3.6}$$

where E is the electric field, $\rho = e(n_p - n_e)$ is the volume charge density, n_p and n_e are the densities of positive ions and electrons. Poisson's equation can be solved in a semi-implicit manner using an explicit predictor step on the electron number densities to predict their values at the next time step. This typically allows for a much larger time step to be taken.

Electron Energy Transport

The electron temperature is computed using an electron energy conservation equation given by

$$\frac{\partial e_e}{\partial t} + \vec{\nabla} \cdot ((e_e + e_p)\vec{u_e} - k_e\vec{\nabla}T_e) = +e\vec{\Gamma_e} \cdot \nabla\phi - e\Sigma_i \Delta E_i^e r_i - \frac{3}{2}k_b n_e \frac{2m_e}{m_{kb}}(T_e - T_g)\bar{v}_{e,kb}. \tag{3.7}$$

Here, the total electron energy is assumed to be approximately equal to the mean electron energy $e_e \approx (3/2)n_e k_B T_e$ while the electron pressure p_e is obtained using the ideal gas law. The right side of the electron energy equation incorporates three source terms: Joule heating, inelastic collisional heating, and elastic collisional heating, respectively. The electron unit charge is e, ΔE_i^e is the energy lost by an electron (in units of eV) in a single collision event involving a reaction i with a rate of progress given by r_i, and $\bar{v}_{e,kb}$ is the species momentum transfer collision frequency.

3.2.2 Plasma Chemistry

Understanding the production mechanism of reactive oxygen and nitrogen species in the N-APPJ is still very much in its early stages, but there have been several decades of research in plasma chemistry exploring how hot electrons interact with neutral components or interactions between neutral components to produce these reactive species. **Table 3.1** shows the plasma chemistry considered in a 2D N-APPJ model consisting of 17 species and 45 reactions [13]. The submechanisms of the helium and helium-air chemistry reactions are taken from Refs. [4,5] and the submechanism of the air chemistry is taken from Ref. [14]. For the reactions which produced N_2 and O_2 in different rotational and vibrational states, their reaction rates are calculated by the cross-section data from Bolsig+ [15]. The more complete reactions set between helium and humid air with over 1000 reactions can be found in Ref. [16].

3.2.3 Photoionization Model

The seed electrons produced by photoionization in front of the plasma bullet play quite an important role in the propagation of plasma jets [8]. The approach for calculating the photoionization of the N-APPJ is the air photoionization source term usually based on the three-term exponential Helmholtz model described by Bourdon et al. [17]. The three-term exponential Helmholtz model is an approximation of a widely used integral model developed in the 1980s by Zheleznyak et al. [18]. The model of Zheleznyak et al. for the photoionization of air serves as the starting point, where the rate of photoionization of oxygen molecules ($O_2 + h\nu \rightarrow e + O_2^+$) at a point $\vec{r}$ due to the emission of ionizing radiation from nitrogen at all surrounding source points $\vec{r}'$ is given by the integral equation,

$$S_{ph}(\vec{r}) = \iiint_{V'} \frac{I(\vec{r}')g(R)}{4\pi R^2} dV'. \tag{3.8}$$

Here, R is the distance between $\vec{r}'$ and $\vec{r}$, $I(\vec{r}')$ is an emission function that characterizes the intensity of radiation from emitting species ($e + N_2 \rightarrow e + N_2^+ + h\nu$ in this case). $h\nu$ represents the ionizing photons emitted by nitrogen whose wavelengths are in the range of 98–102.5 nm (~12.1–12.65 eV). $I(\vec{r}')$ is assumed to be proportional to the rate of ionization $S_i(\vec{r}')$ of the emitting species (N_2) through collisional processes (e.g., electron-impact ionization) multiplied by an efficiency factor ξ(~0.02) and a quenching factor $P_q/(P + P_q)$, where P_q is found from experiments (~30–60 Torr) and P is the gas mixture pressure.

Table 3.1 Helium–Air Chemistry Reactions Used in This Study

Rxn	Reaction	Helium–Air Chemistry (molecules-meters-Kelvin)				
		A	B	C	Activation Energy	References
Helium Chemistry						
R1	$e + He \rightarrow e + He$	BOLSIG+			0	
R2	$e + He \rightarrow e + He^*$	BOLSIG+			19.8	
R3	$e + He \rightarrow 2e + He^+$	BOLSIG+			24.6	
R4	$e + He^* \rightarrow 2e + He^+$	4.661e-16	0.6	4.78	4.78	[8]
R5	$e + He_2^+ \rightarrow 2e + He_2^+$	1.268e-18	0.71	3.4	3.4	[8]
R6	$2He^* \rightarrow e + He + He^+$	4.5e-16	0	0	−15.0	[8]
R7	$e + He_2^+ \rightarrow He^* + He$	5.386e-13	−0.5	0	0	[8]
R8	$He^* + 2He \rightarrow He + He_2^*$	1.3e-45	0	0	0	[8]
R9	$He^+ + 2He \rightarrow He + He_2^+$	1.0e-43	0	0	0	[8]
R10	$e + He^+ \rightarrow He^*$	6.76e-19	−0.5	0	0	[8]

(Continued)

Table 3.1 (*Continued*) Helium–Air Chemistry Reactions Used in This Study

Rxn	Reaction	A	B	C	Activation Energy	References
	Helium–Air Chemistry (molecules-meters-Kelvin)					
R11	$2e + He^+ \rightarrow e + He^*$	6.186e–39	–4.4	0	0	[8]
R12	$e + He^+ + He \rightarrow He^* + He$	6.66e–42	–2	0	0	[8]
R13	$2e + He_2^+ \rightarrow He_2^* + e$	2.8e–32	0	0	0	[14]
R14	$e + He_2^+ + He \rightarrow He^* + 2He$	3.5e–39	0	0	0	[14]
R15	$e + He_2^+ + He \rightarrow He_2^* + He$	1.5e–39	0	0	0	[14]
R16	$2e + He_2^+ \rightarrow He^* + He + e$	2.8e–32	0	0	0	[14]
Air Chemistry						
R17	$e + N_2 \rightarrow e + N_2$	BOLSIG+			0	
R18	$e + N_2 \rightarrow e + N_2 (VIB\ v1)$	BOLSIG+			0.2889	
R19	$e + N_2 \rightarrow e + N_2 (VIB\ 3v1)$	BOLSIG+			0.8559	

(*Continued*)

Table 3.1 (Continued) Helium–Air Chemistry Reactions Used in This Study

Rxn	Reaction	Helium–Air Chemistry (molecules-meters-Kelvin)			Activation Energy	References
		A	B	C		
R20	$e + N_2 \rightarrow e + N_2\,(\text{VIB } 4v1)$	BOLSIG+			1.134	
R21	$e + N_2 \rightarrow e + N_2\,(\text{VIB } 5v1)$	BOLSIG+			1.409	
R22	$e + N_2 \rightarrow e + N_2\,(A)$	BOLSIG+			6.17	
R23	$e + N_2 \rightarrow 2e + N_2^+$	BOLSIG+			15.6	
R24	$e + O_2 \rightarrow e + O_2\,(\text{VIB } 3v1)$	BOLSIG+			0.57	
R25	$e + O_2 \rightarrow e + O_2\,(\text{VIB } 4v1)$	BOLSIG+			0.772	
R26	$e + O_2 \rightarrow e + O_2\,(A1)$	BOLSIG+			0.977	
R27	$e + O_2\,(A1) \rightarrow e + O_2$	BOLSIG+			−0.977	
R28	$e + O_2 \rightarrow e + O_2\,(B1)$	BOLSIG+			1.627	

(Continued)

Table 3.1 (*Continued*) Helium–Air Chemistry Reactions Used in This Study

Rxn	Reaction	Helium–Air Chemistry (molecules-meters-Kelvin)				
		A	B	C	Activation Energy	References
R29	$e + O_2(B1) \rightarrow e + O_2$	BOLSIG+			−1.627	
R30	$e + O_2 \rightarrow e + O_2(EXC)$	BOLSIG+			4.5	
R31	$e + O_2(EXC) \rightarrow e + O_2$	BOLSIG+			−4.5	
R32	$e + O_2 \rightarrow O + O^-$	BOLSIG+			3.6	
R33	$e + O_2 \rightarrow e + 2O$	BOLSIG+			5.58	
R34	$e + O_2 \rightarrow e + O + O(1D)$	BOLSIG+			8.4	
R35	$e + O_2 \rightarrow 2e + O_2^+$	BOLSIG+			12.06	
R36	$2e + N_2^+ \rightarrow e + N_2$	3.165e–30	−0.8	0	0	[14]

(Continued)

Table 3.1 (*Continued*) Helium–Air Chemistry Reactions Used in This Study

Rxn	Reaction	Helium–Air Chemistry (molecules-meters-Kelvin)				
		A	B	C	Activation Energy	References
R37	$e + N_2^+ + N_2 \rightarrow 2N_2$	4.184e-44	-2.5	0	0	[14]
R38	$O^- + O_2^+ \rightarrow O + O_2$	3.488e-14	-0.5	0	0	[8]
R39	$e + 2O_2 \rightarrow O_2^- + O_2$	5.17e-43	-1	0	0	[8]
R40	$O_2^- + O_2^+ \rightarrow 2O_2$	2e-13	0	0	0	[8]
R41	$O_2^- + O_2^+ + M \rightarrow 2O_2 + M$	2e-37	0	0	0	[8]
Helium–Air Interactions						
R42	$He^* + N_2 \rightarrow e + N_2^+ + He$	7.0e-17	0	0	0	[8]
R43	$He_2^* + N_2 \rightarrow e + N_2^+ + 2He$	7.0e-17	0	0	0	[8]
R44	$He_2^+ + N_2 \rightarrow N_2^+ + 2He$	5.0e-16	0	0	0	[8]
R45	$He_2^+ + O_2^- \rightarrow 2He + O_2$	1.0e-13	0	0	0	[8]

Source: Liu, X.Y. et al., *Plasma Sources Sci. Technol.*, 23, 035007, 2014.

$$I(\vec{r}') = \frac{P_q}{P + P_q} \xi S_i(\vec{r}')$$ (3.9)

The photoionization of the absorbing species O_2 also depends on a geometric absorption function $g(R)$, which, in turn, depends on the distance from the photon source and the partial pressure of absorbing species (P_{O2}). The function $g(R)$ can be fit from experimental data in the form

$$\frac{g(R)}{P_{O2}} = \frac{\exp^{-(\chi_{min}P_{O2}R)} - \exp^{-(\chi_{max}P_{O2}R)}}{P_{O2}R\ln(\chi_{max}/\chi_{min})}$$ (3.10)

where χ_{min} and χ_{max} are parameters in the fit. For a finite volume fluid model, this approach requires performing a quadrature for every volume element in order to calculate the photoionization source for each cell. For a domain consisting of n cells, this requires n^2 computations every time the photoionization source terms must be calculated, making this approach computationally prohibitive.

To circumvent the high computational cost of the above integral model, several approximate models have been developed in recent years. A three-term Helmholtz model developed by Bourdon et al. [18] along with relevant fit data and parameters for air was chosen for the low computational cost. It involves splitting the ionization source term into three terms, as below:

$$S_{ph}(\vec{r}) = \sum_{j=1}^{3} S_{ph}^{j}(\vec{r}).$$ (3.11)

Within each term, the absorption function $g(R)/P_{O2}$ is replaced with an exponential function which amounts to approximating the absorption function using a three-term exponential curve fit. The advantage of this method is that the individual photoionization source terms now have the form

$$S_{ph}^{j}(\vec{r}) = \iiint_{v'} \frac{I(\vec{r}')}{4\pi R^2} A_j P_{O2}^2 e^{-\lambda_j P_{O2} R} dv',$$ (3.12)

where the term S_{ph}^{j} is the solution to a Helmholtz equation. The original problem is modified from integrating the Zheleznyak integral n times for n cells to solving three Helmholtz equations over the entire domain; the computational cost is significantly lower than solving the integral equation numerically. The general form of the Helmholtz equation is

$$\nabla^2 S_{ph}^{j}(\vec{r}) - (\lambda_j P_{O2})^2 S_{ph}^{j}(\vec{r}) = -A_j P_{O2}^2 I(\vec{r}),$$ (3.13)

where λ_j and A_j are parameters which are obtained by fitting the three exponential terms in the S_{ph} terms to experimental data obtained for the original

Table 3.2 Air Photoionization Parameters		
	A_j (cm^{-1} Torr^{-1})	λ_j (cm^{-1} Torr^{-1})
S_{ph}^1	0.0067	0.0447
S_{ph}^2	0.0346	0.1121
S_{ph}^3	0.3059	0.5994

Source: Bourdon, A. et al., *Plasma Sources Sci. Technol.*, 16, 656, 2007.

absorption function of air divided by the partial pressure of oxygen $g(R)/P_{O2}$. The values λ_j and A_j, by fitting the three terms, are calculated for air by Bourdon et al. [18] and are tabulated in **Table 3.2**.

For the boundary conditions, a symmetry boundary condition is imposed rather than calculating the exact photoionization term. This results in faster computation times but a less accurate calculation of photoionization near the tube.

Recent Advances in the Physics of N-APPJs

4.1 Study of Basic Phenomena

4.1.1 Conductivity of the Dark Channel

How the energy is coupled to the fast propagating plasma plume is the key issue for N-APPJ. The nanosecond exposure time ICCD images indicate that there is a dark plasma channel between the power electrode and the head of the fast propagating plasma plume. It is not known whether the dark channel has enough conductivity to support the propagation of the plasma bullet (fast-moving head of the plasma plume). Therefore, studies on the optical emission, electron density, electric field, and metastable densities of the dark channel have been carried out.

In order to see whether the dark channel between the jet nozzle and the plasma bullet emits light or not, Xiong et al. took ICCD images of the plasma bullet propagating along the axis of the jet nozzle. The exposure time of the ICCD images is 5 ns. The overexposure problem is avoided by using such a low exposure time. The plasma bullet propagates almost 5 cm over 0.44 μs [19]. **Figure 4.1** suggests that the plasma bullet is connected to the nozzle via a relatively weak optical emission channel. The dark channel is not precisely "dark," which only suggests that the optical emission intensity of the channel is much weaker than that of the bullet. In addition, not only is the optical emission of the dark channel behind the plasma bullet decreased, but also the plasma bullet shrinks during the propagation of the plasma jet.

If the dark channel is not conductive, the increasing applied voltage should not affect the plasma plume when the plasma bullet propagates away from the electrode and the dark channel formed. To verify this, a unipolar square high-voltage pulse in the order of 4.0–7.5 kV is applied to the electrodes with a pulse width of 200 ns and a repetition rate of up to 10 kHz to generate the plasma jet. **Figure 4.2** suggests that for a pulse width of 300 ns and an applied

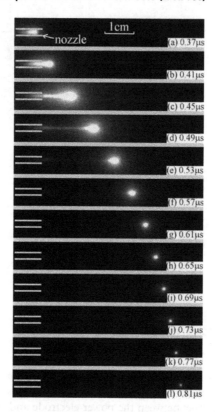

FIGURE 4.1 High-speed photographs of the He plasma plume in open air with exposure time fixed at 5 ns. Photographs (a)–(e) are taken at 40 ns increments. (From Xiong, Q. et al., *J. Appl. Phys.*, 106, 083302, 2009.)

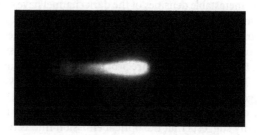

FIGURE 4.2 The plasma bullet propagation showing the channel consisting of emitting species along the ionization channel (applied voltage: 5.5 kV; pulse width: 300 ns; repetition rate: 5 kHz; flow rate: 5.0 L/min). (From Karakas, E., and Laroussi, M., *J. Appl. Phys.*, 108, 063305, 2010.)

voltage of 5.5 kV, the plasma bullet is about 2 cm away from the nozzle before the falling of the applied voltage [20]. The transition point is the point related to the plasma bullet inhibition zones (the secondary discharge ignition zone from low helium mole fraction zone). **Figure 4.3** shows that the transition

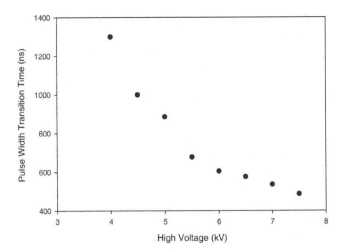

FIGURE 4.3 The pulse width transition time as a function of the applied voltage (repetition rate: 5 kHz; flow rate: 5.0 L/min). (From Karakas, E., and Laroussi, M., *J. Appl. Phys.*, 108, 063305, 2010.)

point shifts from 1300 ns to 425 ns since the plasma bullet is able to arrive to this transition point earlier as the voltage increases from 4 to 7.5 kV. This is attributed to the electric field from the applied voltage. Therefore, the plasma channel is more like a conductive channel rather than an insulating channel. This phenomenon is also confirmed by Lu et al. [21]. They found the length of the plasma plume kept increasing from about 3 cm to more than 6 cm when the pulse width was increased from 500 to 1000 ns. This finding also suggests the existence of a dark ionization channel between the high-voltage electrode and the plasma bullet.

The presence of metastable species is another important characteristic of plasmas. Stefan et al. measured the spatially resolved evolution of metastable densities in a self-pulsing micro-scaled atmospheric pressure plasma jet with the help of tunable diode laser absorption spectroscopy. This microplasma jet has two discharge modes, which are constricted- and α- mode, respectively.

The constricted mode has a high metastable density of 10^{13} cm^{-3} in the region close to the electrode and much lower density of 10^{11} cm^{-3} in the plasma bulk. The map of the development of the metastables suggests an avalanche of metastables is created at the point of the smallest gas gap, and it propagates along the electrodes toward the nozzle with the constricted discharge. The metastable density decreases to 1/e of the maximum value within 100–200 µs, depending on the position in the plasma [22]. In addition, collisions between air molecules and metastables decreased the number density of metastables; therefore, adding only 1% of air decreases the lifetime of metastables from 10 µs (in pure helium) to 39 ns.

Although the nanosecond exposure time image recorded the optical emission from the channel behind the plasma bullet and the metastable density of the plasma channel was measured by diode laser absorption

Nonequilibrium Atmospheric Pressure Plasma Jets (N-APPJs)

spectroscopy, the conductivity of the plasma is still not yet determined because the primary ionization and excitation reactions mainly happen in the head of the plasma bullet. The optical emission of the plasma channel is indeed caused by slow excitation transfer reactions in the afterglow. The measurement and calculation of electron density can deduce the conductivity of the plasma channel. However, the traditional ways to measure the electron density, such as Stark broadening or Thomson scattering, can only measure the electron density of plasma more than 10^{14}/cm^3. The electron density of the plasma channel is estimated to be less than 10^{13}/cm^3; therefore, new measurement techniques are needed.

The electron density of the plasma channel can be obtained through an indirect method, such as measuring the streamer head potential. The streamer head potential (U_h) is measured by a ring (10 mm in diameter, 1mm in height) installed coaxially with the jet and its potential is controlled by a high-voltage power supply. Therefore, the maximum electric field can be estimated from $E_m \approx U_h/2R$. The average electron density n_e can be calculated by $I_s = en_e V_{dr} S$, where I_s is the current following through the streamer channel, $S = \pi r^2$ is the cross-section area of the streamer channel, and e is the charge. Based on the measurements of I_s conducted by Rogowski oil, the electron density is estimated to be about 2×10^{13} cm^{-3} [23]. Another indirect method is measuring the axial electric field by optical emission spectroscopy. The ratio of the nitrogen band ($R_{391\ nm}/_{337\ nm}$) along the jet is calculated by the optical emission intensity at 391 nm (N$_2^+$(B)) and 337 nm (N$_2$(C)) (**Figure 4.4**), then the electric field can be deduced from this ratio (**Figure 4.5**). Finally, the electron density along the jet can be calculated by $I_s = en_e \mu_e ES$ with the known current value and cross-section area of the streamer channel. The electron density of this plasma jet is $10^{11} \sim 10^{13}$ cm^{-3} [24].

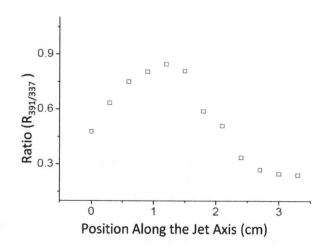

FIGURE 4.4 Intensity ratio of the nitrogen band along the jet axis for an applied voltage of 6.0 kV with a pulse duration 500 ns: the intensity ratio curve has a shape similar to the plasma bullet's velocity curve. (From Begum, A. et al., *AIP Adv.*, 3, 062117, 2013.)

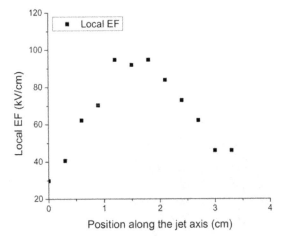

FIGURE 4.5 Local electric field along the jet axis estimated by optical emission spectroscopy. (From Begum, A. et al., *AIP Adv.*, 3, 062117, 2013.)

The electron density obtained by the plasma jet simulation is an efficient way to verify the accuracy of the indirect method mentioned above. The results of a numerical model of a pulsed-DC He plasma jet in open air suggest that the electron density is in the range of $10^{12} \sim 10^{13}$ cm^{-3} (**Figure 4.6**) [13], which is in agreement by an order of magnitude with the results mentioned above. In addition, the tube diameter has a significant effect on the electron density. As the tube diameter decreased from 1 to 0.2 mm, the noble gas flow channel shrank in radial direction; therefore, the plasma channel decreased in radial direction, and the more concentrated ionization increased the electron density from 10^{12} cm^{-3} to 10^{14} cm^{-3} (**Figure 4.7**) [25].

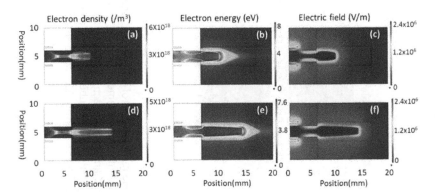

FIGURE 4.6 (a) Electron density, (b) electron energy, and (c) electric field of the pulsed-DC plasma jet at 120 ns (simulation). (d) Electron density, (e) electron energy, and (f) electric field of the plasma jet at 180 ns (simulation). (From Liu, X.Y. et al., *Plasma Sources Sci. Technol.*, 23, 035007, 2014.)

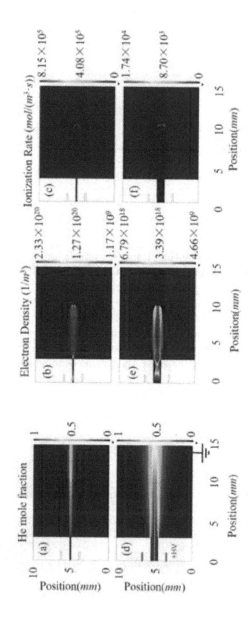

FIGURE 4.7 He mole fraction of (a) 0.2 mm tube diameter and (b) 1 mm tube diameter. Plasma properties 84 ns after the start of pulse excitation of 0.2 mm tube diameter: (b) electron density, (c) ionization rate. Plasma properties 103 ns after the start of pulse excitation of 1 mm tube diameter: (e) electron density, (f) ionization rate. The X axis was the downstream direction of tube. The white dashline in (d) suggests the position of the tube axis. pHV and the small red block in (d) suggest the position of the power electrode. (From Cheng, H. et al., *Plasma Process. Polym.*, 12, 1343–1347, 2015.)

Figures 4.6c,f and **4.7c,f** indicate that electrons are produced by the ionization concentrated in the streamer head with a strong electric field. The comparison between **Figure 4.6a** and **d** indicates that the electron density near the tube nozzle decreased less than 10% over 60 ns. Although the time of plasma decay is typically longer than the time of streamer propagation along the jet, the variation of plasma conductivity along the channel is not significant. Depending on the electron density and streamer radius, the typical resistance of the plasma channel is in the range 10^4–10^6 Ωcm^{-1}, much larger than 10^{-4} Ωcm^{-1}, which is the resistance of copper wire with a diameter of 0.1 mm. Therefore, the channel behind the plasma bullet can be considered as a relatively low conductive channel.

4.1.2 Photoionization and Seed Electrons

The theory attributed to plasma bullet propagation is Dawson's photon-ionization theory [26]. Dawson's original work studied the generation of cathode-directed streamers in air and suggested that photoionization could facilitate streamer tip propagation at very high velocity. The streamer head was assumed to be made up of numerous positive ions. Photons emitted from the streamer tip produce photoelectrons at a short distance in front of the head. The strong electric field between the positive streamer head and the photoelectrons leads to a rapid acceleration of the electrons and the following avalanche. If the avalanche produces enough electrons, the streamer tip is neutralized completely; therefore, a new positive ion region is created at a short distance in front of the original streamer head. This propagation characteristic is illustrated in **Figure 4.8**, which also can explain the plasma bullet propagation with a velocity of $\sim 10^8$ cm/s at a position several cm from the nozzle. The important assumption of Dawson's theory is that electrons produced by photoionization can completely neutralize positive ions. Therefore, the plasma bullet collapses and then regenerates itself at a short distance in front of its previous position [3]. The repetition of generation, collapse, and regeneration occurs so quickly that the plasma jet looks like an illuminant, long, and slim object extending from the nozzle. The nanosecond exposure time image in **Figure 4.1** captured the propagation dynamics of the plasma bullet, which confirms the propagation characteristics caused by photoionization.

In order to understand the role of photoionization, this phenomenon is incorporated in the 2D plasma simulation [8,27]. In this model, photoelectrons are produced through absorption by O_2 of high-energy photons emitted by excited N_2. The seed electron density of this model is assumed to be 10^3 cm^{-3}, which is the background electron density in natural conditions. He plasma jets in air with and without photoionization are carried out by this model. The basic phenomena of streamer propagation along the jet axis are similar for both cases. The difference between these two cases is that photoionization can not only promote plasma bullet propagation speed by 40% but also produce stronger and more uniform ionization in the plasma bullet. It can be concluded that the plasma bullet can propagate at low background electron density. The photoionization creates a weakly ionized cloud ($\sim 10^6$–10^8 cm^{-3}) in front of the streamer head. This ionized cloud significantly increases the streamer propagation speed.

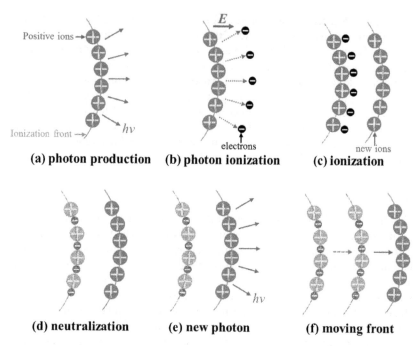

FIGURE 4.8 An illustration of Dawson's theory, with the assumption of complete neutralization of positive ions by photoelectrons. (a) a positively charged front of the plasma bullet emits photons, (b) new electrons produced by photon ionization result in an electric field, (c) photoelectrons move to the original plasma head, leaving behind new positive ions to produce the new plasma head, (d) complete neutralization of ions, (e) photons emitted from the new plasma head, (f) the moving plasma head is produced by this process of generation, disappearance, and regeneration. (From Walsh, J.L. et al., *J. Phys. Appl. Phys.*, 43, 075201, 2010.)

Plasma jet simulations that consider photoionization have also been carried out [28,29]. For plasma jets that propagate through a flexible tube in a Ne/Xe (99.9%/0.1%) gas, photoionization is caused by the absorption by Xe atoms of high-energy photons emitted by excited excimer molecules Ne_2^* [28]. When the applied voltage of the plasma jet is positive, the increasing rates of photoionization ahead of the plasma bullet can increase the speed of the plasma jet. The view angle for photoionizing radiation ahead of the plasma bullet is larger on relatively straight channel sections compared to highly curved channels; hence, the plasma jets is expected to have higher speeds along the channel section with a larger curve radius. For the modelling of positive and negative plasma jets in He with a 1% admixture of air, the photoionization term accounted for ionizing radiation emitted by excited excimer molecules He_2^* and absorbed by air molecules [29]. The plasma jet structure in these two cases are different. The head of the plasma jet has a spherical form at positive polarity, but the shape of the head is like that of a sword at negative voltage, in agreement with observations. In addition, at positive polarity, radiation is emitted mainly from a small region adjacent to the streamer head; at negative polarity, the whole channel is radiating.

Plasma jets without photoionization are also studied in the 2D model, taking into account variations of He-air mixture composition inside the jet. The initial electron density is set to be 10^{10} cm^{-3}. Such a high background electron density makes the role of photoionization negligible. While photoionization is not considered, if the initial electron density is assumed to be less than 10^9 cm^{-3}, the plasma jet is confined inside the tube. The plasma jet cannot propagate in open air when the background electron density is less than 10^8 cm^{-3}.

Photoionization is essential under conditions when the density of seed electrons is low, such as during the initial phase of the plasma jet, especially the very first plasma bullet. Consideration of photoionization is also required for accurate calculation of plasma jet characteristics when the repetition frequency of the plasma bullet is low, because the electron density of the previous plasma bullet decreases to a very low level. As the effect of seed electrons on streamer characteristics is typically not large [29], the requirements for accurate evaluation of the photoionization term are not strict.

We are not aware of any experimental reports on the effect of photoionization on plasma bullet propagation. One of the possible reasons is that, in order to measure the electrons generated by the photoionization ahead of the plasma bullet, diagnostics tools with very high spatial (lm) and temporal (ns) resolution are needed; the density of photoelectrons is also too low to measure using conventional techniques. Another reason is that existing theories suggest that the main source of O_2 ionization are the photons emitted by N_2 with wavelengths in the 98–102.5 nm range. These photons are beyond the range of conventional optical emission spectroscopy, in part due to the very fast absorption by the surrounding air. Another critical factor that affects APNP-Js driven by repetitive kHz voltage waveforms is the density of residual electrons left from previous current pulses (in the stable discharge phase). The density of these electrons could be much higher compared to the seed electrons produced by natural radioactivity; this makes it even more difficult to quantify the contribution of photoionization. However, for the very first discharge (ignition phase), there are no residual electrons due to the previous discharge, while the seed electron density due to natural radioactivity is very low (10^4 cm^3) [30]. Thus, the propagation of the plasma bullet for the very first discharge pulse should be different compared to the stable discharge phase. However, no investigations into the APNP-J ignition phase are presently available, especially to verify the common assumption that the residual electron density within the streamer propagation channel is high.

We thus focus on the ignition phase, where the role of photoionization could be dominant due to the negligible effect of the residual electrons. The base experiment setup is shown in **Figure 4.9a**. A high-voltage (HV) electrode made of a copper wire is inserted into a 10 cm long quartz tube with one end closed. The quartz tube, along with the HV electrode, is inserted into a syringe-like glass tube. The plasma is driven by a pulsed DC power supply. The gas flow rate is controlled by mass flow controllers (MKS Instruments, M100B). The applied

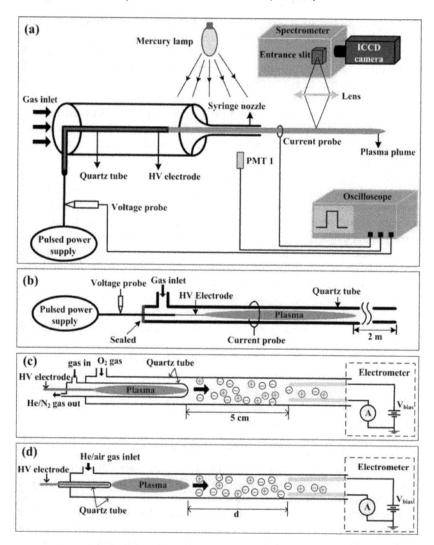

FIGURE 4.9 (a) Schematic of the plasma jet setup for the investigation of the ignition of the very first discharge. (b) The setup for the measurement of the VUV from the plasma plume. (c) and (d) are the schematics of the experiment setup for measurements of charges produced by photoionization and background ionization in APNP-Js, respectively. (From Wu, S. et al., *Phys. Plasmas*, 21, 103508, 2014.)

voltages and currents are measured by a P6015 Tektronix high-voltage probe and a current probe (Pearson model 2877), respectively. The voltage and current waveforms are recorded by a Tektronix DPO7104 broadband digital oscilloscope. A fast intensified charge-coupled device (ICCD) camera (Princeton Instruments, Model: PIMAX2) is used to capture the dynamics of the discharge. The VUV emission spectra are measured by a Princeton Instruments Acton

SpectraHub 2500i spectrometer and detected by the ICCD camera. The light signal is measured by a photomultiplier tube (PMT model: 71D101-CR131) to determine when the plasma is ignited. A mercury lamp (EUV201D) is used to study the illumination effects on the ignition of the very first discharge.

First, it is observed that when high-voltage pulses are applied to the HV electrode and He gas with a flow rate of 1 L/min flows through the syringe, the plasma plume is not generated immediately and is ignited only after many voltage pulses (**Figure 4.10**) when the first light signal is detected by the PMT. The delay time between the first voltage pulse and the first PMT pulse signal is referred to as the ignition delay time Δt_i. An external mercury lamp is used to evaluate whether the illumination of the lamp has any effect on the ignition delay time. The distance between the plasma jet and the mercury lamp is about 30 cm. **Figure 4.11a** is the optical emission spectrum of the mercury lamp. It is found that the illumination of the lamp has a significant effect on the ignition delay time Δt_i as shown in **Figure 4.11b**. When the lamp is off, the delay time Δt_i could be as long as tens of seconds. However, it decreases more than two orders of magnitude when the mercury lamp is turned on. The delay time decreases quickly with the increase of the pulse frequency for both cases.

For the pulse frequency of 2 kHz, when the lamp is off, the ignition delay time Δt_i is about 2000 ms, which corresponds to 4000 voltage pulses before the plasma is ignited. On the contrary, when the lamp is on, the delay time is decreased to less than 5 ms, i.e., the first discharge is ignited only after less than 10 voltage pulses. From **Figure 4.11a**, UV lines at 365 nm and strong visible lines in the 400–600 nm range are detected. But there is no clear optical emission from 100 to 300 nm. Hence, free electrons are likely produced by multi-photon ionization, which shortens Δt_i when the lamp is on. Previous studies show that the plasma plume also emits UV light at 309 and 337 nm wavelengths [31]. Thus, similar to the mercury lamp, plasma UV emissions play a role in photoionization in front of the plasma bullets. As discussed below, plasma also produces VUV emissions, which also affects plasma bullet propagation through the enhanced ionization of oxygen molecules. Second, the dynamics of the plasma bullets generated by the first discharge pulse appear to be random and quite similar to cathode-directed streamers. Indeed, the positions of the plasma bullets change irregularly with time, as opposed to the normal repeatable dynamics of

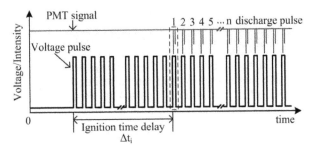

FIGURE 4.10 Schematic of the applied voltage pulses and PMT signals of **Figure 4.9a**. (From Wu, S. et al., *Phys. Plasmas*, 21, 103508, 2014.)

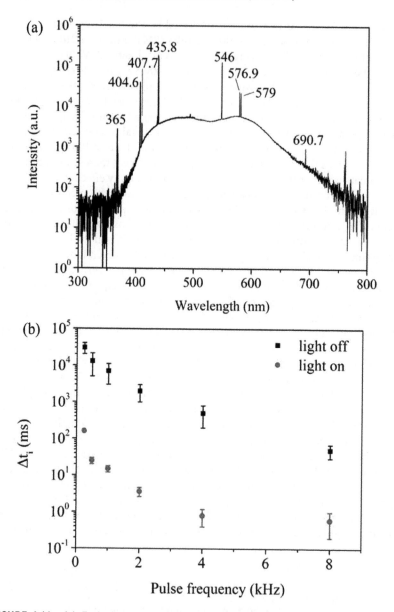

FIGURE 4.11 (a) Emission spectrum of the mercury lamp. (b) The relationship between the ignition delay time Δt_i and the pulse frequency with the mercury lamp turned on or off for the setup of **Figure 4.9a**. Voltage: 6 kV, pulse width: 6 μs, He flow rate: 1 L/min. (From Wu, S. et al., *Phys. Plasmas*, 21, 103508, 2014.)

APNP-Js. Third, measurements by a half-meter spectrometer (the grating and the slit width of the spectrometer are 3600 groove/mm and 100 lm, respectively) and an ICCD camera (exposure time is 20 ns) detected VUV emissions from the plasma in the 94.1–95.3 nm and 101–101.7 nm ranges.

VUV emission was also confirmed in controlled gas environments (**Figure 4.9b**), where the effect of the background gas composition was studied. A HV electrode made of a stainless-steel needle (100 μm tip radius) is inserted into a quartz tube with an inner diameter of 1 mm. The working gas flows into the tube from the left end. To keep the ambient air in the quartz tube as low as possible, a flexible tube with a length of 2 m is connected to the right end. **Figure 4.12** shows the emission from 50 nm to 300 nm of a N-APPJ operating in Ar, Ar/N_2, and Ar/O_2 mixtures. The schematic of the plasma jet device is shown in **Figure 2.12a**. The H30-UVL monochromator is used to measure the spectra from 50 nm to 300 nm. The emission spectrums of the plasma plume, when Ar-N_2/O_2 and He-N_2/O_2 are used as the working gas and discharge gas, respectively, are collected. Small N_2 or O_2 percentages (0.1%–0.5%) are added to the main working gas with a total flow rate of 2 SLM (100%). The comparison of emission spectrums in Ar-N_2/O_2 driven by an AC power supply, with a peak-to-peak value of 16 kV and a frequency of 7 kHz, is shown in **Figure 2.12b** and **c**. The shortest wavelength of spectra peak detected in this experiment is the OII (83.3 nm) transition. The relative intensities of the emission features decrease with the addition of oxygen; however, the addition of nitrogen does not result in significant change. The spectrum is dominated by the strong Ar excimer emission around 126 nm and the OI atomic line at 130 nm. Both the addition of oxygen and nitrogen results in the decrease of these emissions. Besides, as O_2 is added to Ar, emissions of 200–210 nm and 180–190 nm are observed; these are not seen with helium. Besides, the emission lines of NI (149.3 nm, 174.4 nm) are also observed.

Experiments sketched in **Figure 4.9c** were carried out to confirm that electric charges can be produced by photoionization in an APNP-J and to serve as evidence of the role of photoionization in plasma bullet propagation. The HV electrode (same as in **Figure 4.9b**) is inserted into the inner quartz tube with one end closed. The distance between the tip of the HV electrode and the right end of the inner tube is 10 cm. The diameter of the inner quartz tube is 6 mm. A working He + N_2 (2%) gas mixture flows (1 L/min) into the inner quartz tube from the left end. An outer quartz tube is used to guide the O_2 flow (2 L/min). The inner diameter of the outer quartz tube is 12 mm. When high voltage is applied, a plasma plume is generated inside the inner tube. Photons are emitted from the plasma, penetrating the wall of the inner tube and ionizing O_2 inside the outer quartz tube. Then, the charges produced by photoionization move to the charge collector (which consists of two parallel Cu plates spaced 7 mm apart), and they are measured by an electrometer. The distance from the right end of the inner quartz tube to the left edge of the charge collector plates is 5 cm. The photoionization current measurements conducted using an electrometer showed a sharp increase from 0 to 6.5 pA after the plasma plume is turned on, followed by a relatively slow decrease after the plasma plume is turned off. When the repetition frequency of

Nonequilibrium Atmospheric Pressure Plasma Jets (N-APPJs)

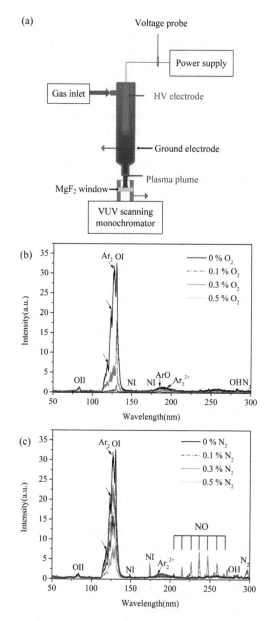

FIGURE 4.12 (a) The schematic of the experiment setup: the entrance slit of the VUV scanning monochromator is 500 μm; the distance between the nozzle and the MaF$_2$ window is about 4 mm and the plasma contacts the MaF$_2$ window immediately. (b and c) Optical emission spectrum patterns of the transitions from 50 nm to 300 nm versus Ar/O$_2$ (b) and Ar/N$_2$ (c) mixtures. (Unpublished)

the DC voltage pulses decreases from 8 to 4 kHz, the corresponding photoionization current decreases from 6.5 to about 3.2 pA. In addition, when the charge collector plates are moved away from the plasma, the time it takes for the photoionization current to start rising becomes longer. This simple experiment further confirms that photoionization does play a role in APNP-J discharges.

These current measurements are used to estimate the charge density. When the bias voltage is increased to more than 900 V, the current saturates and remains unchanged at higher voltages, but it increases linearly with the increase of the gas flow rate from 1 to 3 L/min. Thus, the charge density is independent of the bias electric field and is directly proportional to the gas flow velocity at saturation bias voltages. Given that the mobility for He_2^+, N_2^+, and O_2^-s at atmospheric pressure is about 20.0, 1.7, and 2.4 cm^2/Vs, respectively, the corresponding drift velocities of He_2^+, N_2^+, and O_2^- are about 200, 17, and 24 m/s for an electric field of 1 kV/cm. These values are much higher than the gas flow velocity vg of 0.3 m/s at the flow rate of 2 L/min in the outer quartz tube.

One can thus assume that nearly all the charges flowing through the space between the two plates are collected. The measured current I and charge density n_c are related by $I = en_c v_g S$, where S is the cross-sectional area of the charge collector (~1.1 cm^2). At $v_g = 0.3$ m/s, the calculated charge density n_c per pA of the photoionization current is 1.8×10^5 cm^{-3}. Thus, for a frequency of 8 kHz, the photoionization current of 6.5 pA corresponds to the electron density n_{ph} due to photoionization of about 1.2×10^6 cm^{-3}, which is already within the 10^6–10^8 cm^{-3} range reported previously. Since the VUV transmittance ratio of the quartz tube is very low (yet unknown), the actual n_{ph} value should be much higher.

Existing models predict a relatively high background seed electron density n_b due to previous discharge pulses in the channels where plasma bullets propagate. However, experimental reports that justify these conclusions are still missing. To quantify the role of photoionization in plasma bullet propagation, relative values of n_b and n_{ph} should be known. This is why a dedicated experiment to measure the background electron density in an APNP-J was conducted. The maximum electron and metastable N_2 density appear at the mixing layer where the air molar fraction is about 1% [7]. This is why n_b was measured in pure He and in a He + air (1%) mixture (both at a 1 L/min flow rate), as shown in **Figure 4.1d**. The quartz tube, along with the HV electrode, is inserted into the outer glass tube with an inner diameter of 12 mm. The charges generated by the discharge are measured by the same electrometer as in **Figure 4.1c**. The distance d between the head of the plasma plume and the left edge of the charge collector plates is adjustable. The charge density is estimated from the current measured by the same electrometry approach as above. By varying the distance between the tip of the plasma plume and the charge collector, the decay curve of the charge density versus the distance from the tip of the plasma plume is obtained.

For pure He, the charge density decreases from 10^{11} to 10^6 cm^3 when d is increased from 5 to 50 cm (**Figure 4.13a**). For the He + air (1%) mixture, the charge density decreases from 3.4×10^9 to 2×10^7 cm^{-3} when d is increased from 6 to 20 cm (**Figure 4.13b**). The charge density also decreases nearly linearly with the decrease of the pulse frequency. When the pulse frequency

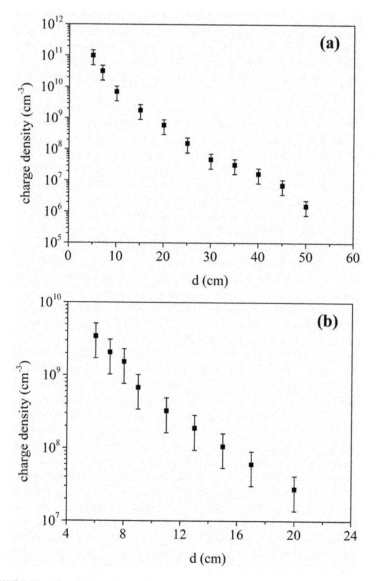

FIGURE 4.13 The decay curves of charge densities with distance for the experiment setup of **Figure 4.9d** for pure He (a) and He + air (1%) gas mixture (b). Voltage: 8 kV, pulse frequency: 8 kHz, pulse width: 400 ns, total gas flow rate: 1 L/min. (From Wu, S. et al., *Phys. Plasmas*, 21, 103508, 2014.)

decreases from 8 to 0.2 kHz, the charge density of the He + air (1%) mixture decreases from 4×10^9 to 1.3×10^8 cm^{-3} at $d = 6$ cm. The above results allow the following conclusions. First, the lifetime of the charges is very long. In our experiment, the gas flow velocity is 0.15 m/s. Therefore, it takes hundreds of milliseconds for the charges to reach the charge collector. In other words, the charges generated by the plasma plume can be blown to the downstream of the gas flow, which is quite surprising.

Second, for pure He gas (with a possible trace amount of air inside the tube due to air diffusion, thus photoionization is still possible), the total background charge density at d¼0 should be larger than 10^{11} cm^{-3} according to the charge decay curve in **Figure 4.13a**, and it is probably even higher in the plasma plume region. On the other hand, the electron density due to photoionization is -10^6-10^8 cm^{-3}. Therefore, photoionization plays a relatively minor role in the propagation of the He plasma plume during its stable phase.

Third, for the He + air (1%) mixture, the total charge density at $d = 0$ is higher than 3.4×10^9 cm^3 (**Figure 4.13b**), and is expected to be even higher in the plasma plume. Although it is still one order of magnitude higher than that due to photoionization, photoionization may still play some role in this case. In addition, this measurement is conducted for the pulse repetition frequency of 8 kHz. As pointed out above, the charge density decreases nearly linearly with the decrease of the pulse frequency. Hence, if the pulse repetition frequency is reduced to 0.2 kHz or less, then the electron density due to photoionization could be close to the background charge density 10^8 cm^{-3}. In other words, photoionization could indeed play a major role under such conditions. This is consistent with previous studies, which show that the dynamics of the plasma becomes unrepeatable when the pulse repetition frequency is reduced to 0.2 kHz [32].

Fourth, it is found that the background electron density plays a key role in the propagation of the plasma bullet in the repeatable or the random mode. The plasma bullet propagates in the repeatable mode when the background electron density is 10^8 cm^{-3} or higher. In this case, the plasma plumes were generated in a gas tube. In applications, plasmas are usually generated in surrounding air, using He as the working gas. Due to strong diffusion, some air is present in the He gas stream, which makes our results for the He + air (1%) mixture most relevant.

Meanwhile, the chemistry of light interactions with the gas may play an important role in the discharge ignition, for example, through vibrational excitation of molecules. This is due to the fact that the energy is largely transferred from the electrons to the molecular vibration states. Several excited molecules, such as N_2, CO, H_2, and CO_2, can maintain their vibrational energy for a relatively long time and without relaxation. This feature is believed to cause selectivity in chemical reactions. In previous studies, the visible light was shown to affect discharge ignition by reducing the breakdown delay time and the breakdown voltage. However, these effects have not been studied systematically for RT-APPJs. Several open questions remain, for example: (1) What is the most effective wavelength range of the visible light? (2) Where is the most effective position for the light to focus on? (3) Does the effect of the visible light depend on the discharge tube size and material? and (4) How does the effect of the visible light depend on the gas composition and pressure? This work focuses on the effects of visible light on the breakdown delay time of an RT-APPJ. The plasma is generated in a discharge tube. The gas in the tube is replaced after each discharge to eliminate the influence of the residual electrons and reactive species. Visible light of different wave-bands is used to illuminate the discharge tube. The ignition delay time is measured to investigate the effect of the visible light on the discharge. Our results suggest that the breakdown delay time decreases with visible light illumination and the most effective light wavelength range is 400–530 nm. This effect is affected by the position of the light source, gas composition, and characteristics of the applied voltage. The details of the experimental results, including the effects of monochromatic light and the 10 nm waveband visible light of different wavelengths on the breakdown delay time, are discussed next.

The experiment setup is shown in **Figure 4.14**. A custom designed vacuum-grade gas handling system is used to ensure the purity of the gases and to avoid the diffusion of the surrounding air as shown in **Figure 4.14a**. The high-voltage (HV) electrode and the ground electrode made of tinfoil are placed on the outside of the tube. The distance between the HV electrode and the ground electrode is 18 mm. It should be pointed out that the system contains no plastic parts, except for the O-ring. The discharge tube is evacuated to 5×10^{-2} Pa and then backfilled with operating gas to a desired pressure. The voltage and current are measured by a HV voltage probe (Tektronix P6015) and a current probe (Tektronix TCP 202), respectively. A photomultiplier tube (PMT, 71D101-CR131) is placed as shown in **Figure 4.14a** to evaluate the ignition of the discharge. **Figure 4.14b** shows a schematic of the sequence of the PMT (photomultiplier tube) signals and the voltage pulses. The PMT signals (in **Figure 4.14b**) are used to identify the breakdown process. The breakdown delay time t_d is the time between the first voltage pulse and the PMT signals shown in **Figure 4.14b**. Three lasers with different wavelengths and a xenon lamp are used as the light sources.

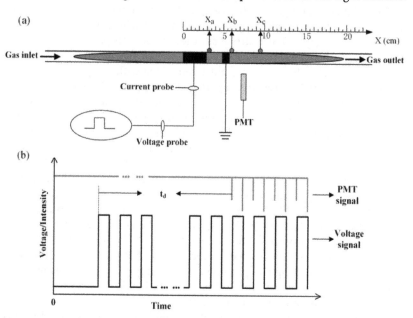

FIGURE 4.14 Schematics of (a) the experiment setup, and (b) the voltage pulses and PMT signals. (From Nie, L. et al., *Phys. Plasmas*, 24, 043502, 2017.)

The three laser sources (with 20 mW output power) provide monochromatic light of 404 nm, 532 nm, and 662 nm wavelengths. The light from the xenon lamp with 30W output power, together with different narrow-band filters, allows light within the selected energy bands to pass through a convex lens and illuminate specific positions along the discharge tube.

It is observed that visible light illumination can reduce the breakdown delay time. In order to better understand this phenomenon, two kinds of visible light sources, namely the three lasers and narrow-bandwidth visible lights, were used to illuminate the gas while the discharge ignition delay time t_d was measured. The optical emission spectra of these light sources (shown in **Figure 4.15**) were measured by a Princeton Instruments Acton Spectra Hub 2500i spectrometer. All the experiments were carried out in a dark room to eliminate the interference of any additional light source. Under the same experiment conditions, the measurements have been repeated 20 times.

First, the effect of monochromatic light on the ignition delay time of a helium plasma jet was studied. In this part, the tube size, material, and gas flow were changed in order to investigate the effect of the monochromatic light on the breakdown process. In addition, the blue laser with a 404 nm wavelength was chosen as the illumination source. Helium with a purity of 99.999% was used as the working gas. The voltage, frequency, and pulse width were fixed at 8 kV, 8 kHz, and 2 ls, respectively. **Figure 4.16a** shows the breakdown delay time of plasma jets with different tube sizes and materials when the monochromatic light is turned on or off. The parameters of the discharge tube in **Figure 4.16a** are listed in **Table 4.1**. **Figure 4.16b** shows the corresponding t_d with different gas flow rates with a No. 5 tube. In **Figure 4.16a**, it can be seen that the 404 nm laser decreases the breakdown delay time effectively with 1–5 orders. It should be mentioned that the t_d of No. 2 without additional light is very difficult to ignite, which means that its t_d is longer than 5 min. The No. 2 jet using a Polytetrafluoroethylene tube is difficult to ignite compared to the No. 1 jet. When comparing No. 1, No. 3, and No. 4 which have the same wall thickness, the t_d values of these tubes are similar without additional light, but the t_d of No. 3

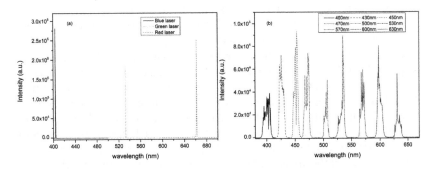

FIGURE 4.15 Emission spectra of (a) the visible light lasers, and (b) the narrow-bandwidth visible light. The grating is 1200 g/mm, and the entrance and exit slits of the spectroscope are fixed at 50 μm. (From Nie, L. et al., *Phys. Plasmas.*, 24, 043502, 2017.)

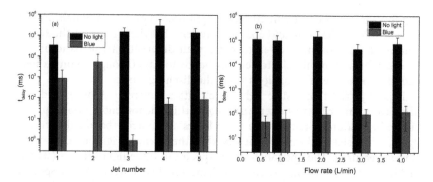

FIGURE 4.16 (a) The delay time of helium with light on or off with different tube sizes and materials according to Table 4.1 (for all cases) and (b) the delay time of helium with light on or off versus gas flow rate. (From Nie, L. et al., *Phys. Plasmas*, 24, 043502, 2017.)

Table 4.1 The Tube Size and Material of RT-APPJs

No.	Inner Diameters (mm)	Outer Diameters (mm)	Material
1	2	4	Quartz
2	2	4	Polytetrafluoroethylene
3	4	6	Quartz
4	1	3	Quartz
5	1.5	3	Quartz

with the 404 nm light turned on is approximately three orders of magnitude less than that of the No. 1 jet. Therefore, the effect of 404 nm light on the ignition delay time can be enhanced with a larger inner diameter. For the No. 4 and No. 5 jets with different wall thickness, the t_d is almost the same with 404 nm monochromatic light turned on. From **Figure 4.16b**, it can be seen that the effect of monochromatic light on the ignition delay time is independent of the gas flow rate. So, monochromatic light can reduce the breakdown delay time of RT-APPJs effectively and the effect can be enhanced with a large inner diameter and the proper wall thickness and material. In addition, the gas flow rate almost has no influence on this effect. Next, the effect of monochromatic light on the ignition delay time of a helium plasma plume was studied. In this section, the effects of monochromatic light at different laser illumination positions and gas pressures were investigated. In addition, the tube was connected with the pump in order to avoid the influence of the air to change the parameters of the gas. The inner and outer diameters of the discharge tube are 1.5 mm and 3 mm, respectively.

Helium with a purity of 99.999% was used as the working gas and the gas is still. The voltage, frequency, and pulse width were fixed at 8 kV, 8 kHz, and 2 μs, respectively. The wavelengths of the monochromatic light in the experiment were 404 nm, 532 nm, and 662 nm. **Figure 4.17a** shows the corresponding t_d when the

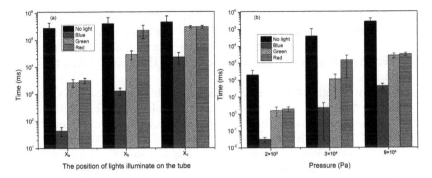

FIGURE 4.17 (a) The values of t_d when the laser illuminates at different positions x_a, x_b, and x_c; gas pressure: 9×10^4 Pa; (b) the delay time of helium with light on or off at pressures 2×10^3 Pa, 3×10^4 Pa, and 9×10^4 Pa. (From Nie, L. et al., *Phys. Plasmas*, 24, 043502, 2017.)

laser illuminates at different positions x_a, x_b, and x_c at a gas pressure of 9×10^4 Pa. **Figure 4.17b** shows the breakdown delay time t_d with (or without) light at pressures of 2×10^3 Pa, 3×10^4 Pa, and 9×10^4 Pa. As can be seen from **Figure 4.17a**, the ignition delay time depends on both the wavelength and the position where the light illuminates. The ignition delay time t_d decreases significantly for all three lasers when the light illuminates position x_a, which is close to the HV electrode. When the blue laser is used, t_d decreases to about 40 ms, which is about four orders of magnitude less than with no light. When the laser illuminates position x_b, which is close to the ground electrode, the t_d decreases by about two orders of magnitude for the blue laser and one order of magnitude for the green laser. For the red laser, the ignition delay time t_d is almost the same as with no light. When the illumination position moves further right to point x_c, which is about 3.5 cm away from the ground electrode, illumination by both the green and red lasers has no effect on the t_d. When the blue laser is used, the t_d decreases by about one order of magnitude. On the other hand, as can been seen from **Figure 4.17b**, the t_d increases when the gas pressure is increased. The blue light laser of 404 nm significantly reduces the discharge delay time by about four orders of magnitude for all three gas pressure conditions. The green and the red lasers are less effective.

As shown in **Figure 4.17**, illumination by all three visible lasers, especially by the blue laser, reduces the ignition delay time effectively. The photon energy E can be calculated from the equation

$$E = hc/\lambda,$$

where h is Planck's constant (4.14×10^{-15} eV), c is the speed of light in a vacuum, and λ is the photon's wavelength. For the lasers with wavelengths of 404 nm, 532 nm, and 662 nm, their photon energy is 3.0688 eV, 2.3305 eV, and 1.8728 eV, respectively. These energies are much lower than the first ionization energy threshold of He, N_2, and O_2, which are 24.6 eV, 15.58 eV, and 12.2 eV, respectively. In other words, the photon energy of the laser light used in the experiment is much lower than the direct ionization energy threshold of the gases. Meanwhile, as the development of the electron avalanche starts from the HV electrode, the seed electrons around the HV electrode can contribute to the development of the electron avalanche. That might be the reason why the effect of the laser on the delay time depends on the position where the laser illuminates.

As shown in **Figure 4.17**, even visible laser light has a significant effect on the breakdown delay time. To further understand the observed effect, light filtered from a xenon lamp was chosen as the light source, as shown in **Figure 4.15b**. The central wavelengths of the optical filter are 400 nm, 430 nm, 450 nm, 470 nm, 500 nm, 530 nm, 570 nm, 610 nm, and 630 nm. In addition, different voltage parameters and gas compositions were chosen to investigate the effect of visible light on the breakdown delay time. The inner and outer diameters of the discharge tube were 1.5 mm and 3 mm, respectively. **Figure 4.18** shows the ignition delay times of discharges illuminated with different center wavelength lights measured at position x_a. Helium of 99.999% purity was used as the working gas at a pressure of 10^4 Pa. The amplitude of the applied voltage was fixed at 8 kV. As can be seen from **Figure 4.18a**, light with a central wavelength from 400 nm to 530 nm can significantly reduce the delay time. However, it was found that light with central wavelengths exceeding 530 nm has no obvious effect. When the frequency is increased to 8 kHz, the ignition delay time decreases several folds in all cases, as shown in **Figure 4.18b**, even without illumination. However, the observed trend remains the same: the ignition delay time is much shorter when the discharge tube is illuminated by light with center wavelengths shorter than 530 nm. A similar trend is observed when the pulse width of the applied voltage is increased to 10 μs. We reiterate that pulsed voltage is used in this work. For each voltage pulse, the actual voltage on time t_{v-on} is very short compared to the pulse period. For a frequency of 2 kHz and a pulse width of 2 μs, the pulse period is 500 μs. Thus, the actual voltage on time is only 0.4%. With the increase of the pulse frequency or the pulse width for the same delay time, the actual duration t_{v-on} of the high voltage applied to the electrode increases. For example, when the frequency is increased from 2 kHz to 8 kHz, the t_{v-on} experiences a four-time increase if the delay time remains the same. Similarly, when the pulse width is increased from 2 μs to 10 μs, the t_{v-on} is increased five times at the same

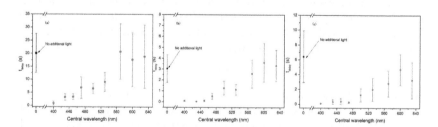

FIGURE 4.18 The ignition delay times of the helium discharge under illumination with narrow-bandwidth light with different center wavelengths: (a) 8 kV, 2 kHz, 2 μs; (b) 8 kV, 8 kHz, 2 μs; and (c) 8 kV, 2 kHz, 10 μs. (From Nie, L. et al., *Phys. Plasmas*, 24, 043502, 2017.)

delay time. Therefore, according to **Figure 4.18a–c**, the ignition delay time is about 20.05 s, 3.05 s, and 6.42 s, respectively, when there is no light illumination at all. According to the discussion above, the actual voltage on time t_{v-on} is about 80.2 ms, 48.8 ms, and 128.4 ms, respectively, for the three cases when there is no light illumination. In other words, when the frequency is 8 kHz, the ignition delay t_{v-on} is the shortest. When the 400 nm light is used, the ignition delay times decrease to about 0.89267 s, 0.11249 s, and 0.0947 s for the three cases, respectively, according to **Figure 4.18a–c**. The corresponding delay times t_{v-on} are 3.57 ms, 1.8 ms, and 1.89 ms, respectively. In other words, high frequencies or long pulses could reduce the observed ignition delays.

Furthermore, the ignition delay time for different gases and gas mixtures was investigated. The ignition delay time for He mixed with one percent of N_2 and O_2 is shown in **Figure 4.19**. Similar results for pure N_2 and O_2 gases are shown in **Figure 4.20**. One can see from **Figure 4.19a** that the delay time for pure helium without light illumination is about 16 times longer compared to that in **Figure 4.5b**. Moreover, light with central wavelengths ranging from 400 nm to 630 nm contributes to the decrease of the delay time. The most effective wavelength range is 400–530 nm, and the effect becomes less obvious when the central wavelength is larger than 530 nm. This phenomenon is almost the same as for the pure helium cases shown in **Figure 4.18b**. For the 1% oxygen admixture in **Figure 4.19b**, the delay time is larger than with 5 min without light. Moreover, the delay decreases to 10–150 s when the central wavelength increases from 400 nm to 470 nm. However, the effect of light with a central wavelength larger than 470 nm appears to be insignificant. Therefore, the effective range of narrow-wavelength visible light is much larger for the 1% nitrogen mixture compared to 1% oxygen. Due to the addition of oxygen, the ignition delay time increases dramatically for all cases. This can be attributed to the electronegativity of oxygen so that oxygen molecules attach to free electrons generated by the extra light illumination. As can be seen in **Figure 4.19**, the ignition delay times for the addition of 1% oxygen reveal some

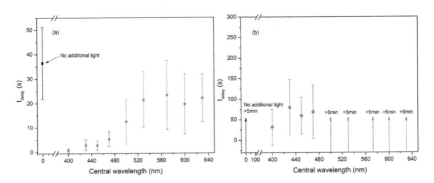

FIGURE 4.19 The ignition delay times for (a) He + 1% N_2 and (b) He + 1% O_2 gas mixtures when light with different central wavelengths is used at a total gas pressure of 10^4 Pa. The voltage, frequency, and pulse duration are 8 kV, 8 kHz, and 2 μs, respectively. (From Nie, L. et al., *Phys. Plasmas*, 24, 043502, 2017.)

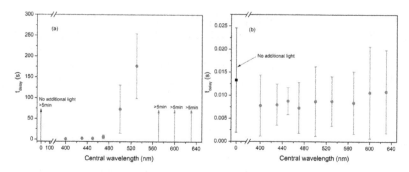

FIGURE 4.20 The ignition delay times for (a) 99.99% N_2 and (b) 99.99% O_2 when light with different central wavelengths is used at a total gas pressure of 10^4 Pa. The other parameters are the same as in **Figure 4.19**. (From Nie, L. et al., *Phys. Plasmas*, 24, 043502, 2017.)

differences. In order to investigate the effect of visible light on nitrogen and oxygen, 99.99% nitrogen was chosen for **Figure 4.20a** and 99.99% oxygen was chosen for **Figure 4.20b**. It is hard to break down pure nitrogen without light, as can be seen from **Figure 4.20a**. Light irradiation with central wavelengths from 400 nm to 530 nm helps decrease the breakdown delay time significantly. However, light with central wavelengths higher than 530 nm plays a less important role in the ignition stage. In **Figure 4.20b**, it is interesting that the delay time is much shorter than for pure He in **Figure 4.18b** and for pure N_2 gases in **Figure 4.20a**. Importantly, light has almost no effect on the delay time for pure oxygen.

Comparing **Figures 4.19b** and **4.20b**, it has been found that the effect of adding O_2 (without illumination) is unexpected. The addition of 1% O_2 to helium leads to the increase of t_d. But in pure O_2, the t_d is several orders less than in He. In order to better understand the effect of O_2 on t_d, the ignition delay time versus oxygen percentage was measured as shown in **Figure 4.21**. As can be seen from **Figure 4.21**, the gas mixture is difficult to ignite with 1%–20% O_2 at a pressure of 10^4 Pa since the ignition delay time is longer than 5 min, but it is much easier to ignite with either less than 1% or more than 20% of O_2. Thus, the reaction process in the plasma system must have changed with the different O_2 percentages. With low oxygen concentration (<1%), Penning processes play a significant role, and the gas can be easily ignited. With 1%–20% oxygen, the electron attachment of oxygen molecules plays the dominant role, which makes the gas difficult to break down, and this is consistent with that of atmospheric pressure [34]. When the O_2 percentage is further increased to more than 20%, the difference in t_d with and without light illumination becomes smaller, which might be explained as follows: the increasing role of the process $(e+O_2 => O_2(1D)+e)$ makes the density of $O_2(1D)$ increase. Then the process $(O^-+O_2(1D) => O_2(1D)+e)$ is enhanced, which supplies enough seed electrons for ignition.

The most common method for producing the electrical breakdown that generates and sustains low-temperature plasma is applying a high electric field

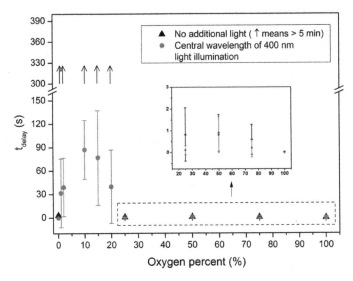

FIGURE 4.21 The ignition delay times versus oxygen percentage with the central wavelength of 400 nm light illuminated on or off (gas pressure of 10^4 Pa; voltage parameters are the same as in **Figure 4.19**). (From Nie, L. et al., *Phys. Plasmas*, 24, 043502, 2017.)

to a neutral gas. Any volume of neutral gas always contains a low density of electrons and ions that are, for example, the result of the interaction of cosmic rays or environmental radiation with the neutral gas. These free charge carriers are accelerated by the electric field, colliding with atoms/molecules in the gas and with the electrode surfaces, and new charged particles may be created. This leads to an avalanche of charged particles and to the initiation of electrical breakdown in the gas. When the electric field in the inter-electrode space E is high enough to create the multiplication of electrons and ions, the avalanche appears. If this multiplication creates a sufficient number of electrons and ions, it will lead to the electrical breakdown. However, if the process of free charge species losses is dominated, the avalanche multiplication will cease. The necessary conditions for gas breakdown are a sufficiently high electric field and enough effective electrons, but how the seed electrons develop into effective electrons is a probabilistic event [35]. As the breakdown delay time is the main focus of this work, we relate the process of gas breakdown to the applied voltage. The breakdown delay time t_d includes the statistical time delay (t_s) and the formation time (t_f), i.e., $t_d = t_s + t_f$. The period of time between the application of voltage and the electron generation is defined as the statistical time delay t_s. Moreover, the breakdown process can only be initiated when the electrons appear in the intense electric field region. That is the reason why the visible light illumination is most effective when it is focused on the HV electrode. When the electrons appear, the avalanche may form in the intense electric field region. This process creates a sufficient number of electrons and ions and it takes a certain amount of time, described as the formation time t_f. Under these conditions, the statistical time delay t_s is much larger than the formation time t_f [35]. Thus, variations in the t_d are mainly affected by the statistical time delay t_s. In this way, the visible light reduces the t_d to supply free electrons during the ignition stage. The next problem is to identify the possible mechanisms that enable visible light to supply free electrons for ignition. There are two distinct possibilities for the generation of free electrons by visible light.

One plausible mechanism is the optically stimulated exoelectron emission (OSEE), which arises from radiation with a wavelength exceeding the photoelectric threshold. OSEE is a particular case of exoelectron emission (EEE), which is a low-temperature emission of electrons from a solid surface subjected to various external effects. These effects are considered preliminary excitations of the electron emitter. In Ref. [36], this excitation was performed by mechanical grinding of the electrode, which leads to the formation of microscopic charged cracks in the oxide layer. Under gas-discharge conditions, the emitter (wall) could be excited as the wall becomes charged upon its contact with the plasma during the tube heating or during the previous breakdown. It is very challenging to measure the EEE current because it is very low, typically ranging from 10^{-18} to 10^{-11} A.

The other possibility is the presence of nitrogen species, which can be vibrationally excited by photons. Though pure helium with 99.999% purity was used, typically at least 5 ppm of nitrogen impurity is present inside the chamber, and the impurity level can be even higher when gas leakage is taken into consideration. The optical emission spectra of the discharges in pure helium, helium with 1% N_2, and helium with 1% O_2 are presented in **Figure 4.22**. From **Figure 4.22**, one can always note N_2 and N_2^+ peaks in pure helium, helium with 1% N_2, and helium with 1% O_2, which confirms the presence of N_2 impurity in pure helium. Because the energy for the nitrogen vibration excitation is about 1.7–3.5 eV, light with central wavelengths of 400–630 nm facilitates the vibration excitation as the corresponding energies range from 3.105 eV to 1.97 eV. Nitrogen molecules are vibrationally excited by the narrow visible light, which makes the electron impact ionization easier afterward.

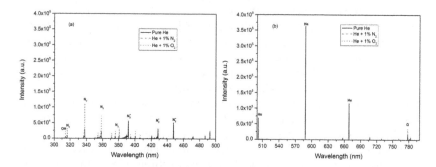

FIGURE 4.22 Optical emission spectra of pure He and He + 1% N_2 and He + 1% O_2 gas mixtures. (a,b) are two parts of OES. (From Nie, L. et al., *Phys. Plasmas*, 24, 043502, 2017.)

4.1.3 Penning Ionization and Donut Shape

One interesting phenomena of an APPJ is how its cross-section image indicates the donut shape of the plasma plume in a radial direction (**Figure 4.23**) [1]. Therefore, experiments and models have been carried out by several groups to study the mechanism that forms the donut shape and to determine whether the donut shape is related to Penning ionization or not.

The donut shape represents the structure of the optical emission of the plasma over a broad spectral range. The optical emission spectroscopy measurement indicates that the light emissions of the plasma bullet are dominated by the excited N_2, He, and N_2^+ [37]. The spatially resolved optical emission of the N_2 second positive system (SPS) at 337 nm for different axial positions is shown in **Figure 4.24** [38]. Abel inverted profiles at 40 mm and

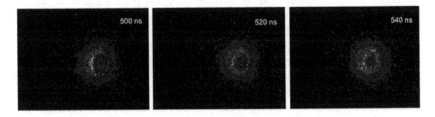

FIGURE 4.23 Photographs of the bullets illustrating their donut shape. (From Mericam-Bourdet, N. et al., *J. Phys. Appl. Phys.*, 42, 055207, 2009.)

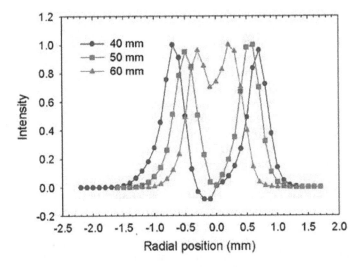

FIGURE 4.24 Radial profile of the peak intensity of molecular nitrogen SPS at different axial positions (HV pulse amplitude: 8 kV). At 40 mm and 50 mm the emission peaks are at the edge of the bullet indicating a ring-shaped emission profile. At 60 mm, the emission from the center increases, indicating contraction for the bullet. (From Jarrige, J. et al., *Plasma Sources Sci. Technol.*, 19, 065005, 2010.)

50 mm from the nozzle suggest two emission peaks on the left and right edges, while the intensity is quite low in the center of the jet. The two emission peaks tend to shift to the center as the bullet propagates away from the nozzle. Two peaks almost merge as a result of bullet contraction at 60 mm.

The spatial distribution of He metastable 2^3S_1 atoms (He_m) of a He plasma plume driven by a bipolar pulse voltage was measured by laser absorption spectroscopy (LAS) [39]. The result indicates that the He_m density has a ring-like, hollow shape at the center too. The radius of the ring and the peak density decreases gradually with the distance from the tube exit. In addition, the radius of the ring increases as the tube diameter increases. It seems like the counterdiffusion of He and the surrounding air is responsible for the formation of the donut shape. If this is true, the donut shape should disappear when N_2 is added to the working gas, He. High-speed photographs of the plasma bullet in the surrounding air taken head on with respect to the plasma plume shown in **Figure 4.25** confirm this assumption. This figure clearly shows that the plasma bullet appears as a solid disk shape rather than a donut shape when 1.5% of N_2 is added into the working gas He [40]. In addition, **Figure 4.26** indicates that the diffusion of the ambient air has a significant effect on the propagation of the plasma bullet. As we know, O_2 plays an important role in the energy loss of electrons due to the electron attachment effect; therefore, the acceleration behavior of the plasma bullet is probably due to the diffusion of N_2 from the surrounding air. When He/N_2 is used as the working gas, **Figure 4.26a** and **c** shows similar results. Therefore, the diffusion of N_2 is probably responsible for the acceleration of the plasma bullet in open air because N_2 can be further ionized by He metastable states.

The simulation results based on a one-way coupled model between neutral gas flow and plasma dynamics analyze the development of a ring-shaped emission profile in a plasma bullet. In the early stage of the plasma jet, the discharge is substantially sustained by the electron impact ionization of helium. The density of electrons and ions shows a relatively broad peak along the axis of the jet nozzle though the applied electric field is uniform. The position of the peak of Penning ionization between nitrogen and helium metastables is only at the position of 0.4 mm from the axis. The peak of helium metastable density is at the position of 0.5 mm from the axis. The neutral gas flow calculated by the total mass continuity equation, the momentum continuity equation, and the mass continuity equation suggests nitrogen density near the tube axis is significantly low, whereas the nitrogen density far away from the axis is so high that electron energy is quenched. In the simulation domain, the distribution of electron impact excitation/ionization is nonuniform. Penning ionization plays only a minor role in sustaining the plasma during the early stage of the discharge. However, the off-axis Penning ionization peak eventually generates an off-axis peak of electron density in the late stage of the discharge [4].

In addition, to understand the role of Penning ionization in the propagation and the shape of the plasma bullet, a special experiment was conducted [98] to investigate the effect of Penning ionization on the ignition of the plasma plume. They found out that, when He is used as the working gas, the discharge actually is first ignited in a region with a high air content ratio rather than

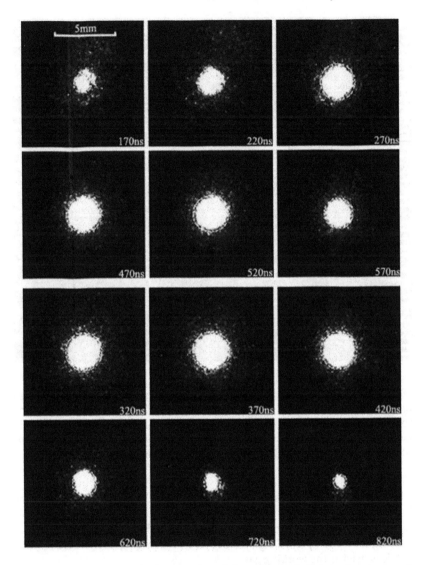

FIGURE 4.25 High-speed photographs of the plasma bullet in the surrounding air taken head on with respect to the plasma plume. The exposure time is 5 ns. The internal diameter of the nozzle is 2.5 mm. The working gas is a He/N$_2$ (He: 1 L/min; N$_2$: 0.015 L/min) mixture. (From Wu, S. et al., *IEEE Trans. Plasma Sci.*, 39, 2286–2287, 2011.)

a region with a low air content ratio as the applied voltage increases. In other words, the presence of air is favorable for the ignition of the discharge. A similar phenomenon is observed when Ne is used. But when Kr is used as the working gas, the discharge is first ignited in a region with a low air content ratio rather than a region with a high air content ratio. They believe that this

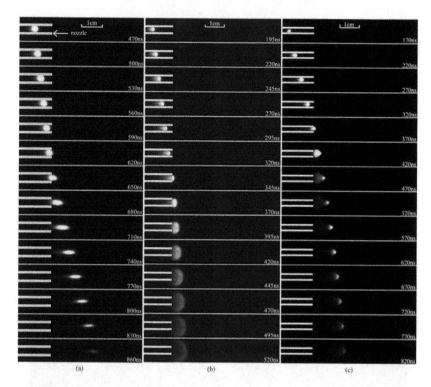

FIGURE 4.26 High-speed photographs of the plasma plume. The working gas for (a) and (c) is a He/N$_2$ (He: 1 L/min; N$_2$: 0.015 L/min) mixture, and the working gas for (b) is He (1 L/min). (a) No vacuum chamber is connected. (b) and (c) Vacuum chamber connected. (From Wu, S. et al., *IEEE Trans. Plasma Sci.*, 39, 2286–2287, 2011.)

is because the metastable states of Kr (9.9 eV, 10.6 eV) could not sustain the direct Penning ionization with N$_2$ molecules, and the reaction responsible for Kr APPJ is mainly direct impact ionization. Thus, they concluded that Penning ionization plays a significant role in the propagation of APPJs when He/Ne is used as the working gas.

Although the experimental and numerical results shown above support the assumption that the donut shape is due to the air diffusion into the helium flow and the consequent Penning ionization, the following simulation results suggest the opposite. The simulation results captured a plasma bullet with a ring-shaped structure, and the ring radius decreased as the plasma bullet propagated further from the nozzle. Once the Penning ionization was shut down, the ring-shaped structure did not change. The only difference was the peak electron density and excited N$_2$(C) decreased by 13.3% and 33%, respectively. Therefore, according to the simulation results, the effect of Penning ionization on the streamer propagation is weak, which can slightly change the peak density value of several species [7].

These results are consistent with numerical results verified by the experiment results [13]. This numerical work is for the 2D axisymmetric geometry shown in **Figure 4.27a**. Pure helium flows through a dielectric tube with a radius of 1 mm. The gas flow rate considered in the simulation is 1.1 standard liters per minute (SLM). The simulation region extends to 1.5 cm from the tube exit. The mole fraction of helium decreases in the radial direction because of convection and the diffusion of air (**Figure 4.27b**); as a result, the mole fraction of air increases in the radial direction (**Figure 4.27c**). The mixing layer defines the region where chemical reactions between helium and air species occur, which is crucial for the propagation of the plasma jet.

The experimental image of the plasma jet is shown in **Figure 4.28a**. For the part of the plasma plume outside the tube, its optical emission along the tube axis extension is weaker. This emission pattern is the signature of a ring-shaped streamer head. For the part of the plasma plume inside the tube, the optical emissions shrinks in the radial direction. The plasma plume shown in the long exposure time image is an accumulation of optical emissions from the streamer head moving at a very high speed. The luminous streamer head is followed by the dark plasma channel, which is full of electrons that maintain the current continuity between the electrode and the streamer head. Therefore, this long exposure time image of the plasma actually shows the distribution of the dark plasma channel. The electron density distribution shown in **Figure 4.28b** indicates that the plasma channel has the same pattern as in the experiment.

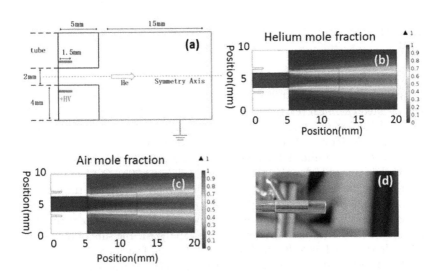

FIGURE 4.27 (a) Simulation geometry. The pulsed DC voltage is applied to the inner electrode. The outer electrode is grounded. (b) The mole fraction of helium. (c) The mole fraction of air. (d) The plasma jet device employed in the experiment. (From Liu, X.Y. et al., *Plasma Sources Sci. Technol.*, 23, 035007, 2014.)

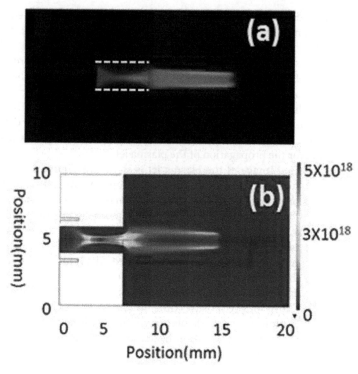

FIGURE 4.28 (a) The image of a plasma jet with an exposure time of 1 s. (b) The electron density distribution at 180 ns (simulation result). (From Liu, X.Y. et al., *Plasma Sources Sci. Technol.*, 23, 035007, 2014.)

Although the plasma channel is electrically neutral, it is interesting to analyze the density distribution of different positive ions because of the complicated chemical reactions in the mixing layer. For the helium species, He^+ and He_2^+ were concentrated in the streamer head and inside the tube, respectively (**Figure 4.29a and d**). The distribution of N_2^+ in the plasma channel was more uniform than the other species, and its peak value was even higher than that of He^+ in the streamer head (**Figure 4.29b**). O_2^+ had the same distribution profile as N_2^+, and its density peak value was almost the same as N_2^+; the slight difference was that most of them were confined in the region closer to the streamer head (**Figure 4.29e**). Because of the high oxidative stress of O and O(1d), their distributions were also analyzed. The sum of the O and O(1d) densities is 1.58×10^{18} m^{-3} (**Figure 4.29c and f**), which is higher than the N_2^+ density and is the key reason for the plasma jet's reactivity. The high-density regions of O and O(1d) suggest the proper distance between the tube exit and the object for applications.

To understand the different distributions of positive ions and to examine the role of reactions in streamer propagation, **Figure 4.30** shows the distribution of ionization reactions for each species. The electron impact ionization of He

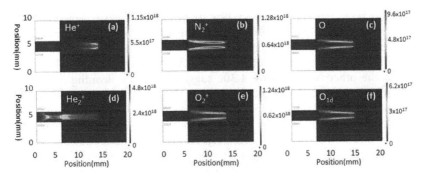

FIGURE 4.29 Density distribution of plasma species (m^{-3}) at 180 ns. (a) He$^+$, (b) N$_2^+$, (c) O, (d) He$_2^+$, (e) O$_2^+$, (f) O$_{1d}$. (From Liu, X.Y. et al., *Plasma Sources Sci. Technol.*, 23, 035007, 2014.)

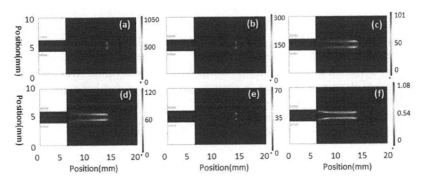

FIGURE 4.30 Ionization reactions rate (mol m^{-3}) distribution at 180 ns. (a) Helium related ((R3)–(R5) in Table 4.1), (b) ionization of N$_2$ by electron (R23), (c) Penning ionization of O$_2$ by He* (R46), (d) Penning ionization of N$_2$ by He* and He$_2^*$ ((R42) and (R43)), (e) ionization of O$_2$ by electron (R35) and (f) Penning ionization of O$_2$ by He$_2^*$ (R47). (From Liu, X.Y. et al., *Plasma Sources Sci. Technol.*, 23, 035007, 2014.)

and He* was concentrated in the streamer head (**Figure 4.30a**), and the peak ionization rate occurred off axis at a radial position of ~0.5 mm from the axis. Their ionization rate is much higher than that of the other cases; therefore, He$^+$ dominated the ion species in the streamer head. The peak ionization rates of N$_2$ and O$_2$ are only 28.6% and 7% of that of He (**Figure 4.30b** and e). Because the electron energy distribution function is assumed to be Maxwellian, the ionization rates of He, N$_2$ and O$_2$ by electrons are functions of the density of each species, n_e and ε. For the same distribution of electron energy density ($n_e \times \varepsilon$), the electron impact ionization is a function of the density of He, N$_2$, and O$_2$. Although the ionization energy of helium is 24.6 eV, which is higher than the 15.6 eV of N$_2$ and the 12.6 eV of O$_2$, the mole fractions of He, N$_2$, and O$_2$ are 0.902, 0.08, and 0.018, respectively, which means the ionization of helium is more probable. In addition, the peak value of the rotational and vibrational reaction rates of N_2 is 1.39×10^5 mol m^{-3} s^{-1}, which is much higher than 300 mol m^{-3} s^{-1}; therefore, most of the applied power is consumed by these rotational and vibrational reactions ((R17)–(R22) and (R24)–(R29), Table 3.1) even if they only need less than 5 eV of electron energy [13].

On the other hand, **Figure 4.30e** suggests that the Penning ionization rate of N_2 by He* and He$_2^*$ ((R42) and (R43), Table 3.1) decreases slowly along the channel behind the streamer head. The Penning ionization rate of O_2 by He* has the same distribution (**Figure 4.30c**). This is attributed to the long lifetime and high density of He* and He$_2^*$. Although the peak value of Penning ionization is only 11.2% of the peak electron impact ionization rate of He and He*, the space-averaged value shown in **Figure 4.30** suggests that the contribution of Penning ionization to the production of positive ions is higher than the sum of electron impact ionization at 180 ns. Therefore, N_2^+ is the dominant positive ion in the plasma channel. Also, the electrons produced by this Penning ionization increase the electric conductivity of the plasma channel and facilitate the ring-shaped streamer propagation. A comparison between the space-averaged ionization rate at 120 ns and that at 180 ns (**Figure 4.31**) indicates more reactions in the mixing layer gradually cause the N_2 and O_2 species to dominate the plasma during the streamer advancement [13]. To better understand of the factors governing the streamer structure and the role of Penning ionization in various gas mixtures, further experiments and simulations are needed.

The effects of the surface discharge, the surrounding gas diffusion, and Penning ionization on the plasma jet ring shape is investigated in Ref. [41].

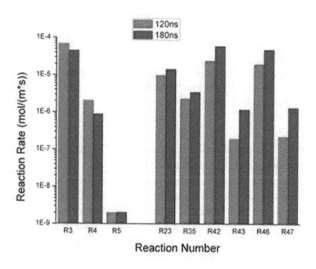

FIGURE 4.31 Space-averaged ionization rate of helium (electron impact ionization, (R3)–(R5) in Table 4.1), N_2 (electron impact ionization, (R23)), O_2 (electron impact ionization, (R35)), N_2 (Penning ionization, (R42) and (R43)) and O_2 (Penning ionization, (R46) and (R47)). (From Liu, X.Y. et al., *Plasma Sources Sci. Technol.*, 23, 035007, 2014.)

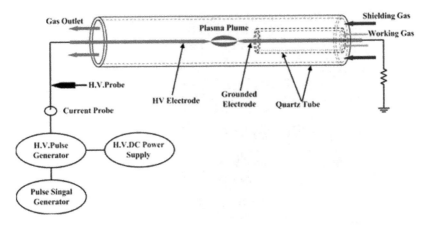

FIGURE 4.32 Schematic of the experiment setup. (From Xian, Y. et al., *Plasma Process. Polym.*, 11, 1169–1174, 2014.)

Figure 4.32 depicts the schematic of the experiment setup. The setup basically includes two quartz tubes and two electrodes. The outer tube has an inner diameter of 10 mm, and the inner tube has an inner diameter of 4 mm and an outer diameter of 6 mm. The two tubes are coaxial. The distance between the left ends of the two tubes is 65 mm. Two stainless-steel needle electrodes are placed on the axial. The diameter of the needle electrodes is 0.4 mm. The right electrode is connected to the ground through a 200 kV resistor while the left electrode is connected to a pulsed DC power supply. By moving the inner tube horizontally, the left tip of the right electrode could be in or out of the inner tube. The gap between the two electrodes is about 27 mm. The working gas (He, Ar, or Kr) flows from the right end of the inner tube at a flow rate of 4 L/min and the shielding gas flows through the outer tube at 2 L/min. The shielding gases are air, N_2, O_2, Ar, He, and Kr for different cases. The applied voltage is adjustable up to 10 kV. The frequency and the pulse width of the pulse voltage are fixed at 4 kHz and 1 ms, respectively. **Figure 4.33** shows a photograph of plasma plumes that use He as the working gas and different types of gas as the shielding gas. It shows clearly that the plasma plumes have a ring-shaped structure for all the shielding gases. They have a spindle shape and concentrate on the two electrodes at both ends. Such a result illustrates that the diffusion between the working gas and the ambient gas definitely plays an important role in the formation of the ring-shaped structure of the plasma bullet. Since the radius of the plasma jets is limited by the surrounding gas, the structure of the plasma jets could also be affected by the radius limitation effect. On the other hand, when Ar and Kr are used as the working gas, as shown in **Figures 4.34** and **4.35**, the discharge between the two electrodes is filamentary because the energy of the metastable states of Ar and Kr is lower than the ionization potential of the shield gas. Thus, Penning ionization plays no role in these cases.

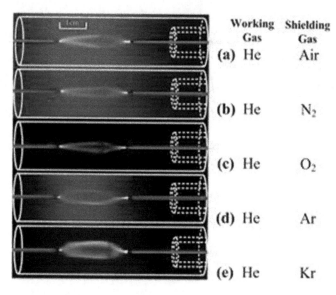

	Working Gas	Shielding Gas
(a)	He	Air
(b)	He	N_2
(c)	He	O_2
(d)	He	Ar
(e)	He	Kr

FIGURE 4.33 Photographs of plasma plumes. The working gas is He, and the shielding gases are air, N_2, O_2, Ar, Kr for (a–e), respectively. (From Xian, Y. et al., *Plasma Process. Polym.*, 11, 1169–1174, 2014.)

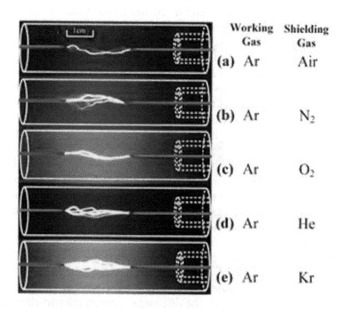

	Working Gas	Shielding Gas
(a)	Ar	Air
(b)	Ar	N_2
(c)	Ar	O_2
(d)	Ar	He
(e)	Ar	Kr

FIGURE 4.34 Photographs of plasma plumes. The working gas is Ar, and the shielding gases are air, N_2, O_2, He, Kr for (a–e), respectively. (From Xian, Y. et al., *Plasma Process. Polym.*, 11, 1169–1174, 2014.)

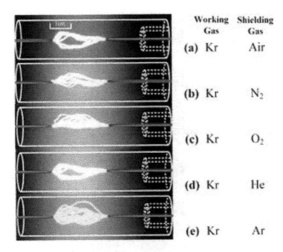

FIGURE 4.35 Photographs of plasma plumes. The working gas is Kr, and the shielding gases are air, N_2, O_2, He, Ar for (a–e), respectively. (From Xian, Y. et al., *Plasma Process. Polym.*, 11, 1169–1174, 2014.)

When He mixed with 2.5% of N_2, O_2, He, Ar, or Kr is used, the ring-shaped structure disappears, as shown in **Figure 4.36**. This is because the atoms or molecules of the added gases have a lower ionization energy than the energy potential of the metastable state He, so they could be ionized by the metastable state He. As a result, the ring-shaped structure disappears. This result further demonstrates that the diffusion of the surrounding gas and Penning ionization play an important role in the formation of the ring-shaped structure. When the main working gas is Ar or Kr mixed with a small amount of another gas, as shown in **Figures 4.37** and **4.38**, the discharge is always filamentary.

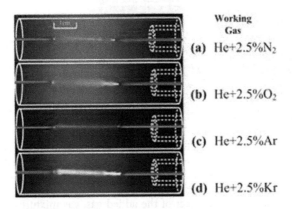

FIGURE 4.36 Photographs of plasma plumes for He mixed with 2.5% of (a) N_2, (b) O_2, (c) Ar and (d) Kr, respectively. The shielding gas is air for all the photographs. (From Xian, Y. et al., *Plasma Process. Polym.*, 11, 1169–1174, 2014.)

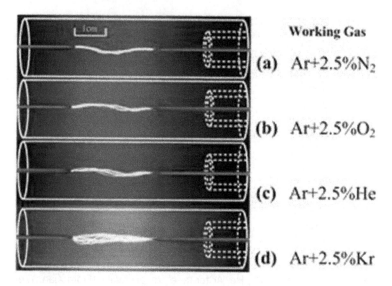

FIGURE 4.37 Photographs of plasma plumes for Ar mixed with 2.5% of (a) N$_2$, (b) O$_2$, (c) He, and (d) Kr, respectively. The shielding gas is air for all the photographs. (From Xian, Y. et al., *Plasma Process. Polym.*, 11, 1169–1174, 2014.)

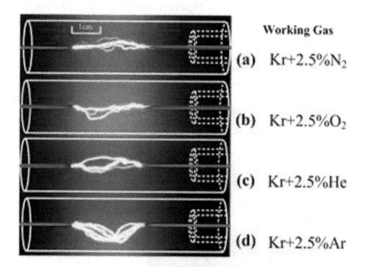

FIGURE 4.38 Photographs of plasma plumes for Kr mixed with 2.5% of (a) N$_2$, (b) O$_2$, (c) He, and (d) Ar, respectively. The shielding gas is air for all the photographs. (From Xian, Y. et al., *Plasma Process. Polym.*, 11, 1169–1174, 2014.)

To better understand the effect of the added gas, gas mixtures of He mixed with different percentages of N$_2$, O$_2$, and Kr were used. The shielding gas was air. When 2%, 3.5%, and 6.5% of N$_2$ or Kr is added to the working gas He, respectively, as shown in **Figures 4.39** and **4.40**, the plasmas are always solid.

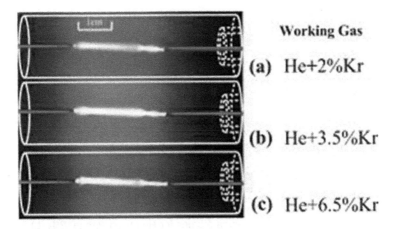

FIGURE 4.39 Photographs of plasma plumes for He mixed with (a) 2%, (b) 3.5%, and (c) 6.5% of Kr, respectively. The shielding gas is air. (From Xian, Y. et al., *Plasma Process. Polym.*, 11, 1169–1174, 2014.)

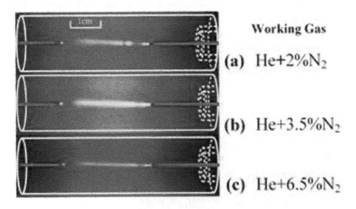

FIGURE 4.40 Photographs of plasma plumes for He mixed with (a) 2%, (b) 3.5%, and (c) 6.5% of N_2, respectively. The shielding gas is air. (From Xian, Y. et al., *Plasma Process. Polym.*, 11, 1169–1174, 2014.)

On the other hand, when 2% or 3.5% of O_2 is added to the working gas He, the plasma is still solid. However, when the percentage of O_2 is increased to 6.5%, the plasma has ring-shaped structure, as shown in **Figure 4.41**. This is not understood at the moment. It may be because the high concentration of O_2 in the core zone of the plasma plume results in the strong attachment of the electrons, thus causing low electron density in the core zone.

The effect of the dielectric tube, and thus the effect of surface discharge on the ring-shaped structure of the plasma bullet, was also investigated. To eliminate the influence of air diffusion, 2.5% of a different gas is added to the main working gas He. The right electrode is placed in the inner tube with its tip

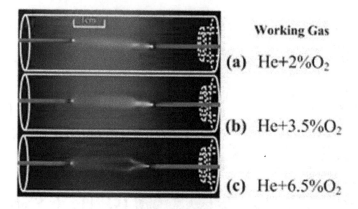

FIGURE 4.41 Photographs of plasma plumes for He mixed with (a) 2%, (b) 3.5%, and (c) 6.5% of O_2, respectively. The shielding gas is air. (From Xian, Y. et al., *Plasma Process. Polym.*, 11, 1169–1174, 2014.)

1 cm away from the left end of the inner tube. The applied voltage is 4 kV. As shown in **Figure 4.42**, the plasma plumes can be divided into three sections. The first section is in front of the right electrode; the second section is the surface discharge on the wall of the quartz tube; and the third section is between the exit of the tube and the left electrode. We can see that the last section has a ring-shaped structure. Since the influence of gas diffusion is eliminated by adding a small amount of the atom and molecule gases, the ring-shaped structure that appears in this case must be due to the effect of the inner tube, i.e., the surface discharge.

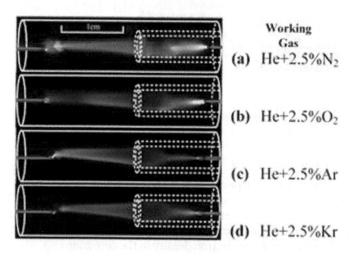

FIGURE 4.42 Photographs of plasma plumes with the right electrode placed in the inner tube. He mixes with 2.5% of (a) N_2, (b) O_2, (c) Ar, and (d) Kr. The tip of the right electrode is 1 cm away from the exit. The applied voltage is 4 kV. (From Xian, Y. et al., *Plasma Process. Polym.*, 11, 1169–1174, 2014.)

The reactive species produced by plasma jets are quite important for the application. To investigate the OH distribution and production in APPJs, the temporally and spatially resolved LIF technology is adopted. Two-dimensional OH-LIF intensities are measured in the APPJ without a ground electrode and with a ground electrode, independently. The results indicate that OH distributes as a donut shape in the plasma plume for the first several high-voltage (HV) pulses. However, with more HV pulses applied, the core of the plasma plume is filled with high-density OH and the donut shape disappears. Further analysis suggests that the OH radicals outside the plasma jet nozzle are generated from two parts: (1) the plasma plume outside the tube, which results in the donut-shape distribution, and (2) the intense reactions inside the tube, which lead to the disappearance of the donut shape [42].

Figure 4.43a and 1b shows the image of a plasma jet with an exposure time of 100 ms and the schematics of a laser diagnostics system, respectively. For the setup of the plasma jet, as shown in **Figure 4.43b**, there are two dielectric tubes. The inner tube (inner diameter 2 mm and outer diameter 4 mm) is used to

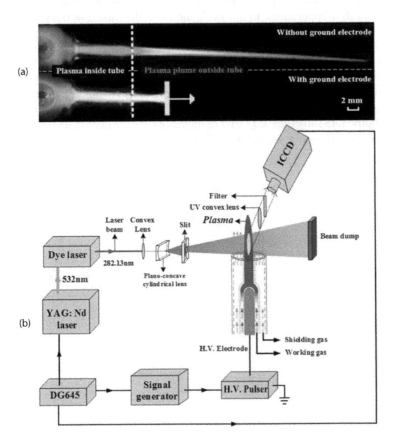

FIGURE 4.43 (a) The photograph of the plasma jet (b). The schematic of the OH LIF experimental set up. (From Yue, Y. et al., *J. Appl. Phys.*, 121, 033302, 2017.)

generate the plasma jet with the working gas (helium) flowing at a rate of 2 L/min (~10.6 m/s). The outer tube (inner diameter 18 mm) is applied to control the humidity of the surrounding gas, whose fluctuation may have a great effect on OH production. In addition, the shielding gas (80% N_2, 20% O_2) is introduced coaxially with a flow rate of 4 L/min corresponding to ~0.26 m/s. A pinlike electrode for generating the plasma is covered by a dielectric tube (thickness 1 mm), and it is placed in the center of the inner tube with the tip 20 mm away from the nozzle. The radius of curvature of the electrode tip is 1.5 mm. The plasma jet is driven by a high-voltage pulse generator with a repetitive frequency of 8 kHz, a duration time of 1 ls, and an applied voltage of 8 kV. It is operated under two conditions: (1) without the ground electrode; and (2) with the ground electrode. A ground-connected metal plate, which is not shown in **Figure 4.43**, serves as the ground electrode and is placed 5 mm from the exit of the tube. The voltage and current waveforms are also illustrated in **Figure 4.44**. The consumed power for producing the plasma is less than 0.1 W and ~1.8 W for conditions (1) and (2), respectively. It is well known that the OH intensity increases dramatically when adding tens to hundreds ppm of H_2O into the working gas, which is beneficial to many kinds of applications. In this work, therefore, the chosen working gas is He with seed gas 40 ppm H_2O monitored by a Dewpoint Tramsmitter (DPT-500). As for the humidity in the shielding gas, it is estimated that the H_2O concentration is at ~20,000 ppm based on the temperature and relative humidity in our atmosphere (temperature 25°C and relative humidity 65%). Thus, the humidity in our shielding gas is set at 20,000 ppm. The regulation of humidity is achieved by means of water bubble vapor.

Figure 4.45 shows the 2D OH-LIF intensity near the plasma jet nozzle with different experimental conditions. All the fluorescence images are captured at

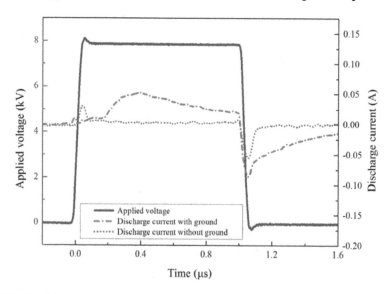

FIGURE 4.44 The waveforms of applied voltage and discharge current. (From Yue, Y. et al., *J. Appl. Phys.*, 121, 033302, 2017.)

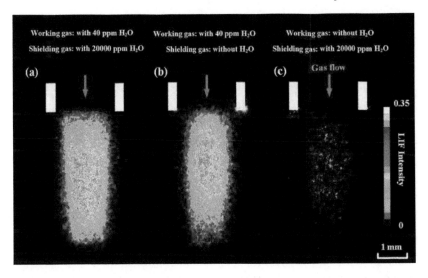

FIGURE 4.45 Two-dimensional OH-LIF intensity without using the ground electrode when mixing H_2O into (a) both the working gas and the shielding gas (b) the working gas only and (c) the shielding gas only. (From Yue, Y. et al., *J. Appl. Phys.*, 121, 033302, 2017.)

5 ls after the rising edge of the HV pulse, and the plasma jet is operated with continuous HV pulses. In **Figure 4.45**, H_2O is added into: (a) both the working gas and the shielding gas; (b) the working gas only; and (c) the shielding gas only. It is obvious that the OH distributes as a solid disk in the radial direction when adding H_2O into the working gas. The humidity in the shielding gas shows little difference in the shape of OH density compared to **Figure 4.45a** and **b**.

What is more, when the H_2O is mixed into only the shielding gas, weak fluorescence signals are detected in the core of the plasma jet, which may be due to the working gas impurity.

In helium atmospheric-pressure plasma jets, the lifetime of OH lasts several ms [43]. Within its lifetime, the OH generated by new pulses can be overlapped by accumulated OH, which was produced by previous pulses. Thus, to further understand the OH production mechanism, the trigger pulses for the HV power supply were modulated so that 100 pulses were artificially deleted for every 800 pulses by a signal generator. The same method was used in our previous study [44]. In this way, thus, OH has enough time (12.5 ms) to decay by diffusion and termination reactions. When the decay is checked, it is observed that the OH-LIF intensity reaches background emission intensity at the end of 100 deleted pulses. Under this condition, an investigation of the time-resolved 2D LIF is conducted to study the OH production during the first several pulses after the decay, as shown in **Figure 4.46**. The plasma jet is operated with 40 ppm H_2O in the working gas and 20,000 ppm H_2O in the shielding gas similar to **Figure 4.45a**. The fluorescence is captured at 5 μs after the rising edge of (a) the 1st pulse, (b) the 2nd pulse, and (c) the 3rd pulse. Obviously, it is found that the OH intensity distributes as a donut shape. Specifically, OH intensity increases gradually at the margin of the effluent

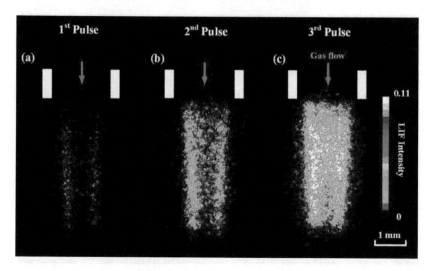

FIGURE 4.46 Time-resolved OH intensity generated by the first three pulses without using the ground electrode. The fluorescence is captured at 5 μs after the rising edge of (a) the 1st pulse, (b) the 2nd pulse, and (c) the 3rd pulse. (From Yue, Y. et al., *J. Appl. Phys.*, 121, 033302, 2017.)

with the first three pulses. With more HV pulses applied, the center will be filled with OH and the structure changes to a solid disk as in **Figure 4.45a**. It is observed that the OH density generated by the first three pulses is much lower than that produced by continuous pulses, as shown in **Figure 4.45a** and **b**.

In the second case, the ground electrode was placed 5 mm from the exit of the tube. First, OH distribution was investigated with the continuous HV pulses. As shown in **Figure 4.47**, the plasma jet is operated with (a) H_2O in both the working gas and the shielding gas, (b) H_2O in the working gas only, and (c) H_2O in the shielding gas only. The fluorescence is also captured at 5 ls after the rising edge of the pulse. It is obvious in **Figure 4.47a** and **b** that OH also distributes like a solid disk when adding H_2O into the working gas in the radial direction, as shown in **Figure 4.45a** and **b**. Moreover, the difference in LIF intensity can be neglected for the case of with and without humidity in the shielding gas, which is similar to that in **Figure 4.45**. Compared to **Figure 4.45**, however, it is found that the OH intensity is almost three times higher. And when the humidity exists in only the shielding gas, OH density in the margin is enhanced and distributes as a donut shape compared to the weak fluorescence in **Figure 4.45c**. The difference may be attributed to the dissociations by higher electron density and its ring-shape distributions in the effluent when introducing the ground electrode.

Similar to **Figure 4.46**, **Figure 4.48** investigates the production of OH by time-resolved LIF using modulated HV pulses. Except for the use of the ground electrode, the experimental conditions are the same as in **Figure 4.46**. Again, the decay is checked at the end of 100 deleted pulses. The phenomenon of the donut shape–distributed OH is even more obvious than in **Figure 4.46**. From the 1st pulse to the 3rd pulse, OH intensity is fortified

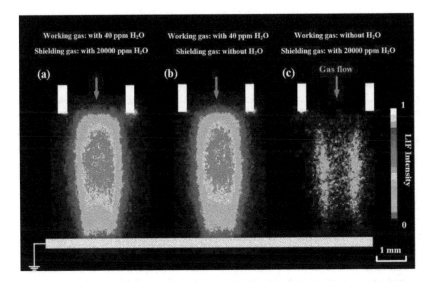

FIGURE 4.47 Two-dimensional OH-LIF intensity while using the ground electrode when mixing H_2O into (a) the working gas and the shielding gas, (b) the working gas only, and (c) the shielding gas only. (From Yue, Y. et al., *J. Appl. Phys.*, 121, 033302, 2017.)

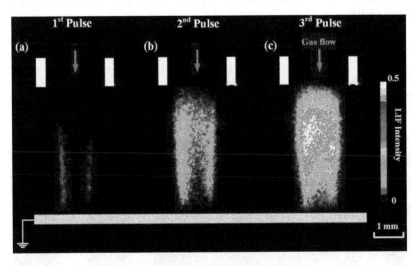

FIGURE 4.48 Time-resolved OH intensity generated by the first three pulses using the ground electrode. (a) the 1st pulse, (b) the 2nd pulse, and (c) the 3rd pulse. (From Yue, Y. et al., *J. Appl. Phys.*, 121, 033302, 2017.)

at the edge of the effluent. Simultaneously, OH-LIF intensity increases in the core of the plasma jet, as illustrated in **Figure 4.48c**. With more HV pulses applied, the OH density is able to develop a solid-disk structure, which is shown in **Figure 4.47a**.

Illustrated in **Figures 4.46** and **4.48,** the OH density distributes as a donut shape for the first several HV pulses. Then, the OH density in the core of the effluent increases gradually and finally distributes as a solid disk with more HV pulses applied. The phenomenon is especially obvious in **Figure 4.48b–c.** In order to understand the filling process in the core, an investigation of the time-resolved LIF is conducted from the 2nd pulse to the 3rd pulse in **Figure 4.48,** which is shown in **Figure 4.49.** The fluorescence is captured at different times after the rising edge of the 2nd pulse. The other experimental conditions are the same as in **Figure 4.48.** It is clear that OH distributes as a donut shape to a solid disk from 5 to 120 μs after the 2nd pulse and before the 3rd pulse. With the 3rd pulse applied, the OH density at the margin is enhanced and distributes as a transitional structure from the donut shape to the solid disk. To better understand the OH distribution, it is necessary to discuss the kinetics of OH. As we all know, the reaction pathways to produce OH are strongly dependent on the gas temperature (T_g), electron temperature (T_e), and electron density (n_e). For our plasma jet, the gas temperature is about 300 K; electron temperature is in a range of a few eV; electron density is in the order of 10^{12} cm^3. Thus, according to the literature, several main reaction pathways to generate the OH radicals can be concluded. The principal reaction is considered to be electron dissociation of H_2O:

$$e + H_2O \rightarrow H + OH + e. \tag{4.1}$$

The dissociation by radicals and metastable state N_2 also enhances OH density, as follows:

$$O(^1D) + H_2O \rightarrow 2OH, \tag{4.2}$$

$$N_2(A^3\Sigma^+) + H_2O \rightarrow N_2 + OH + H, \tag{4.3}$$

where $O(^1D)$ is a singlet oxygen generated by a collisional dissociation and $N_2(A^3\Sigma^3)$ is a metastable nitrogen molecule generated by electron impact excitation.

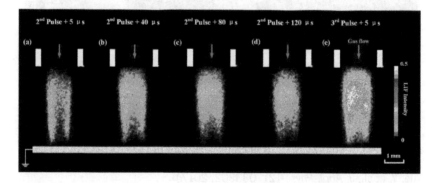

FIGURE 4.49 OH-LIF intensity from the 2nd pulse to the 3rd pulse using ground electrode. (a) 5 μs after the rising edge of the 2nd pulse, (b) 40 μs after the rising edge of the 2nd pulse, (c) 80 μs after the rising edge of the 2nd pulse, (d) 120 μs after the rising edge of the 2nd pulse, (e) 5 μs after the rising edge of the 3rd pulse. (From Yue, Y. et al., *J. Appl. Phys.*, 121, 033302, 2017.)

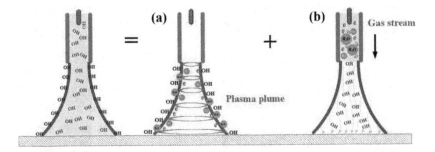

FIGURE 4.50 The schematic of the OH production mechanism. (a) the OH produced in the plasma plume outside the tube. (b) the OH produced inside the tube.

When the electron energy is low (<2 eV), metastables are also considered to be beneficial to OH production by an indirect pathway:

$$He_m + H_2O \rightarrow He + H_2O^+ + e, \tag{4.4}$$

$$e + H_2O^+ \rightarrow OH + H. \tag{4.5}$$

Based on the phenomenon in **Figure 4.49**, it is suggested that total OH radicals are generated from two parts. The first part is produced in the plasma plume with a donut-shape structure outside the tube. The other part is generated inside the tube with a solid-disk structure. The schematic is shown in **Figure 4.50**.

For the first part, it is well known that the plasma plume, which looks continuous to the naked eye, are actually bullet-like volumes of plasma traveling at a speed of 10^4 m/s [45]. Further investigation shows that the plasma bullet propagates as a donut-shape structure in the plasma plume; namely, the optical emission intensity reaches peaks at the margin of the effluent. Moreover, it is also found that the electric field intensity and electron density also distribute as a donut shape in the plasma plume. In order to further study OH production by plasma bullets in our work, a simulation using COMSOL was conducted to investigate the distribution of electron density, which relates to the principal pathway of OH generation (Reaction 4.1). **Figure 4.51** shows the 2D electron density at 120 ns after the rising edge when (a) no ground electrode is applied and (b) a ground electrode is used. The rising time and voltage are 50 ns and 2 kV, respectively; the other conditions are the same as in **Figures 4.46** and **4.48**. The core solvers, based on our previous work, comprise a couple sets of governing equations for charged and neutral species continuity, electron energy transport, and self-consistent electrostatic potential. The ICCD images, shown in **Figure 4.51**, show the plasma bullets (after the rising edge) traveling to about 3 mm from the exit of the tube. It is clear that the plasma bullet propagates as a donut-shape structure whether or not the ground electrode is used. Furthermore, from **Figure 4.51**, the electron density also distributes as a ring-shaped structure. The donut shape–distributed electron density contributes to the principal pathway of OH production (Reaction 4.1). It also enhances the production of O(1D), $N_2(A^3\Sigma^+)$, and He_m, which are beneficial to OH generation according to Reactions 4.2 through 4.5.

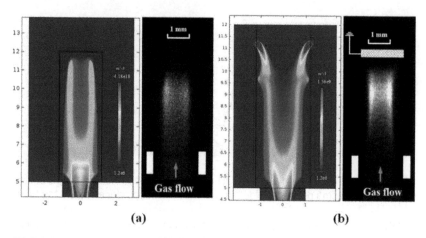

FIGURE 4.51 Two-dimensional electron density and ICCD image of plasma bullets. (a) no ground electrode is applied and (b) a ground electrode is used. (From Yue, Y. et al., *J. Appl. Phys.*, 121, 033302, 2017.)

Therefore, for OH produced by a plasma bullet, the distribution suggests a donut shape, which is shown in **Figures 4.46** and **4.48.**

Comparing **Figure 4.51a** and **b,** the electron density when using the ground electrode is much higher than without using the ground electrode, which promotes OH generation. As expected, the OH intensity is three times higher (refer also to **Figure 4.47** and **Figure 4.45**). Moreover, due to the higher electron density in the margin, the OH density shows a donut-shape distribution for dehumidified working gas in **Figure 4.47c.** It is observed that there are two discharges in one pulse period, specifically at the rising edge and the falling edge. Considering the optical emission intensity and consumed power, it is suggested that the main production of OH is from the discharge at the rising edge [46]. Because of this suggestion, the simulation and ICCD image in **Figure 4.51** were both obtained after the rising edge. In this work, the size of the plasma bullet is in mm. Some investigations show that the properties of the plasma are significantly changed when the size is further reduced to the nanometer domain. Furthermore, the nanoplasma, defined by Ostrikov et al., has great application prospects in nanoscale synthesis and processing [47].

With more HV pulses applied, the OH distribution will transition from a donut shape to a solid disk, as shown in **Figure 4.49.** It is suggested that OH generated inside the tube contributes to it, which is the other part of the OH source, as shown in **Figure 4.50.** For the other part, the high density of OH radicals is generated inside the tube because of the impact of high electron density (10^{12}–10^{13} cm^3) and high electric field intensity (20 Td around the electrode) on OH production according to Reactions 4.1, 4.2, 4.4, and 4.5. With the gas stream, the generated OH is carried downstream to the effluent. The structure of this part of OH is a solid disk. As shown in **Figure 4.52,** in the 2nd cycle, the center of the effluent is filled over time with OH, which may be derived from upstream inside the tube. To verify this, the investigation in **Figure 4.52** was conducted.

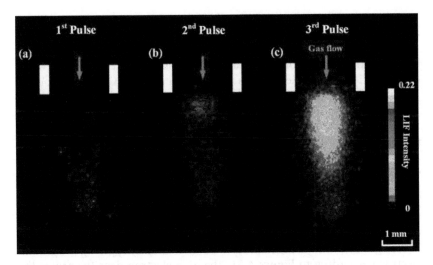

FIGURE 4.52 OH-LIF density by blocking the propagating of plasma bullets. (a) the 1st pulse, (b) the 2nd pulse, and (c) the 3rd pulse. (From Yue, Y. et al., J. Appl. Phys., 121, 033302, 2017.)

The experimental conditions are basically the same in **Figures 4.51** and **4.52**. The only difference is that a thin wire with a diameter of 0.1 mm, which is connected to the ground through a 1-MX resistor, is placed at the exit of the tube to block the propagation of plasma bullets. The fluorescence is also captured at 5 ls after the rising edge of the 1st pulse, the 2nd pulse, and the 3rd pulse. It is presented that OH radicals propagate from the exit of the tube to the effluent in the axial direction from the 2nd pulse to the 3rd pulse with a solid-disk structure. According to the OH traveling distance, the velocity is estimated to be about 16 m/s, which approaches the gas velocity (~10.6 m/s). It further confirms that the ground state OH radicals are able to propagate with the gas stream within their lifetime. Moreover, referring to **Figure 4.49**, which shows the transition from a donut shape to a solid disk in the 2nd cycle, it is proven that the additional OH in the core of the effluent comes from upstream in the tube. Moreover, the disappearance of the donut shape is achieved by this filling process with the gas flow effect.

Whether the ground electrode is applied, total OH radicals are produced from two parts: the plasma plume outside the tube and the intense reactions inside the tube. Through the dissociation of the plasma bullet outside the tube, OH at the margin is enhanced and forms a ring-shaped structure. According to the simulation results, it may be due to the donut shape–distributed electron density. With more HV pulses applied, the ring-shaped structure disappears. It is a result of the gas phase–driven OH propagating from upstream inside the tube to downstream outside the tube. As the OH fills the core of the effluent, the donut shape disappears and a solid-disk structure is detected. It is worth noting how the OH distributes as a donut shape in **Figure 4.47c** as well as a solid disk in **Figure 4.45c**. It may be due to the higher electron density in the margin of the effluent when the applying the ground electrode according to **Figure 4.51**.

In **Figure 4.45c**, although OH and electrons in plasma bullet distribute as a donut shape, OH can still be produced by intense reactions inside the tube because of the gas impurity. In addition, the donut shape is overlapped by OH transported from upstream. It is also suggested that the water concentration in the shielding gas may have an effect on OH generation. As analyzed above, the humidity in shielding gas only functions through the plasma bullet process outside the tube. Avoiding the severe OH overlapping from the upstream, the investigation of the content of H_2O in shielding gas on OH production with dehumidified working gas is conducted in **Figure 4.53**. The other conditions are the same as in **Figures 4.45c** and **4.47c**. The integration window is w2 × h5 mm^2 outside of the tube to collect OH-LIF signals. The inset images show the OH-LIF maps in the effluent. The OH intensity increases with the humidity in the shielding gas when the ground electrode is applied. However, the OH density remains almost constant, with small fluctuations when the ground electrode is not applied. This might be due to the enhanced electron density by applying the ground electrode according to **Figure 4.51**. The high electron density ensures the forward process of the principal reaction and promotes OH generation. Note that at 0 ppm, the LIF intensity is not at zero, which is probably caused by gas impurity.

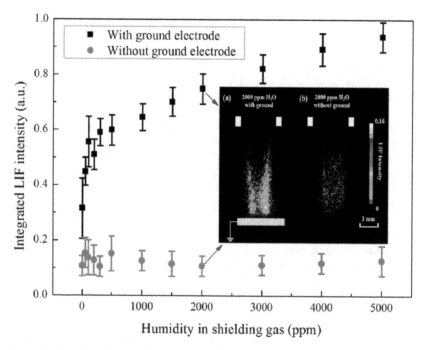

FIGURE 4.53 Relation of OH intensity and humidity in the shielding gas. (From Yue, Y. et al., *J. Appl. Phys.*, 121, 033302, 2017.)

4.1.4 Repeatability of Plasma Bullet

The streamer propagation in uniform media is known to be not repetitive and always reproducible. The streamers do not propagate along the same path and over the same distance during the same time interval. For the plasma jet, the plasma plume propagation is thought to behave in a repetitive and reproducible mode: it consistently propagates along the same distance during the same time interval. However, recent studies suggest this phenomenon does not happen under certain conditions [3,32,48]. The plasma bullet does not propagate along the same path or same distance for different shots under specific conditions. For the pulsed-DC plasma jet, when the applied voltage is 8 kV, **Figure 4.54** clearly indicates that the locations of plasma bullets are irregular at the same time, and the discharge is rather unstable. As the applied voltage increases to 9 kV, **Figure 4.55** indicates that the plasma bullet appears repeatable [48]. A similar chaotic mode with nonperiodic current oscillations is also captured by average power spectra for the kHz AC driving plasma jet [3].

Besides the applied voltage, decreasing the pulse frequency to a certain level can also lead to significant changes in the plasma bullet propagation dynamics. The dynamics of the plasma bullet remain similar as the pulse repetition

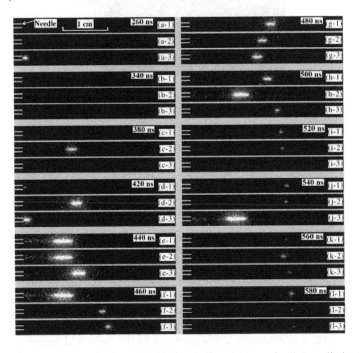

FIGURE 4.54 High-speed photographs of the plasma plume for the applied voltage of 8 kV. Due to the randomness of the discharges, three photographs are taken for every delay time. The plasma plume at (a1-3) 260 ns, (b1-3) 340 ns, (c1-3) 380 ns, (d1-3) 420 ns, (e1-3) 440 ns, (f1-3) 460 ns, (g1-3) 480 ns, (h1-3) 500 ns, (i1-3) 520 ns, (j1-3) 540 ns, (k1-3) 560 ns, (l1-3) 580 ns. (From Xian, Y. et al., *IEEE Trans. Plasma Sci.*, 37, 2068–2073, 2009.)

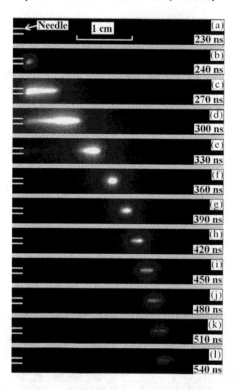

FIGURE 4.55 High-speed photographs of the plasma plume for the applied voltage of 9 kV. The plasma plume at (a) 230 ns, (b) 240 ns, (c) 270 ns, (d) 300 ns, (e) 330 ns, (f) 360 ns, (g) 390 ns, (h) 420 ns, (i) 450 ns, (j) 480 ns, (k) 510 ns, and (l) 540 ns. (From Xian, Y. et al., *IEEE Trans. Plasma Sci.*, 37, 2068–2073, 2009.)

frequency decreases from 10 kHz to 0.25 kHz (**Figure 4.56**) [33]; the location of plasma bullets was always the same for the same delay time. When the pulse frequency is further decreased to 0.2 kHz, although the visual appearance of the plasma plume does not change, the ICCD images indicate that the position of the plasma bullet is not the same for the same delay time for different shots (**Figure 4.57**). The electron attachment to oxygen and electron-ion recombination are thought to be the possible reasons. The duty-off time increases as the pulse frequency increases; therefore, the seed electron density decreases just before the next pulse. This may cause the observed randomness of the delay time of plasma bullet formation for the low pulse repetition frequency. Another possible cause of the random delay times of the discharge ignition at low frequencies is the stochastic appearance of initial electrons in the high-field region near the electrode. Till the frequency is sufficiently high, residual electrons produced in the preceding discharge pulse are present near the electrode. At low frequencies, the residual electrons are swept away by the gas flow before the beginning of the following pulse. This argument is supported by the observation that the transition frequency separating the regimes of regular and random formation of plasma bullets increases almost linearly at higher gas flow rates.

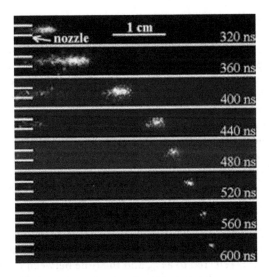

FIGURE 4.56 High-speed photographs of the plasma plume for the frequency of 0.25 kHz. Voltage: 7 kV, pulse width: 800 ns. Exposure time of ICCD camera: 20 ns. Single shot for each image. The time labeled on the photographs corresponds to the rise edge of the applied voltage. (From Wu, S. et al., *Curr. Appl. Phys.*, 13, S1–S5, 2013.)

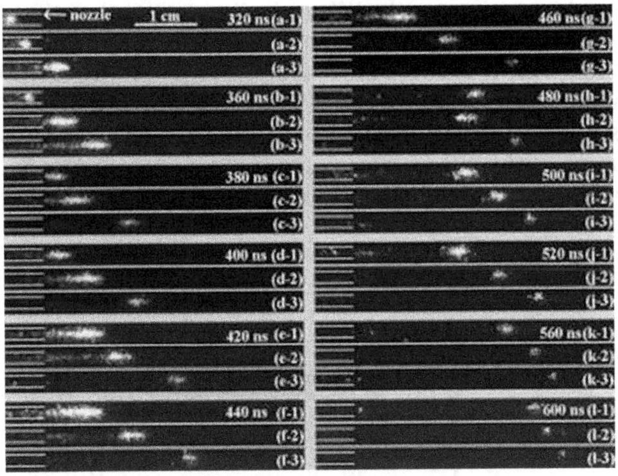

FIGURE 4.57 High-speed photographs of a discharge with the frequency of 0.2 kHz. Voltage: 7 kV, pulse width: 800 ns. Exposure time of ICCD camera: 20 ns. Single shot for each image. The time labeled on the photographs corresponds to the rise edge of the applied voltage. Due to the randomness of the discharges, three photographs are taken for every delay time. The plasma plume at (a1-3) 320 ns, (b1-3) 360 ns, (c1-3) 380 ns, (d1-3) 400 ns, (e1-3) 420 ns, (f1-3) 440 ns, (g1-3) 460 ns, (h1-3) 480 ns, (i1-3) 500 ns, (j1-3) 520 ns, (k1-3) 560 ns, and (l1-3) 600 ns. (From Wu, S. et al., *Curr. Appl. Phys.*, 13, S1–S5, 2013.)

The effects of the seed electrons n_{seed} on the repeatability of the plasma bullets were also investigated. The ignition delay and the propagation velocity of the plasma bullets are measured by two photomultiplier tubes (PMT). By adjusting the percentage of the gas mixture, the frequency of the applied voltages, and the pressure of the gas, the threshold seed electron density n_{seed} for plasma bullets propagating in repeatable mode was evaluated [49].

The schematic of the experiment setup is presented in **Figure 4.58**. The inner and outer diameters of the discharge tube are 6 mm and 8 mm, respectively. The left side of the quartz tube is terminated with a high-voltage electrode covered with a quartz tube. The high-voltage copper wire electrode is inserted into the concave part with an outer diameter of 3 mm. The applied voltage is unipolar square wave pulse with an amplitude of 8 kV and a pulse width of 2 μs fixed throughout the paper. The frequency of the applied voltage is adjustable up to 10 kHz.

The typical voltage and PMT waveforms are exhibited in **Figure 4.59**. The time between the peak of PMT-1 and the rising edge of the voltage (50% of its maximum) is referred to as t_d, and the time between the two PMTs is referred to as t_y. Since the distance l between the two PMTs is 2.5 cm, the propagation velocity of the plasma bullet can be calculated by $v = l/t_y$. If v is the same for the same discharge conditions, it is reasonable to assume that the propagation velocity of the plasma bullet from the electrode to the position of PMT-1 is also a constant. Under such conditions, the ignition delay time t_d can be used to judge the mode of discharge, i.e., repeatable or unrepeatable/random mode. When the jitter of t_d is less than a certain value, the plasma bullet propagation is considered to be in repeatable mode. If not, then it is considered to be in unrepeatable/random mode. In this manuscript, the chosen threshold of the jitter of t_d is 2 ns. If the jitter t_d is larger than 2 ns, the plasma bullet propagation is considered to be in unrepeatable mode. The reason that we chose 2 ns is because the propagation velocity of the plasma bullet is on the order of 10^5 m/s. Thus, it propagates about 0.2 mm in 2 ns, which is about the resolution of an ICCD camera [49].

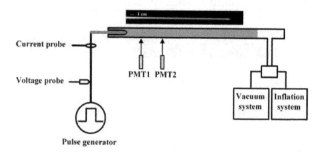

FIGURE 4.58 The schematic of the experiment setup. (From Nie, L. et al., *Phys. Plasmas*, 23, 093518, 2016.)

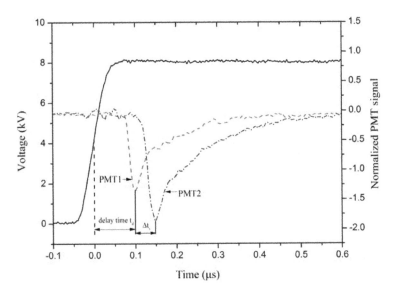

FIGURE 4.59 Typical applied voltage and the two PMT waveforms for an applied voltage of 8 kV, a frequency of 1 kHz and a working gas of He mixed with 2.5% O_2. Total gas pressure: 0.2 atmospheric pressure. (From Nie, L. et al., *Phys. Plasmas*, 23, 093518, 2016.)

Figure 4.60 shows the velocity of the plasma bullet for different frequencies with the O_2 percentages 0.5% and 10%. Each of the measurements is repeated 20 times, and the mean error bar is also presented in the figure. As we can see the value of the error bar is very small for all the experiment conditions, which means that under such conditions, the plasma propagation velocities are constant. In other words, if the plasma bullet propagation is in unrepeatable mode under such conditions, then it must be due to the inconsistency of the ignition time of the plasma bullet. When 0.5% of O_2 is used, the propagation velocity of the plasma bullet increases dramatically when the frequency of the applied voltage increases to more than 250 Hz. On the other hand, when 10% of O_2 is used, there are no dramatic changes to the propagation velocity for the ranges of frequency studied, and it is actually almost a constant. It should be pointed out that as the O_2 content increases, the required minimum frequency to ignite the plasma becomes higher, which is about 1 kHz for 10% of O_2. **Figure 4.61** shows the propagation velocities of the plasma bullet for a different percentage of O_2 at a fixed frequency of 2 kHz. It shows that the velocity increases with the increase of O_2 from 0% to 2.5%, and then it decreases if the O_2 percentage is increased further. It is also worth emphasizing that the error bar of the velocities is small for all data points. In other words, the plasma bullet propagation velocities are stable for all the conditions studied.

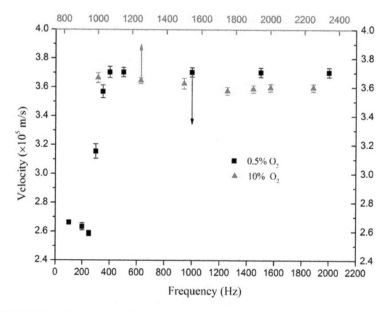

FIGURE 4.60 The propagation velocity of the plasma bullet vs. frequency for 0.5% and 10% of O_2. Total gas pressure: 0.2 atmospheric pressure. (From Nie, L. et al., *Phys. Plasmas*, 23, 093518, 2016.)

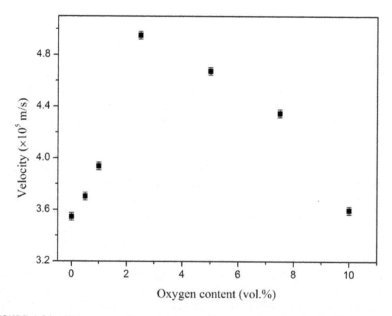

FIGURE 4.61 The propagation velocity vs. O_2 percentage for the fixed frequency of 2 kHz. Total gas pressure: 0.2 atmospheric pressure. (From Nie, L. et al., *Phys. Plasmas*, 23, 093518, 2016.)

Figure 4.62 shows the ignition delay time t_d for different O_2 percentages and different frequencies. It can be seen from **Figure 4.62** that both the jitter (error bar) of delay time t_d and the delay time t_d itself decrease with the increase of the frequency. As mentioned before, the plasma bullet propagation is considered to be in repeatable mode when the jitter of the t_d is less than 2 ns. The discharge mode transitions from random mode to repeatable mode with the increase of the frequency. This can be explained as follows: With the increase of the pulse frequency, the pulse period becomes shorter, and thus the pulse off time between pulses becomes shorter and the decay time of the electrons becomes shorter, which results in the relatively high seed electron density n_{seed}; so, the plasma propagation mode changes from random mode to repeatable mode. To understand more about the effect of N_2 on the repeatability of the plasma bullet, the ignition delay time t_d was studied for different frequencies with the N_2 percentages of 1% and 2.5%.

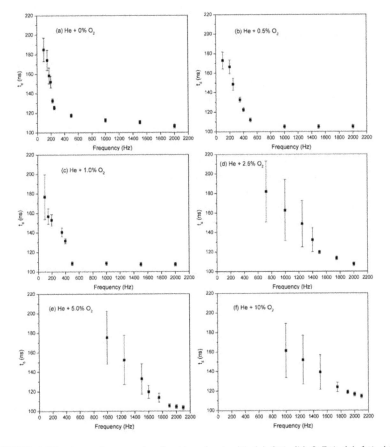

FIGURE 4.62 t_d vs. frequencies for He mixed with (a) 0%, (b) 0.5%, (c) 1%, (d) 2.5%, (e) 5%, and (f) 10% O_2. Total gas pressure: 0.2 atmospheric pressure. (From Nie, L. et al., *Phys. Plasmas*, 23, 093518, 2016.)

It should be pointed out that the discharge is difficult to ignite when a higher N_2 percentage is used (Nie et al. 2016). As described above in **Figure 4.62a**, f_{cri} is about 250 Hz for pure He gas.

On the other hand, it remains at about 250 Hz when 1% of N_2 is added, as shown in **Figure 4.63a**, and it decreases slightly to 190 Hz when 2.5% of N_2 is added (**Figure 4.63b**). It is worth pointing out that the jitter of the ignition delay time t_d is much larger when N_2 is added for the unrepeatable mode, as can be seen in **Figures 4.62a** and **4.63**. This might be due to three effects. First, when N_2 is added, due to the rotational and vibrational states of N_2, it could result in the decrease of the electron temperature. Second, compared to He, N_2 has a relatively low ionization potential, which could result in a relatively high ionization rate and thus high electron density. Third, as mentioned before, some of the metastable states of N_2 could contribute to the ionization process. Thus, when the frequency of the applied voltage is higher than the critical frequency f_{cri}, the first effect is compensated by the other two effects, which results in a critical frequency similar to pure He. However, when the frequency is lower than the f_{cri}, due to the long pulse off time, the contribution from the third effect becomes weaker, and thus the n_{seed} is less than that of pure He, which finally results in the larger jitter of the ignition delay time [49].

As we know, the diffusion of electrons to the wall could contribute to the loss of seed electrons for low-pressure plasma. To see if it also plays some role in our case, the t_d jitter for a total gas pressure of 4×10^3 Pa with a working gas of He mixed with a different percentage of O_2 was also investigated. It can be seen from **Figure 4.64** that the f_{cri} are 250 Hz, 700 Hz, 800 Hz, and 1200 Hz for pure He, He + 5% O_2, He + 10% O_2, and He + 25% O_2, respectively, the increasing tendency of which is similar with what we obtained from **Figure 4.62** for the gas pressure of 0.2 atmospheric pressure.

In this section, the plasma propagation appears in two modes, one of which is random mode and the other one is repeatable mode. Further investigations

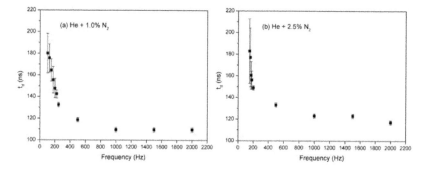

FIGURE 4.63 t_d vs. frequencies for the mixture of He with (a) 1% and (b) 2.5% N_2. Total gas pressure: 0.2 atmospheric pressure. (From Nie, L. et al., *Phys. Plasmas*, 23, 093518, 2016.)

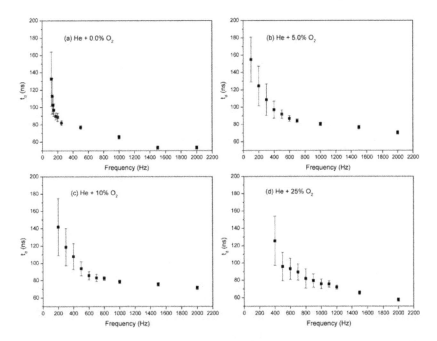

FIGURE 4.64 t_d vs. frequencies for (a) He, (b) He + 5% O_2, (c) He + 10% O_2, and (d) He + 25% O_2. Total gas pressure is 4×10^3 Pa for all the cases. (From Nie, L. et al., *Phys. Plasmas*, 23, 093518, 2016.)

show that the ignition time appears randomly in the random mode, while the propagation velocities are always repeatable for both the random and the repeatable modes. In addition, the jitter of the ignition delay time decreases with the increase of the frequency of the applied voltages. This is believed to be due to the seed electrons. With the increase of the frequency of the applied voltages, the time between the nearby pulse becomes shorter, so the seed electron density increases. When the seed electron density increases to a certain point, the propagation of the plasma transitions from the random mode to the repeatable mode. The crucial frequency f_{cri} for the mode transition increases with the increase of the O_2 percentage, but it changes slightly when the N_2 percentage is increased.

Furthermore, the critical frequency f_{cri} for the mode transition depends on the O_2 percentage. When O_2 is added, electrons attach to O_2 quickly and form O_2^-. Before the ignition of the plasma, the electric field in front of the electrode is high, which results in the detachment of the O_2^- and thus contributes to the seed electrons. The seed electron density is contributed to two pathways, i.e., the residual electrons n_{res} from the previous discharges and the electrons n_{det} due to the detachment of O_2^-. After further estimation, it is found that, in a repeatable mode, the minimum seed electron density is about 10^8 cm^3 for all

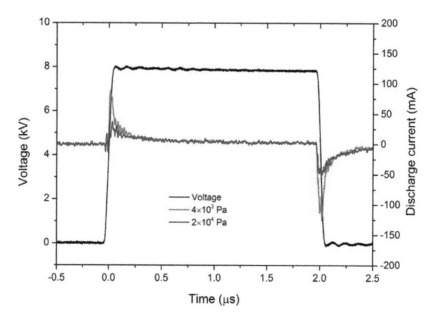

FIGURE 4.65 The applied voltage and discharge current vs. time. (From Nie, L. et al., *Phys. Plasmas*, 23, 093518, 2016.)

the different percentages of O_2 and He mixtures investigated in this work at the total gas pressure of 2×10^4 Pa. On the other hand, when the gas pressure is reduced to 4×10^3 Pa, in order to propagate in repeatable mode, the minimum seed electron density is about 10^7 cm^{-3} (**Figure 4.65**).

For better applications of plasma jets in complicated conditions, the propagation of plasma streamers in a tube with a part of it covered by a conductor was investigated. It was found that the propagation of the plasma is inhibited by the surrounding conductor, and a secondary plasma streamer begins to propagate at the downstream end of the conductor [50].

Figure 4.66 shows the schematic of the discharge device. Atmospheric-pressure plasma streamers are generated in a glass tube. The inner diameter of the tube is 1 mm. The wall thickness is 0.25 mm. The high-voltage electrode is made of a steel needle with a radius of 100 μm, which is inserted into the right end of the tube. A 30 mm section of the tube is covered by a saline layer of 1 mm thickness. The distance between the tip of the steel needle and the left end of the saline is 15 mm. The saline layer is floating without connecting to the ground or a HV power supply. To minimize the ambient air contamination, another 3 m long tube is connected to the end of the tube. When high pulsed DC voltage (amplitude: 8 kV, pulse repetition frequency: 8 kHz, pulse width: 800 ns) is applied to the HV electrode and helium gas is fed into the tube with a flow rate of

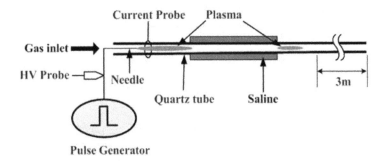

FIGURE 4.66 Schematic of the discharge device. (From Xian, Y.B. et al., *Phys. Plasmas*, 22, 063507, 2015.)

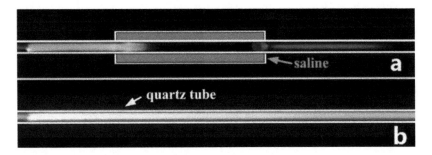

FIGURE 4.67 Photos of the plasmas in tubes: (a) with a saline layer outside of the tube, (b) without a saline layer. Applied voltage: 8 kV, pulse repetition frequency: 8 kHz, pulse width: 1 μs, helium gas flow rate: 1 L/min. (From Xian, Y.B. et al., *Phys. Plasmas*, 22, 063507, 2015.)

1 L/min, plasma streamers are generated inside the tube. As can be seen in **Figure 4.67a**, when there is a saline layer outside of the tube, the plasma streamer is divided into three sections: the primary plasma streamer at the right side of the HV electrode, the dark section in the saline layer, and the secondary plasma streamer at the downstream end of the saline layer. The total length of these three sections is much shorter than the plasma streamer propagating in a tube without a saline layer. It is clear that the primary plasma streamer is restrained by the saline layer. However, the primary streamer could enter the region covered by saline for about 5 mm, and a secondary streamer is generated. **Figure 4.68** shows the discharge current and voltage waveforms for the plasma streamer propagating in a tube with a saline layer outside the tube. Two distinct current pulses are observed for one single voltage pulse, which correspond to the rising and falling edge of the pulse voltage, respectively. **Figure 4.69** shows the dynamics of the

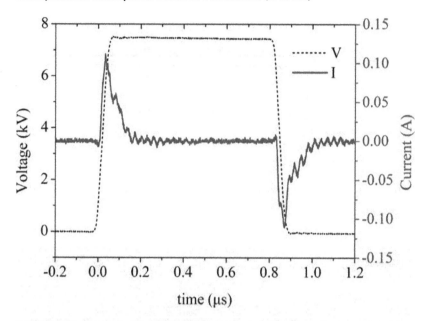

FIGURE 4.68 The discharge current and applied voltage waveforms *vs.* time. The discharge parameters are the same as that in **Figure 4.67a**. (From Xian, Y.B. et al., *Phys. Plasmas*, 22, 063507, 2015.)

discharge in the tube covered by a 30 mm long saline layer. It shows that the plasma bullet arrived at the area covered by the saline layer at 110 ns. Then, the plasma became weaker and weaker after it entered the saline layer and extinguished at 350 ns. At 415 ns, a secondary plasma bullet was generated at the right end of the saline layer. It should be noted that, with some other setups, the secondary plasma bullet could be ignited before the primary plasma bullet extinguishes [50].

Since the saline layer used above could be replaced by tinfoil without affecting the phenomenon, the experiment in this section is carried out with tinfoil rather than a saline layer for an easier experimental process. The experiment setup is shown in **Figure 4.70a**. A steel needle with a radius of 100 μm is inserted into a glass tube. A 3 cm long tinfoil ring is pasted outside the tube at 2 cm away from the needle. A slot of 0.5 mm is made on the tinfoil ring for convenient observation. Another 4 mm long tinfoil ring is pasted at 4 cm away from the first one. These two rings are connected by a 1 mm wide tinfoil strip. When a pulsed high voltage is applied to the needle electrode, a secondary streamer is generated beside the right ring. First, pulsed HV is applied to the needle, and a secondary streamer is generated beside the right ring. Second, pulsed HV is applied to the rings, and the

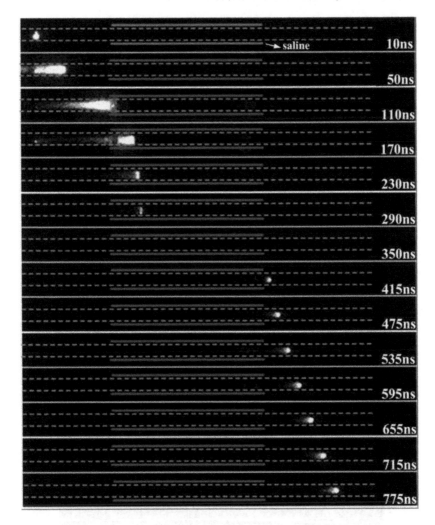

FIGURE 4.69 High-speed images of the discharges in the tube covered by a 30 mm long saline layer. The exposure time is fixed at 5 ns. Each image is an integrated picture of 20 shots with the same delay time. The time labeled on each image corresponds to **Figure 4.68**. The discharge parameters are the same as those in **Figure 4.67a**. (From Xian, Y.B. et al., *Phys. Plasmas*, 22, 063507, 2015.)

needle is afloat. The right ring will also generate a streamer. The voltage is changed to get the streamer generated by the right ring to be as long as the secondary streamer generated in the first step. Then, the voltage applied to the rings in the second step could be regarded as equal to the electric potential of the tinfoil rings in the first step.

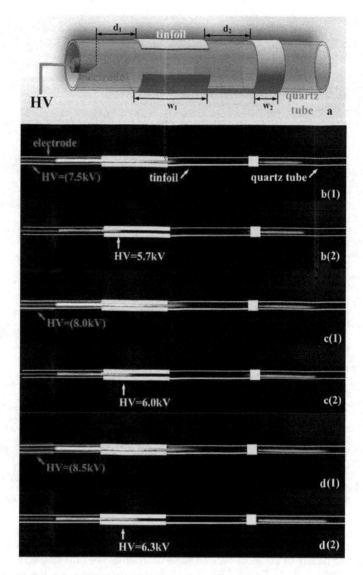

FIGURE 4.70 The method of equal effects used to measure the electric potential of the conductor. The high voltage is applied to the needle in b(1), c(1), and d(1), and the conductor surrounding the tube is floating. In contrast, the high voltage is applied to the conductor in b(2), c(2), and d(2), and the needle is floating. (From Xian, Y.B. et al., *Phys. Plasmas*, 22, 063507, 2015.)

With this method, we measured the electric potentials of the tinfoil rings when the voltage applied to the needle was 7.5 kV, 8.0 kV, and 8.5 kV, respectively. As shown in **Figure 4.71**, the electric potentials were 5.7 kV, 6.0 kV, and 6.3 kV, respectively. Although the measurement is not accurate, this result indicates that the electric potential of the conductor is high enough

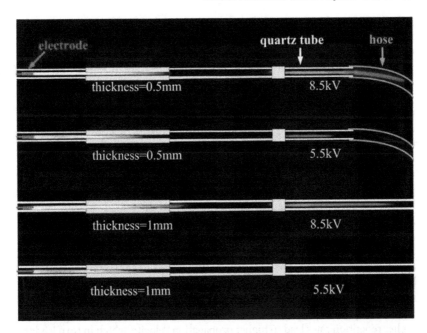

FIGURE 4.71 Photographs of plasma streamers in tubes of different thickness. The thickness of the tubes and the applied voltage are labeled in every picture. (From Xian, Y.B. et al., *Phys. Plasmas*, 22, 063507, 2015.)

to sustain the propagation of a plasma streamer. It explains the reason why a secondary streamer could be generated. It also demonstrates that a large amount of positive surface charge accumulates on the inner surface of the tube, especially in the section covered by the conductor. Since it is agreed that the electric field in front of a positive streamer is positive, the positive surface charge will significantly reduce the electric field in front the of the streamer. As a result, the plasma bullet will slow down and extinguish [50].

4.2 Effect of Discharge Parameters

4.2.1 Pulse Repetition Rate

The reactive species in a plasma that are produced by repetitive streamers, including electrons, ions, radicals, and metastable neutrals, decay after each discharge pulse. When the pulse repetition rate increases, the decay time between successive streamers shortens; the concentrations of reactive species rise immediately before the start of the next streamer, which affects the subsequent discharge.

For streamers propagating inside dielectric tubes, the variation of repetition frequency also affects the distribution of residual surface charge accumulated on the tube walls. However, the results reported by different groups concerning the effect of repetition frequency on plasma characteristics are not consistent for plasma propagating either in a dielectric tube or in the surrounding air. For plasma propagating in a dielectric tube, the effect of the pulse repetition frequency on the propagation of the plasma plume, driven by frequencies from a few Hz to hundreds of Hz, was studied [51–53]. The results clearly indicate that the plasma length first decreases from around 26 cm to 13 cm as the pulse repetition rate is increased from 1 Hz to 75 Hz, and then it remains constant from 75 Hz to 350 Hz [52].

The effect of the pulse frequency repetition rate f from 0.1 kHz to 10 kHz on the length of the plasma L_{pla} was also studied. The variation of this parameter did not affect the APPJ length noticeably. **Figure 4.72** shows the temporal dynamics of the applied voltage and the discharge current for the pulse frequencies of 0.1 kHz, 1.0 kHz, and 10 kHz [53]. One can notice a slight increase in the peak current value when f is increased. It is worth stressing that the primary discharge current appears much earlier when the pulse frequency is higher. This may be related to the accumulation of excited species, such as $He(2^1S_1)$, N_2W^3u, $a^1\Pi g$, and $W^1\Delta u$, which have lifetimes between 0.1 ms and 10 ms, which are comparable to the duration of the pulse off cycle. When the pulse frequency is increased, the excitation of such species becomes more effective. Consequently, larger amounts of these species are involved in the gas breakdown and discharge maintenance processes, which ultimately leads to noticeably earlier detection of the discharge current (**Figure 4.72**).

For the plasma propagation outside the tube in the surrounding air [54], higher repetition rates lead to higher propagation velocity, which in turn leads to an increase of the plasma plume length. This is quite different from the observations [55], which show that the variation of the repetition frequency from 0.4 kHz to 10 kHz did not affect the streamer velocity and the propagation length of the

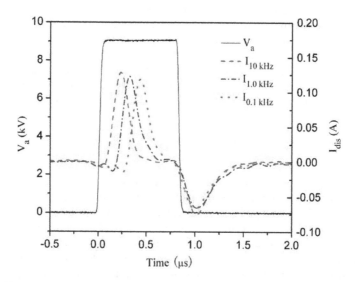

FIGURE 4.72 Temporal dynamics of the applied voltage and the actual discharge current I_{dis} for different values of V_a for different pulse frequencies f. Applied voltage $V_a = 9$ kV, pulse width $t_{pw} = 800$ ns, and He flow rate $q = 0.5$ L/min. (From Xiong, Q. et al., *Phys. Plasmas*, 16, 043505, 2009.)

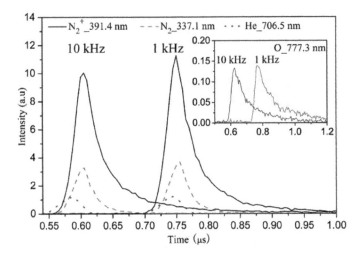

FIGURE 4.73 The time evolution of the emission for the N_2 337.1 nm dashed, N_2^+ 391.4 nm solid, He 706.5 nm dotted, and O 777.3 nm inset image lines of the plasma jet for the pulse frequencies of 1 kHz and 10 kHz, with $V_a = 8$ kV and $t_{pw} = 800$ ns. (From Xiong, Q. et al., *J. Appl. Phys.*, 107, 073302, 2010.)

plasma in the surrounding air. **Figure 4.73** shows the effect of the pulse frequency f (repetition rate) on the temporal behaviors of the emission intensities of the four spectral lines. Because the emission intensity of the O 777.3 nm line is much weaker than that of the other three lines, the temporal evolution of the O emission for a different f is shown in the inset figure in **Figure 4.5**. The pulse frequency varies from 1 kHz to 10 kHz. As can be seen, the pulse frequency does not affect the emission intensities significantly. On the other hand, it is worth noting that the emission of the four lines appears much earlier for higher frequencies. This is consistent with the temporal behavior of the bullet velocity for different f, as shown in **Figure 4.74a**. The bullet propagates out from the nozzle much earlier for high pulse repetition rates. But the bullet almost propagates to the same distance in the open air and accelerates to a peak velocity of about 2×10^5 m/s at the same position when f varies from 0.1 kHz to 10 kHz, as shown in **Figure 4.74b**.

For the pulse-modulated radio-frequency discharges, **Figure 4.75** gives the measured optical emission spectroscopy at 706 nm as a function of the duty cycle for various modulation frequencies [56]. Because high-energy electrons with $\varepsilon > 22$ eV are required to produce $He(^3S_1)$, optical emission spectroscopy at 706 nm ($He(^3S_1) \rightarrow He(^3P_2) + h\lambda$) is used to trace the high-energy electrons in the plasma. In **Figure 4.75a**, a significant increase in 706 nm emission intensity suggests efficient electron heating by the RF electric field. Similar evolution trends of electron temperature are captured by the numerical results in **Figure 4.75b**. At a given modulation frequency, as the duty cycle increases, the electron temperature initially increases significantly, then for a large enough duty cycle, it increases slowly. For example, at 6.25 kHz the 706 nm intensity increases significantly for a duty cycle smaller than 60%, then changes

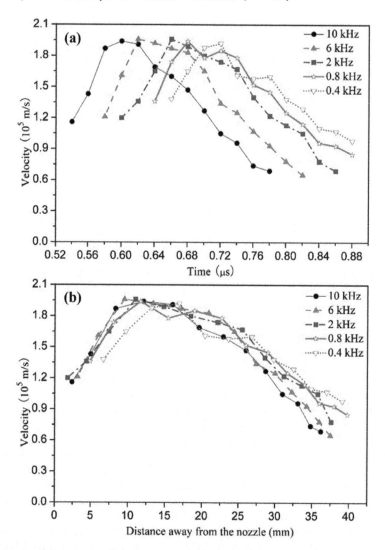

FIGURE 4.74 The (a) temporal and (b) spatial evolution of the bullet velocity for different pulse frequencies f, with $V_a = 8$ kV and $t_{pw} = 800$ ns. (From Xiong, Q. et al., *J. Appl. Phys.*, 107, 073302, 2010.)

evenly afterward. Thus, to improve the proportion of high-energy electrons or the electron temperature in the discharge region, a relatively large duty cycle (>30%) should be used, but a very high duty cycle (>60%) cannot contribute to a further increase in electron temperature, as seen in **Figure 4.75b**. The experimental and computational data also show that by increasing the modulation frequency at a given duty cycle, the electron temperature decreases mainly due to fewer RF cycles during the power-on phase (**Figure 4.75**). The gas temperature in the discharge region shown in **Figure 4.75a**, which was estimated by

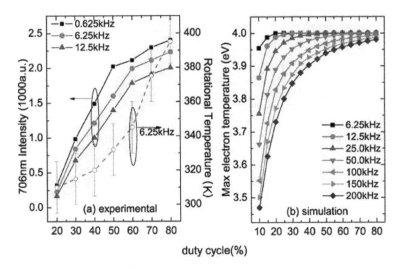

FIGURE 4.75 (a) Measured 706 nm intensity and rotational temperature and (b) simulated maximum electron temperature as a function of duty cycle for various modulation frequencies. (From He, J. et al., *Plasma Sources Sci. Technol.*, 22, 035008, 2013.)

comparing measured and calculated optical emission spectra of OH around 309 nm, increases sharply when the duty cycle is larger than 60% at 6.25 kHz. Usually a very high gas temperature is not acceptable in many temperature-sensitive applications, particular the interaction of plasmas with living tissue. Hence, a duty cycle smaller than 60% and larger than 30% should be applied in many applications to effectively lower the gas heating, while a high electron temperature can still be obtained.

The different observations discussed above are not clearly understood. The difference between the results from inside the dielectric tube compared to the results from outside the dielectric tube in the surrounding air may be because of the surface charges when the plasma is confined inside the tube. For streamers propagating inside tubes, the major effect is probably related to surface charges that determine the electric field distribution. The difference in the results obtained at different frequencies [54,55] can be related to the different electrode configurations used in these studies. As mentioned in the previous section discussing photoionization and seed electrons, the frequency variation affects the initial electron density n_0, which could affect both the streamer velocity and the propagation length. On the other hand, as discussed in the section discussing photoionization and seed electrons, the variation of n_0 has a rather weak effect on the streamer parameters. For better understanding of the frequency effects, additional experimental and computational studies are required.

4.2.2 Rise Time of Pulse Voltage

The short rise time of nanosecond pulse voltage is thought to result in over-voltage breakdown. More energetic electrons produced by the overvoltage

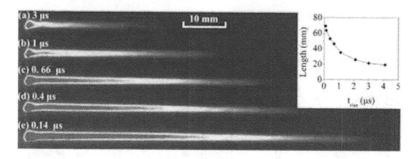

FIGURE 4.76 Photographs of the plasma plume for different pulse rising times. The plasma plume at (a) 3 μs, (b) 1 μs, (c) 0.66 μs, (d) 0.4 μs, and (e) 0.14 μs. (From Wu, S. et al., *Plasma Process. Polym.*, 10, 136–140, 2013.)

breakdown can enhance ionization and excitation processes. However, only a few studies have been carried out to confirm this assumption. These results are all consistent with the case when the pulse rise time is about 50 ns or longer [6,51,57]. These studies indicate that the plasma plume propagates faster and longer as a result of the shorter pulse rising time.

Figure 4.76 shows photographs of the plasma plume for different pulse rising times [57]. It shows clearly that the length of the plasma plume increases drastically with the reduction of the pulse rising time. Its length is less than 20 mm for a pulse rising time of 4 ms, and it increases to 70 mm when the pulse rising time is reduced to 140 ns. This observation indicates that the pulse rising time has a significant effect on the plasma. To investigate how the pulse rising time affects the electrical characteristics of the discharge, the current and voltage waveforms of the discharge for different pulse rising times were measured, as shown in **Figure 4.77**. **Figure 4.77a** shows that the shorter the pulse rising time is, the higher the peak discharge current is. It should be pointed out that the currents shown in this figure are the actual discharge currents. Displacement current has been subtracted from the total current. **Figure 4.77b** gives a close look of the current and voltage waveforms for a pulse rising time of 1 ms. It shows that the discharge current has a peak value of about 0.5 A and it lasts for about 10 ns. To have a more detailed discussion about the relationship between the discharge current and the pulse rising time, the peak value of the discharge current I_{peak} versus the pulse rising time is plotted in **Figure 4.78**. The breakdown voltage on the needle V_{dis} versus the pulse rising time is also plotted in **Figure 4.78**. It shows that both the I_{peak} and the V_{dis} increase with the reduction of the pulse rising time. When t_{rise} is reduced from 4 ms to 140 ns, the peak value of the discharge current I_{peak} increases from about 0.2 A to 1.3 A and the breakdown voltage V_{dis} increases from less than 4 kV to about 6 kV. These results confirm that the reduction of the pulse rising time could result in the overvoltage breakdown.

Figure 4.79 shows the results of the electron temperature. It shows that the electron temperature for a pulse rising time of 140 ns is about 1.55 eV. It decreases to less than 1.25 eV when the pulse rising time is increased to 4 ms. These results confirmed for the first time that plasma driven by voltage pulses

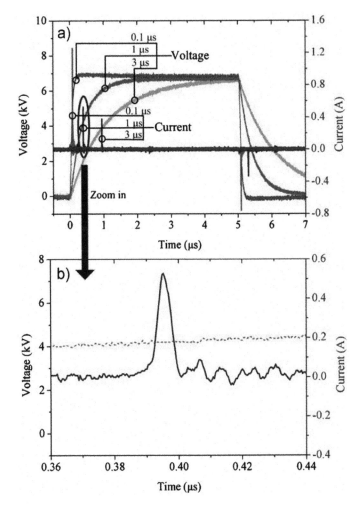

FIGURE 4.77 (a) Typical current and voltage waveforms for three different pulse rising times. (b) close look of the current and voltage waveforms for a pulse rising time of 1 ms. (From Wu, S. et al., *Plasma Process. Polym.*, 10, 136–140, 2013.)

with shorter rising times has a higher electron temperature than those driven by longer rising times. Plasma bullet behavior, one of the most interesting phenomena of APNP jets, has attracted lots of attention recently. In the next section, the effect of the pulse rising times on the plasma propagation velocity is also investigated. First, the dynamics of plasma is captured by a fast ICCD camera. The exposure time of the camera is fixed at 5 ns for all the photographs. Then the plasma bullet propagation velocities are evaluated according to the photographs. **Figure 4.80** shows the photographs taken at various delay time for different pulse rising times. Each picture is an integrated image of over 100 shots with the same delay time. The figure clearly indicates that the plasma bullet propagates faster with a pulse rising time of 140 ns. To calculate

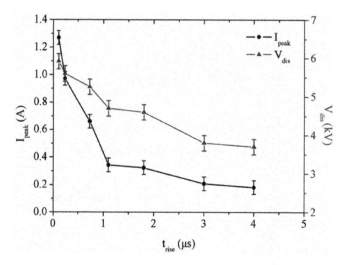

FIGURE 4.78 The peak value of the current pulse I_{peak} and breakdown voltage V_{dis} versus pulse rising time. (From Wu, S. et al., *Plasma Process. Polym.*, 10, 136–140, 2013.)

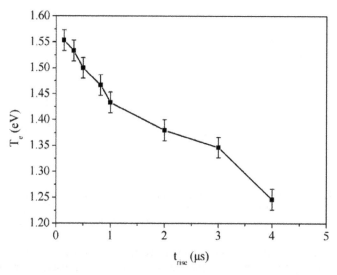

FIGURE 4.79 Estimated electron temperature T_e for different pulse rising times. (From Wu, S. et al., *Plasma Process. Polym.*, 10, 136–140, 2013.)

the velocities, the location of the plasma bullet is determined by observing the head of the plasma bullet. The details of the propagation velocities of the plasma are plotted in **Figure 4.81**. As can be seen from **Figure 4.81**, the peak value of the plasma propagation velocities increases with the reduction of the pulse rising time. It reaches about 7×10^4 m/s for a pulse rising time of 140 ns, which is more than three times faster than for a pulse rising time of 3 ms.

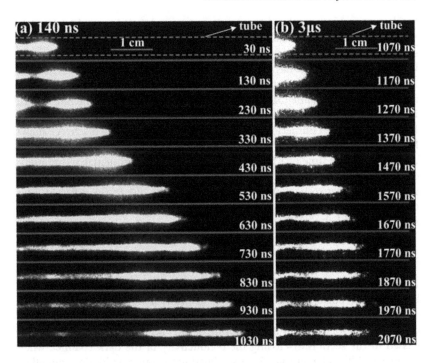

FIGURE 4.80 High-speed photographs of the plasma plume for different pulse rising times: (a) 140 ns, (b) 3 ms. The exposure time is fixed at 5 ns. (From Wu, S. et al., *Plasma Process. Polym.*, 10, 136–140, 2013.)

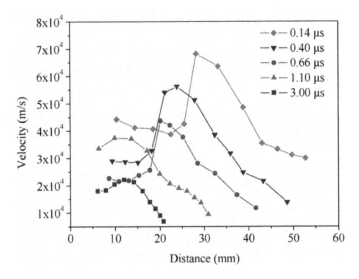

FIGURE 4.81 Plasma bullet propagation velocities versus distance from the ground electrode for different pulse rising times. (From Wu, S. et al., *Plasma Process. Polym.*, 10, 136–140, 2013.)

The experiment results shown above suggest the shorter pulse rise time makes the plasma plume longer and the plasma more reactive. Is this true for even shorter pulses? This is still not clear because of the lack of sufficient studies. However, this effect has been studied for dielectric barrier discharges. The Ar metastable generation efficiency in DBD for a pulse rising time of 150 ns and 10 ns were the same [58], even when dielectric losses were included. That is, there is no advantage to reducing the pulse rise time from 150 ns to 10 ns in a DBD. In addition, a numerical study on the capacitive coupled discharge shows that the maximum electron energy is achieved with a pulse rise time of 40 ns rather than a rise time of 1 ns, 10 ns, or 20 ns. The experimental and numerical studies on the effect of pulse rise time on APPJs and capacitively coupled discharges appear to be contradictory. This may be due to the different ways of coupling the power to the plasma in these two types of discharges. Further studies are needed.

4.2.3 Pulse Width

Pulse width is one of the important parameters that can affect plasma plume behavior. For pulse widths in the ~100 ns scale, increasing the pulse width results in higher discharge currents, and consequently, an increase of the energy coupled to the plasma and the extending of the plasma plume [52,59]. As discussed in the section discussing the conductivity of the dark channel, when the pulse width is larger than 500 ns, adjusting the pulse width does not affect the shape and the peak value of the discharge current pulses [21]. A further increase of the pulse width up to 1000 ns results in the increase of the length of the plasma plume. This is partly due to the longer duration of the external electric field, which affects the plasma bullet propagation. A further increase of the pulse width has no obvious effect on plasma plume behavior.

However, when the pulse-off-time (interruption) between consecutive pulses is very short, totally different plasma plumes are observed [59]. The atmospheric plasma jet device employed in this study is shown in **Figure 4.82a**. **Figure 4.82b** shows a single 8 kV DC voltage pulse of a 2 ms duration (and 1 ms repetition rate) and the current it produces during the initial rise phase and the following fall phase. A continuous 8 kV DC voltage was repeatedly interrupted for a duration of 2 ms every 1 ms, as shown in **Figure 4.82c**. The interruption time was also reduced to 900 ns, retaining the same repetition rate (**Figure 4.82d**). During the interrupted DC operation, the fall phase preceded the rise phase as shown in **Figure 4.82c** and **d**. In all cases, the discharges were only produced during the rise or fall phases, as evidenced by the current spikes in **Figure 4.82b–d**. All three waveforms in **Figure 4.82b–d** produced three totally different plasma plumes, depicted in **Figure 4.83a–c**. The 2 ms pulse has generated a common plume with a large dim area indicative of a propagating bullet (**Figure 4.83a**). Importantly, this type of plasma plume was generated using DC pulses with durations up to 990 ms at the same repetition rate of 1 ms. However, the other two interrupted DC waveforms produced exotic plumes with two and three distinctive areas labeled in **Figure 4.83b** and **c**, respectively. When the break between

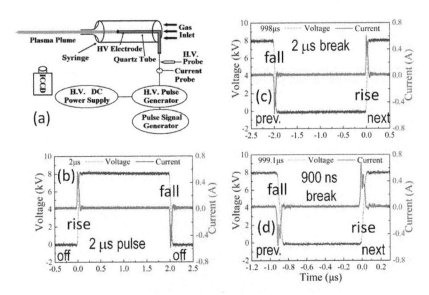

FIGURE 4.82 (a) Schematic of the atmospheric plasma jet device, gas feed, power supply, and plasma diagnostics. Current-voltage pulsed DC waveforms corresponding to pulse widths of (b) 2 μs, one single pulse; (c) 998 μs, with a 2 μs interruption between two pulses; and (d) 999.1 μs, with a 900 ns interruption between two pulses is shown. Breakdown occurs only during the voltage rise and fall phases. (From Xian, Y. et al., *Sci. Rep.*, 3, 1599, 2013.)

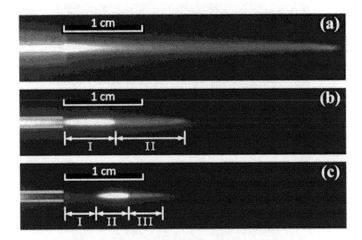

FIGURE 4.83 Photographs of three distinctive plasma plume types generated using (a) a2 μs pulse (Mostly a dim area, typical of a plasma bullet, is seen.); (b) a2 μs interruption (Bright [I] and dim [II] areas are seen.); and (c) a900 ns interruption (Dark [I], bright [II], and dim [III] areas are seen.). Images (a)–(c) were produced using a 1 μs exposure from the beginning of the voltage rise phases, labeled "0" in **Figure 4.82(b)–1(d)**, respectively. (From Xian, Y. et al., *Sci. Rep.*, 3, 1599, 2013.)

the pulses was 2 ms (**Figure 4.83b**), the plume appeared as a bright channel (I) and a dim (likely bullet propagation) area (II). This type of structure was revealed for the 2–10 ms interruptions of 8 kV DC pulses. When the duration of the DC interruption was reduced to less than 1 ms, a very different plume structure was discovered (**Figure 4.83c**). One can clearly see the dark area (I), the bright area (II), and the dim (likely bullet propagation) area (III) in the plume generated using a 900 ns interruption (**Figure 4.82d**).

When a short 2 ms pulse is used, the plume dynamics (**Figure 4.84**) are typical of the plasma bullet propagation reported previously. The bright plasma channel originates within the nozzle and starts exiting it at 528 ns

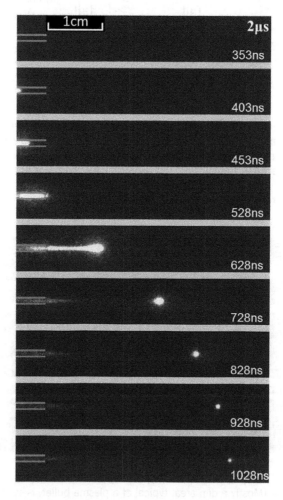

FIGURE 4.84 High-speed photographs of the plasma plume generated during the voltage rise phase of the 2 µs pulse. The exposure time is 5 ns. The time labeled on each photograph corresponds to the time in **Figure 4.82b**. (From Xian, Y. et al., *Sci. Rep.*, 3, 1599, 2013.)

into the pulse. At 628 ns, the plume elongates to 1 cm, forming a bright bullet at its head. One hundred nanoseconds later, the bullet separates from the rest of the plume. Other images in **Figure 4.84** show that the bullet propagates further, with the speed of the ionization front, 10^5 m/s (note that the speed of the gas flow is much lower, typically on the order of 10 m/s), and eventually it slows down and shrinks in size. The rest of the plasma plume darkens significantly while the bullet propagates. **Figure 4.85** shows the very different dynamics of a plasma plume generated by a short 2 ms interruption. The plume brightens and exits the nozzle much faster, forming a bright, non-propagating channel at 278 ns, which then fully exits from the nozzle 100 ns later. The length of this channel is approximately 6 mm. At 428 ns, the channel brightens, and the plasma bullet forms and tends to detach from the channel. Fifty nanoseconds later, the bullet is fully detached and a dark channel forms between the head of the bright plasma channel and the small bullet-like plasma. As the bullet propagates further, its speed and size decrease while the luminosity of the bright channel (area I in **Figure 4.83b**) decreases. The speed and length of the bullet propagation are noticeably smaller than in **Figure 4.84**, thereby leading to a shorter plasma plume in **Figure 4.83b** compared to **Figure 4.83a**. For example, the propagation speed of the bullet is approximately three times smaller than in the 2 ms pulse case. In order to elucidate whether the bright plasma channel (area I in **Figure 4.83b**) is formed locally or is due to the integrated effect of a fast-moving bullet, a dedicated experiment using the ICCD camera with an exposure time of 2 ns and a step of 2 ns was performed. This high-speed imaging confirms that the bright plasma channel at 378 ns is clearly not due to the integrated effect of a fast-moving bullet. If the break between the voltage fall and rise phases is further reduced, the bright plasma channel starts forming within the nozzle and then moves forward much faster. This can be seen in **Figure 4.86**, which corresponds to the waveform in **Figure 4.82d**. Already at 173 ns, the bright channel clearly separates from the nozzle and leaves an approximately 4 mm dark channel behind. The length of the dark channel between the nozzle and the left end of the bright plasma channel becomes smaller when the voltage interruption time decreases. At about 273 ns, a plasma bullet is formed at the right end of the bright plasma channel and starts propagating forward. One hundred nanoseconds later, a clear gap between the bullet and the channel is visible. At this moment, the channel becomes dimmer and this tendency continues thereafter. Interestingly, the length of the dark space (area I in **Figure 4.83c**) between the nozzle and the left end of the bright channel does not noticeably change with time. As the bullet propagates, it also reduces in size and slows down, similar to the two other cases. The propagation speed of the bullet in **Figure 4.86** is somewhat smaller than in **Figure 4.85**.

The exotic plumes are thought to be related to the seed electron density, which is affected by the pulse width. The spatial and temporal resolution measurements of electron density are needed to confirm this assumption. But the seed electron density is probably quite low, so traditional diagnostic methods such as Stark broadening can not be used. Thus, the development of new diagnostic methods for the measurement of such low

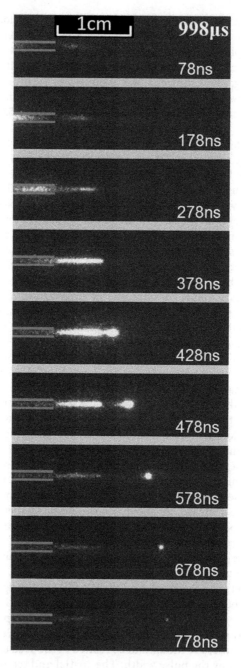

FIGURE 4.85 Same as in **Figure 4.84** for the 2 μs DC interruption. The time labeled on each photograph corresponds to the time in **Figure 4.82c**. (From Xian, Y. et al., *Sci. Rep.*, 3, 1599, 2013.)

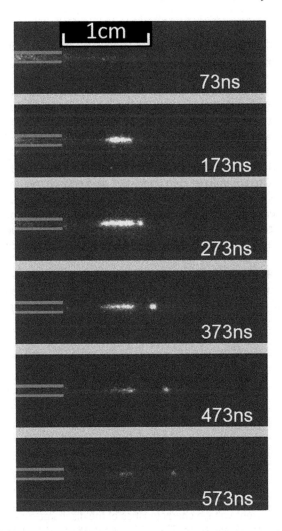

FIGURE 4.86 Same as in **Figure 4.84** for the 900 ns DC interruption. The time labeled on each photograph corresponds to the time in **Figure 4.82d**. (From Xian, Y. et al., *Sci. Rep.*, 3, 1599, 2013.)

electron density in the plasma plumes is needed. In addition, electron density can be significantly affected by reactive O_2 species. A disappearing, multisegmented plasma plume is expected when O_2 is added to the working gas. Therefore, a related experiment should be carried out to confirm whether the multisegmented plasma plume is related to the seed electron density.

4.2.4 Voltage Polarity

A plasma jet driven by a positive pulse is believed to propagate much faster and much longer than a plasma jet driven by a negative pulse [60–62]. Studies on the dynamics of the plasma plume suggest plasma bullets travel at a constant speed of about 30 km/s before they exit the nozzle for both polarities. It has been revealed that the plasma bullet structures for both polarities are quite different [61]. For both the positive pulse and the negative pulse, the pulse widths and pulse frequency were fixed at 800 ns and 8 kHz, respectively. The applied voltages were fixed at 8 kV and −8 kV for the positive pluse and the negative pulse, respectively. The helium flow rate was maintained at 2 L/min. The current-voltage characteristics of the discharge for the positive pluse and the negative pulse is the total current. The total current is the sum of the displacement current and the actual discharge current, where the actual discharge current is much smaller than the displacement current, as shown in **Figure 4.87a** and **b**, respectively. It should be pointed out that current I is the total current.

Figure 4.88a and **b** are the photographs of the plasma plume generated by the positive pulse and the negative pulse, respectively. It can be seen that the plasma plume generated by the positive pulse is much longer than the plasma

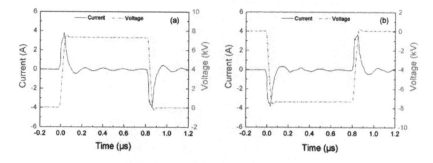

FIGURE 4.87 Current-voltage characteristics of the discharge for (a) positive and (b) negative pulses. (From Xiong, Z. et al., *J. Appl. Phys.*, 108, 103303, 2010.)

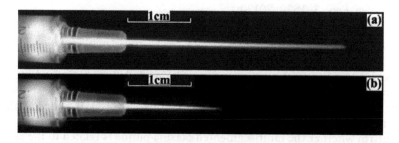

FIGURE 4.88 Photographs of the plasma for (a) positive and (b) negative pulses. (From Xiong, Z. et al., *J. Appl. Phys.*, 108, 103303, 2010.)

plume generated by the negative pulse. To further understand this phenomenon, a high-speed ICCD camera was used to capture the dynamics of the plasma plumes. **Figures 4.89** and **4.90** show the high-speed photographs of the discharges taken at different delay times for the positive pulse and the negative pulse, respectively. The time labeled on each photograph corresponds to the times in **Figure 4.87**. As shown in **Figures 4.89** and **4.90**, the plasma travels like a bullet before it exits the nozzle. For the positive pulse, the plasma becomes much brighter as soon as it exits the nozzle. On the other hand, for the negative pulse, the plasma also becomes brighter when it exits the nozzle, but it is not as obvious as it was for the positive pulse. It is interesting to note that the shape of the plasma head for the positive pulse is different from that of the negative pulse. The plasma head has a spherical shape for the positive pulse, whereas the plasma head is more shaped like a sword for the negative pulse. The reason behind the shape difference is not clear at the moment.

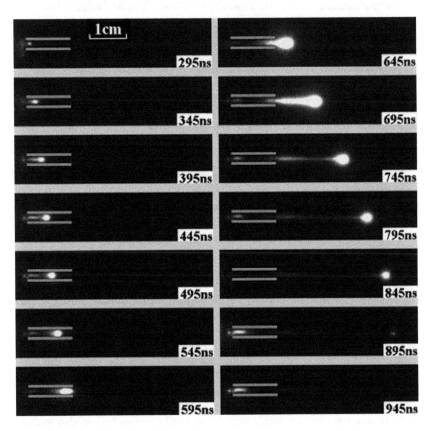

FIGURE 4.89 High-speed photograph of the plasma plume for the positive pulse. The exposure time is fixed at 5 ns. The time labeled on each photograph corresponds to the time in **Figure 4.87a**. (From Xiong, Z. et al., *J. Appl. Phys.*, 108, 103303, 2010.)

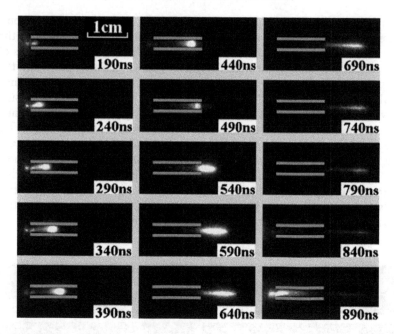

FIGURE 4.90 High-speed photograph of the plasma plume for the negative pulse. The exposure time is fixed at 5 ns. The time labeled on each photograph corresponds to the time in **Figure 4.87b**. (From Xiong, Z. et al., *J. Appl. Phys.*, 108, 103303, 2010.)

The optical emission spectrum was used to better understand the phenomenon mentioned above. For both the positive pulse and the negative pulse, the optical emission spectra of the plasma bullets inside and outside the nozzle were measured. When the positive pulse was used, **Figure 4.91a** and **c** shows that the emission intensity of N_2^+ increased significantly by more than one order of magnitude when the plasma exited the nozzle. Thus, it is reasonable to assume that the N_2^+ charge density increases when the plasma exited the nozzle. On the other hand, the higher the N_2^+ charge density, the stronger the local electric field induced by the plasma plume head, and the higher the plasma plume propagation velocity. This may explain why the propagation velocity of the plasma bullet increases when the plasma exits the nozzle. As to why the N_2^+ charge density increases when the plasma exits the nozzle, it can be explained as follows. Before the plasma exits the nozzle, the concentration of N_2 is very low inside the nozzle and most of the N_2 is ionized. When the plasma propagates out of the nozzle, due to diffusion, the concentration of N_2 in the plasma plume increases significantly. Due to the high collision frequency, the metastable state He not only excites N_2 but also ionizes N_2 and then further excites it. As a result, the concentration of N_2^+ increases significantly. This assumption is consistent with our following experimental observation. As can be seen from **Figure 4.91(2)**, before the plasma exits the nozzle, the emission intensities of both N_2 and N_2^+ are very low. The emission

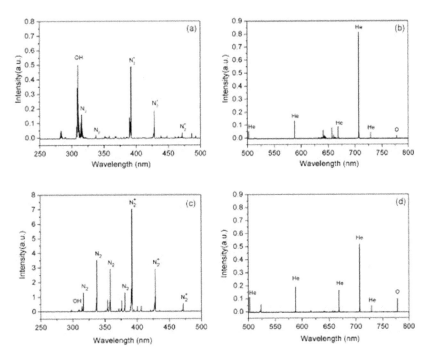

FIGURE 4.91 Emission spectra of the plasma for the positive pulse. (a) and (b) are the emission of the plasma inside the nozzle. (c) and (d) are the emission of the plasma outside the nozzle (5 mm away from the nozzle). (From Xiong, Z. et al., *J. Appl. Phys.*, 108, 103303, 2010.)

of N_2 can only just be detected. Thus, it is reasonable to say that most of the N_2 is ionized. When the plasma exits the nozzle, **Figure 4.91c** shows that the emission intensities of both the N_2 and N_2^+ increase dramatically. Therefore, it is reasonable to say that the concentration of N_2^+ increases when the plasma exits the nozzle. When the negative pulse is applied, a similar behavior is observed, as shown in **Figure 4.92**. However, for the negative pulse, the peak velocity achieved is much lower than that for the positive pulse as mentioned above. This is similar to the streamer discharge. The head of the plasma plume is like a plasma bullet containing N_2^+. Moreover, the electric field due to the external applied voltage plays an important role in the propagation of the plasma bullet. Therefore, when the positive pulse is used, the local electric field induced by the plasma bullet and the electric field from the external applied voltage have the same direction, so the total electric field is enhanced, which results in the high peak velocity of the plasma bullet. On the other hand, when the negative pulse is used, the two electric fields mentioned above have opposite directions, so the total electric field is weakened, which is the reason why the peak velocity achieved by the positive pulse is higher than that resulting from the negative pulse [61].

The behavior of the plasma bullets for the conditions of the experiment above was simulated in [29]. The calculated image for both the shape of the

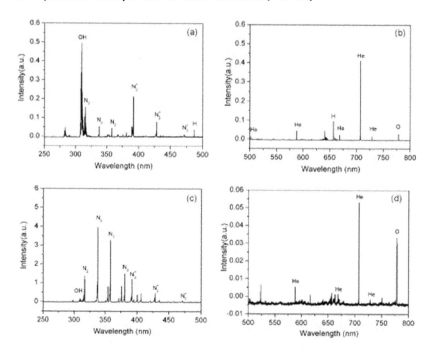

FIGURE 4.92 Emission spectra of the plasma for the negative pulse. (a) and (b) are the emission of the plasma inside the nozzle. (c) and (d) are the emission of the plasma outside the nozzle (5 mm away from the nozzle). (From Xiong, Z. et al., *J. Appl. Phys.*, 108, 103303, 2010.)

streamer head and the structure of the regions of strong optical emission are very similar to the experimental observations [60]. The characteristics of the streamer propagation were also reproduced in simulations; initially, both the positive streamer and the negative streamer move with close velocities, but later the velocity of the negative streamer decreases steeply while the positive streamer continues propagating. The effect of the voltage polarity on the simulated streamer dynamics and structure is directly related to the difference between the profiles of self-consisted electric fields in positive and negative streamers. The calculated profiles of the electric field at the streamer axis for both polarities suggest the maximum electric field values in streamer heads for the positive streamer are much higher than those for the negative one. On the contrary, the electric field in the channel of the negative streamer is much higher than that in the positive one. The generation of radiating $N_2(C^3\Pi)$ molecules in the positive streamer channel, where the electric field is small, appears to be insignificant; that is why these species are produced mostly in the streamer head region. Since the lifetime of $N_2(C^3\Pi)$ is much smaller than the time of streamer propagation, $N_2(C–B)$ radiation is localized in a small region near the head. At the negative polarity, the electric field in the channel is much higher, so $N_2(C^3\Pi)$ molecules are generated effectively in the whole channel. That is the reason why, at the positive polarity, photons are emitted mainly from a small region adjacent to the streamer head, while at the negative polarity the whole channel produces optical emission.

The electric field in the plasma bullet for the positive and negative voltage was measured [63]. The obtained electric field values (up to 30 kV cm^{-1}) for positive streamers appeared to be an order of magnitude higher than those measured for the negative streamers (less than 4 kV cm^{-1}), which is in good qualitative agreement with the numerical results [60]. The effect of voltage polarity on the plasma plume propagation inside capillaries was investigated [51]; for a peak value of 12 kV, the average velocity for the negative polarity is almost two times higher than the average velocity for the positive polarity. The reason for this difference may be related to the specific boundary conditions of the capillary channels, and it represents an opportunity for future research.

4.2.5 Diameter, Dielectric Constant, and Shape of the Tube

It is typical for the plasma jet devices that the plasma is first generated inside the tube. Thus, the characteristics of the tube will also affect the dynamics of the plasma plumes. The effect of two of the dielectric tube parameters (i.e., the diameter and the dielectric constant) on the plasma characteristics have been investigated by several groups.

The numerical results of the effect of tube diameter on a microplasma jet were obtained for the 2D axisymmetric geometry, as shown in **Figure 4.93**. The working gas (helium) flows through a dielectric tube with an inner diameter from 0.1 to 1 mm. The flow rate considered in this simulation is 200 cm^3/min. Power is applied to a thin electrode embedded within the dielectric tube. The vertical distance between electrode and inner surface of dielectric tube is

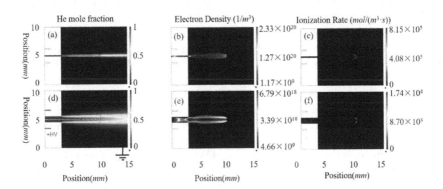

FIGURE 4.93 He mole fraction of (a) 0.2 mm tube diameter and (d) 1 mm tube diameter. Plasma properties 84 ns after the start of pulse excitation of 0.2 mm tube diameter: (b) electron density, (c) ionization rate. Plasma properties 103 ns after the start of pulse excitation of 1 mm tube diameter: (e) electron density, (f) ionization rate. X axis was the downstream direction of tube. The white dashline in (d) suggests the position of tube axis. +HV and small black block in (d) suggested the position of power electrode. (From Cheng, H. et al., *Plasma Process. Polym.*, 12, 1343–1347, 2015.)

1 mm. The horizontal distance between the power electrode and the nozzle of the tube is 2 mm. A positive voltage from 2.6 kV to 4.3 kV with a rising phase of 50 ns is applied to the power electrode to initiate the plasma jet. The electrostatic potential for the far-field boundaries of the ambient air region is grounded for all simulations presented (**Figure 4.93d**) [25].

The dependence of the mole fraction of helium on the tube diameter is shown in **Figure 4.93**. The mole fraction of helium decreased in the radial direction because of convection and the diffusion of air [8,64]. The diameter of a mixing layer with a helium mole fraction of 10% is 0.36 mm for a 0.2 mm tube diameter at a horizontal position 5 mm from the nozzle, and the ratio between the two diameters is $0.36/0.2 = 1.8$ (**Figure 4.93a**). For a 1 mm tube diameter, the diameter of a mixing layer with a helium concentration of 10% is 1.5 mm, and its ratio to the tube diameter is $1.5/1 = 1.5$. The larger ratio of 0.2 mm is attributed to the faster gas velocity of the smaller tube diameter at the same gas flow rate. The reactions between the helium and the air species in the mixing layer have significant effects on the propagation of the micro plasma jet. The ionization of He, N_2, and O_2 by energetic electrons dominated the ionization reactions in the plasma bullet. For the 0.2 mm tube diameter, **Figure 4.93c** suggests the position of the peak ionization rate (the sum of the ionization of He, N_2, and O_2 by energetic electrons) is at 0.13 mm from the the the tube axis, where the concentration of helium is 45.5%. The ionization of He, N_2, and O_2 are 1.64×10^5 mol/(m³s), 5.36×10^5 mol/(m³s) and 1.15×10^5 mol/(m³s), respectively. For a 1 mm tube diameter, the position of the peak ionization rate is at 0.35 mm from the tube axis (**Figure 4.93f**), where the concentration of helium is 67.5%. The ionization of He, N_2, and O_2 are 8.9×10^3 mol/(m³s), 7.5×10^3 mol/(m³s), and 1.6×10^3 mol/(m³s), respectively, which are nearly two orders of magnitude less than that of the 0.2 mm tube diameter. The dominant ionization reaction changed from electron impact ionization of He to electron impact ionization of N_2 as the tube diameter decreased. The off-axis ionization results in the ring shape of the plasma bullet observed in the experiment. The plasma channel filled with electrons, located behind the fast-moving plasma bullet, is also ring-shaped (**Figure 4.93b** and **e**). The boundary of the plasma channel in the radial direction is assumed to be the place where the electron density decreased to 1×10^{18}/m³. The width of the plasma channel decreased from 1.5 mm to 0.95 mm as the tube diameter decreased from 1 mm to 0.2 mm. A five-fold decrease of the tube diameter only led to a 33% decrease of the plasma channel width. The higher ionization of air species with the smaller tube diameter is the reason for that. In addition as the tube diameter decreased from 1 mm to 0.2 mm, the peak electron density inside the tube increased from 6.79×10^{18}/m³ to 2.33×10^{20}/m³, and the peak electron density in the plasma bullet increased from 5.9×10^{18}/m³ to 3.6×10^{19}/m³ (compare **Figure 4.93b** and **e**). The six-fold increase of electron density outside the tube is similar to the four-fold increase measured in the experiment.

The slowly decreasing width of the plasma channel and the increasing electron density suggest that more power is needed to sustain the plasma jet as the tube diameter decreases. The pulsed DC plasma jet usually arose at the end of the rising phase of the pulsed DC voltage in the experiment; therefore, in the simulation, the sustaining voltage is considered to be the pulsed DC voltage, which can drive the plasma jet at the same time. In order to analyze the reason why the sustaining voltage increased with decreasing diameter, the discharge dynamics inside the tube are shown in **Figures 4.94** and **4.95**. **Figure 4.94a** shows the distribution of electron density for a 0.2 mm diameter when the applied voltage was in the rising phase and its value was 2100 V. The discharge did not happen at this time. The distribution of electrons is used to describe the pre-avalanche situation. Electrons were concentrated at the nozzle and had a peak density value of $1.96 \times 10^{12}/m^3$, but the electron density inside the tube decreased from $1 \times 10^{11}/m^3$ (the initial background value) to $2.5 \times 10^7/m^3$. The characteristic time of diffusion to the wall for the electrons $\tau_{e,D} = (R/2.4)^2/D_e$ (R is the radius of the tube, D_e is the diffusion coefficient of the electrons) is 66 ns.

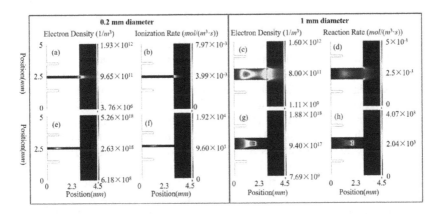

FIGURE 4.94 Electron density and ionization rate of a 0.2 mm tube diameter with an applied voltage of 2100 V (a) and (b), and 3700 V (e) and (f). Electron density and ionization rate of a 1 mm tube diameter with an applied voltage of 2100 V (c) and (d), and 2600 V (g) and (h). (From Cheng, H. et al., *Plasma Process. Polym.*, 12, 1343–1347, 2015.)

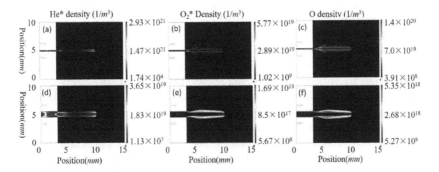

FIGURE 4.95 Plasma properties (a, d) He* density, (b, e) O_2* density, (c, f) O density for a 0.2 mm tube diameter and a 1 mm tube diameter. (From Cheng, H. et al., *Plasma Process. Polym.*, 12, 1343–1347, 2015.)

The characteristic time of drift movement across the tube $\tau_{e,\mu} = R/E\mu_e$ (E is the electric field, 6.2×10^5V/m, μ_e is the mobility of the electrons, 1132 cm²/Vs is 1.4 ns. Therefore, the drift movement of the electrons to the inner wall led to the depletion of electrons. Simultaneously, the electric field (5.3×10^5V/m) at the nozzle accelerated electrons from open space toward the tube, and the subsequent ionization reactions further increased the electron density at the nozzle (**Figure 4.94b**). The high-electron density region at the nozzle provided the seed electrons necessary to initiate the avalanche inside the tube, which is similar to the ejection of electrons from the cathode of the Townsend breakdown. The stronger electric field in the remaining rising phase of the applied voltage accelerated these seed electrons toward the electrode, so the electron impact ionization rate increased to 1.92×10^4 mol/m³s (**Figure 4.94d**) and the electron density increased to 5.26×10^{18}/m³ (**Figure 4.94c**) near the electrode when the applied voltage was 3700 V at 50 ns. The rapidly increasing electron density near the electrode eventually facilitated the avalanche-to-streamer transition, and the ionization front moved out of the tube at 58 ns.

The depletion of electrons inside the tube is not observed for the 1 mm diameter (**Figure 4.95a**). The characteristic time of electron diffusion $\tau_{e,D}$ and drift movement $\tau_{e,\mu}$ increased to 1650 ns and 12.7 ns, respectively, compared to the 0.2 mm diameter. A five-fold increase of the tube diameter and an electric field with the strength of 3.4×10^5 V/m are the reasons for the increasing time of drift movement across the tube. A more than nine-fold increase of drift movement time led to a slight decrease of electron density in the small arc region near the inner surface of the tube. The electron impact ionization reaction rate is a function of electron density. The higher electron density inside the tube compared to the 0.2 mm diameter case results in more electron impact ionization reactions (compare **Figures 4.94b** and **4.95b**), so the electron density increased further to 5.2×10^{12}/m³. The accumulation of electrons at the nozzle also happened, and its peak electron density was 5.05×10^{12}/m³. The high electron density throughout the tube facilitated the propagation of the ionization wave. As the voltage increased to 2600 V at 50 ns, the position of the ionization front is 0.9 mm closer to the nozzle than that of the 0.2 mm diameter case (compare **Figures 4.94d** and **4.95d**). The ionization front eventually moved out of the tube at 54 ns [25].

The comparison of the development of discharge in the rising phase of the applied voltage between the 0.2 mm diameter case and the 1 mm diameter case suggests the pre-avalanche electron density is crucial for the avalanche-to-streamer transition. **Figure 4.96a** shows the pre-avalanche electron density inside the tube as a function of tube diameter. The pre-avalanche electron density was less than the initial background value (1×10^{11}/m³) when the tube diameter was less than 0.3 mm, and it especially decreased to 3.4×10^6/m³ when the tube diameter was 0.1 mm. When the initial background electron density decreased to the natural background electron density of 1×10^9/m³, the evolution of pre-avalanche electron density was similar to 1×10^{11}/m³ cases. Therefore, the sustaining voltage has to be increased to accelerate seed electrons from the high electron density region at the nozzle and produce discharge in the region near the electrode. The sustaining voltage increased

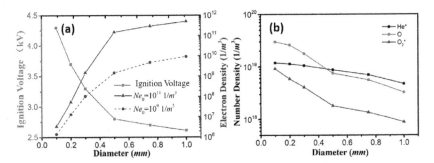

FIGURE 4.96 (a) The ignition voltage and the pre-avalanche electron density for varying tube diameters. The pre-avalanche electron densities inside the tube were obtained when the applied voltage was 2100 V for varying tube diameters. Two initial background electron densities 1×10^{11} and $1 \times 10^9/m^3$ were used to analyze the effect of tube diameter on pre-avalanche electron density. (b) The peak densities of He^*, O_2^*, and O at the position 7 mm from the nozzle for varying tube diameters. (From Cheng, H. et al., *Plasma Process. Polym.*, 12, 1343–1347, 2015.)

from 2800 V to 4300 V, as the tube diameter decreased from 0.5 mm to 0.2 mm. The 1.53 times increase of the sustaining voltage is close to the 1.86 times increase of the sustaining voltage observed in the experiment. The peak densities of He^*, O_2^* and O at the position 7 mm downstream of the nozzle as a function of the tube diameter are shown in **Figure 4.96b**. The peak density of O at the 7 mm position increased from $3.2 \times 10^{18}/m^3$ to $2.5 \times 10^{19}/m^3$ as the tube diameter decreased from 1 mm to 0.2 mm. Its increasing amplitude between 0.5 mm and 0.1 mm is larger than the amplitude between 1 mm and 0.5 mm, which is consistent with the increasing amplitude of the sustaining voltage between the varying tube diameters shown in **Figure 4.96b**. The O_2^* case is similar to the O case. The increasing amplitude of He^* was less than that of O_2^* and O because more He^* was consumed by reactions with air species as the tube diameter decreased [25].

A microplasma plume generated in a quartz tube, with its inner diameter decreased from 245 μm to 6 μm, was also studied. The maximum electron density of the plasma is around 10^{16} cm^{-3} based on the Stark broadening of H_α line and electric current measurement. When the discharge ignition voltage increases from 11 kV to 40 kV, the plasma is able to reach the positions within the tube where the inner diameters are 245 μm and 10 μm, respectively. This also leads to the corresponding increase in the gas temperature from about 420 K to 540 K. Further analysis shows that, if this trend holds true while the diameter further decreases to 1 μm, the density of the plasma could reach as high as ~10^{18} cm^{-3}.

The schematic of the experiment setup is shown in **Figure 4.97a**. A high-voltage (HV) electrode made of a stainless-steel needle is inserted into the left end of the quartz tube. (The radius of the needle tip is 50 μm.) The inner diameter of the tube gradually decreases from 245 μm to 6 μm.

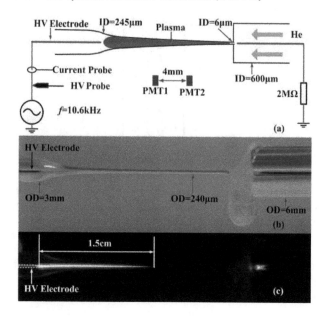

FIGURE 4.97 (a) The schematic of the experiment setup. (b) The photograph of the discharge tube. (c) The photograph of the plasma. (From Gou, J. et al., *Phys. Plasmas*, 23, 053508, 2016.)

The right end of the quartz tube is connected to another quartz tube with an inner diameter of 600 μm. The two quartz tubes are connected and sealed by glue, as shown in **Figure 4.97b**. The plasma is driven by an AC power supply with a frequency of 10.6 kHz. When the AC voltage (peak-to-peak amplitude up to 40 kV) is applied to the HV electrode and the working gas (98% He and 2% H_2) is supplied, a plasma plume with a length of around 1.5 cm is generated in the quartz tube, as shown in **Figure 4.97c**. The small amount of H_2 is introduced to facilitate the electron density measurements. With the help of a microscope (Olympus CX341), the inner diameter of the quartz tube vs. the position is obtained, as shown in **Figure 4.98**. (The end of the needle tip is considered position 0.) The inner diameter of the tube at the 1.5 cm position is around 10 μm.

The current and voltage waveforms of the plasma are shown in **Figure 4.99**. There are current pulses both at the rising edge and at the falling edge of the voltage. However, the number of current pulses and the onset time of the discharge at the rising or falling edge of the voltage are not constant. The peak of the discharge current can be as high as ~60 mA. It is observed that the plasma plume can reach the position with the inner diameter down to 10 μm; thus, the current density can be estimated to be as high as 7.6×10^4 A cm^{-2}. **Figure 4.100** shows the gas temperature of the plasma at different positions. It indicates that the gas temperature of the plasma first increases with the increase in the distance from the end of the needle tip to the plasma and then saturates at a constant of around 540 K.

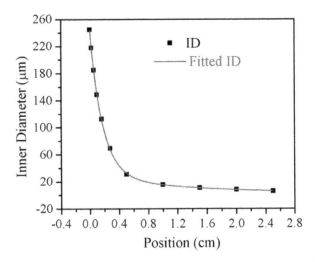

FIGURE 4.98 The inner diameter of the quartz tube vs. the position. (The end of the needle tip is considered position 0.) (From Gou, J. et al., *Phys. Plasmas*, 23, 053508, 2016.)

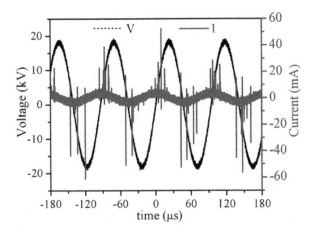

FIGURE 4.99 Typical current-voltage waveforms of the plasma. (From Gou, J. et al., *Phys. Plasmas*, 23, 053508, 2016.)

The gas pressure and the flow velocity at different positions can also be obtained, as shown in **Figure 4.101a**. The pressure difference from the reference position to 1.5 cm is small. The main pressure change occurs in the region between 1.5 cm and 2.5 cm, where the pressure increases from 1.2 bars to 3 bars. The spatially resolved Knudsen and Reynolds numbers can be calculated and are shown in **Figure 4.101b**. As the figure indicates, Re is smaller than 0.28 in the entire region of 0 cm to 2.5 cm. Thus, the gas flow is laminar in the entire region. Meanwhile, the gas in the region between

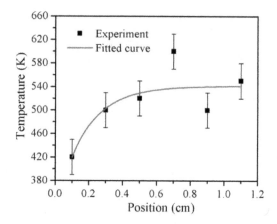

FIGURE 4.100 The gas temperature of the plasma at different positions. (The end of the needle tip is considered position 0.) (From Gou, J. et al., *Phys. Plasmas*, 23, 053508, 2016.)

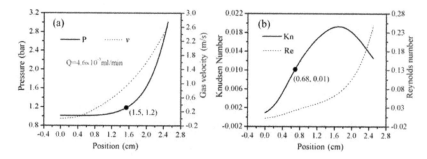

FIGURE 4.101 (a) Gas pressure and velocity vs. position. (b) Knudsen number and Reynolds number vs. position. (From Gou, J. et al., *Phys. Plasmas*, 23, 053508, 2016.)

0 cm and 0.68 cm is in the continuum flow regime, whereas the gas in the region between 0.68 cm and 2.5 cm is in the slip flow regime. Similar to the slip flow regime, continuum conservation equations can still be used to describe the bulk flow, and the Poiseuille law is applied to our experiment to acquire the gas characteristics. The electron density of the plasma plume at different positions, obtained by the same method, is shown in **Figure 4.102**. The figure indicates that the electron density of the plasma increases with the decrease of the diameter of the tube. As shown in **Figure 4.103**, the electron density obtained from H_α Stark broadening and the electron density estimated from the discharge current show a similar trend. The difference between the electron densities deduced from the two methods becomes larger and larger as the distance from the end of the needle tip to the plasma decreases. The plasma does not fill the whole cross section of the tube until the diameter of the tube is reduced to about 30 μm, as shown in **Figure 4.104**.

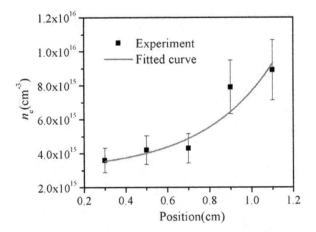

FIGURE 4.102 Electron density of the plasma plume at different positions. (From Gou, J. et al., *Phys. Plasmas*, 23, 053508, 2016.)

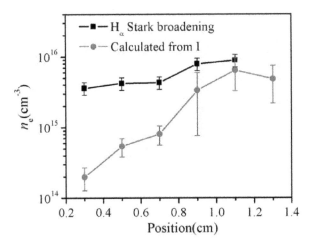

FIGURE 4.103 The electron density at different positions obtained from H_α Stark broadening and estimated from the discharge current. (From Gou, J. et al., *Phys. Plasmas*, 23, 053508, 2016.)

This results in a significant underestimation of the electron density of the plasma. Therefore, the electron density obtained from H_α Stark broadening is closer to the real value of the electron density of the plasma. The length of the plasma plume vs. the applied voltage is shown in **Figure 4.105a**. The corresponding ignition voltage vs. the inner diameter of the discharge tube is shown in **Figure 4.105b**. From the simple assumption given above, we can estimate that in order to ignite the plasma in a tube with an inner diameter of 1 µm, an applied voltage (peak to peak) of 65 kV is needed [65].

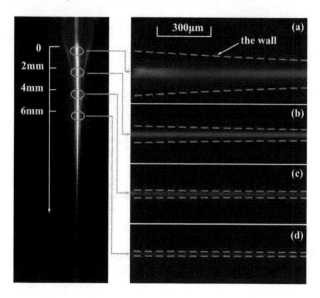

FIGURE 4.104 Photograph of the plasma in different areas of the plume: (a) 0–1 mm; (b) 2–3 mm; (c) 4–5 mm; and (d) 6–7 mm. (From Gou, J. et al., *Phys. Plasmas*, 23, 053508, 2016.)

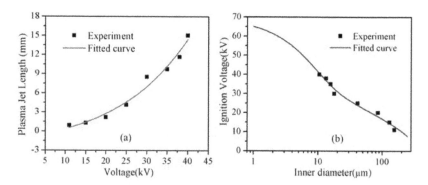

FIGURE 4.105 (a) The length of the plasma plume for different applied voltages and (b) the ignition voltage for different inner diameters of the discharge tube. (From Gou, J. et al., *Phys. Plasmas*, 23, 053508, 2016.)

An atmospheric pressure nonequilibrium Ar microplasma generated inside a micro-tube with a plasma radius of 3 μm and a length of 2.7 cm (**Figure 4.106**) was also studied [66]. The electron density of the plasma plume estimated from the broadening of the Ar emission line reaches as high as 3×10^{16} cm^{-3}. The electron temperature obtained from the CR model is 1.5 ev while the gas temperature of the plasma estimated from the N_2 rotational spectrum is close to room temperature. The sheath thickness of the plasma could be close to the radius of the plasma. The ignition voltages of

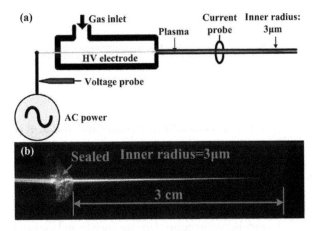

FIGURE 4.106 (a) The schematic of the experiment setup. (b) The photograph of the Ar plasma plume. (From Lu, X. et al., *Sci. Rep.*, 4, 7488, 2014.)

the plasma increase one order of magnitude when the radius of the dielectric tube is decreased from 1 mm to 3 μm [66] (**Figure 4.107**).

To better understand plasma bullet behavior in dielectric tubes, the propagation of plasma streams in a specially designed U-shaped tube was also investigated. The length of plasma stream decreases with the reduction of the distance between the bended tubes. Interestingly, when a secondary discharge occurs near the high-voltage (HV) electrode at the falling edge of the pulsed DC voltage, the plasma bullet in the bottom tube starts to speed up.

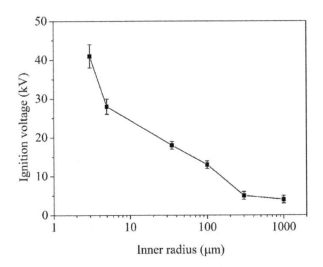

FIGURE 4.107 The ignition voltage of the Ar plasma plume vs. the inner radius of the glass tube. (From Lu, X. et al., *Sci. Rep.*, 4, 7488, 2014.)

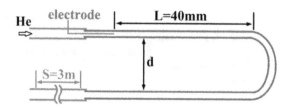

FIGURE 4.108 Schematic of the discharge device. (From Wu, S. et al., *AIP Adv.*, 5, 027110, 2015.)

Figure 4.108 shows the schematic of the discharge device. Atmospheric-pressure plasma streams are generated in the U-shaped glass tube. The inner diameter of the tube is 1 mm. The wall thickness is 1 mm. The high-voltage electrode is made of a steel needle with a radius of 100 μm, which is inserted into the end of the upper tube. The distance between the tip of the steel needle and the bending point is 40 mm. To minimize the ambient air contamination, another 3 m long tube is connected to the end of the bottom tube. When high pulsed-DC voltage (amplitude: 8 kV, pulse repetition frequency: 8 kHz, pulse width: 1 μs fixed throughout the paper) is applied to the HV electrode and helium gas is fed into the U-shaped tube with flow rate of 1 L/min, a plasma stream is generated inside the tube.

Figure 4.109a shows the photo of typical discharges in a straight tube. The inner diameter of the straight tube is 1 mm. The experimental conditions are the

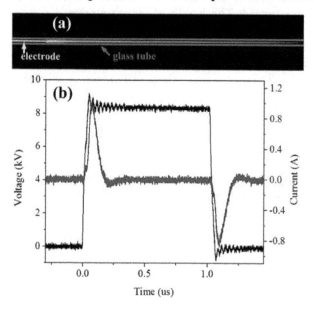

FIGURE 4.109 (a) The photo of the discharge in a straight tube. (b) The discharge current and applied voltage waveforms vs. time for the straight tube. The current probe is located at 15 mm away from the tip of the HV electrode. Applied voltage: 8 kV, pulse repetition frequency: 8 kHz, pulse width: 1 μs, helium gas flow rate: 1 L/min. (From Wu, S. et al., *AIP Adv.*, 5, 027110, 2015.)

same as those in the U-shaped tube. To measure the discharge current and applied voltage of the discharge, a Tektronix current probe (TCP312) and Tektronix voltage probe (P6015A) were used. The diameter of the probe jaw is 3.8 mm, which is bigger than the outer diameter of the glass tube. It allows us to measure the stream current by letting the plasma stream across the probe jaw. The discharge current is obtained by subtracting the displacement current from the total current. **Figure 4.109b** shows the discharge current and voltage waveforms for the straight plasma stream. Two distinct current pulses are observed for one single voltage pulse, which correspond to the rising and falling edge of the pulse voltage, respectively. The secondary discharge is probably related to the charges deposited on the inner surface of the tube during the primary discharge [67].

Figure 4.110 shows that the plasma streams can pass through the curve of the tube and keep propagating in the opposite direction in the bottom part of the tube. As can be seen in **Figure 4.110**, the smaller the d is, the

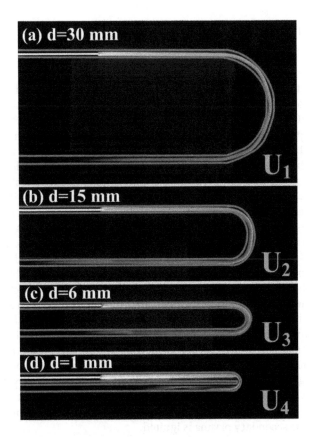

FIGURE 4.110 Photos of the plasmas in U-shaped tubes for different distances (d). (a) $d = 30$ mm, (b) $d = 15$ mm, (c) $d = 6$ mm, (d) $d = 1$ mm. Applied voltage: 8 kV, pulse repetition frequency: 8 kHz, pulse width: 1 μs, helium gas flow rate: 1 L/min. (From Wu, S. et al., *AIP Adv.*, 5, 027110, 2015.)

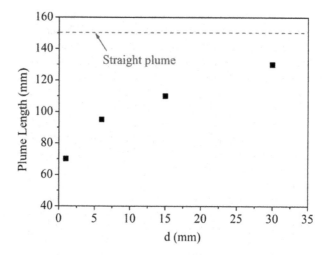

FIGURE 4.111 The stream length vs. *d*. (From Wu, S. et al., *AIP Adv.*, 5, 027110, 2015.)

shorter the total length of the plasma stream is. To have a detailed discussion about the length of the plasma stream, the length of the plasma stream vs. *d* is plotted in **Figure 4.111**. When *d* increases from 1 mm to 30 mm, the stream length increases from 70 mm to 130 mm. It is worth pointing out that, in **Figure 4.110c and d**, the upper inner surface propagation of the plasma stream is observed after the U turn rather than all along the whole tube uniformly. To understand how the plasma stream propagates in the U-shaped tube and why the stream propagates along the upper inner surface of the tube when *d* is small, an intensified charged-coupled device (ICCD) camera (Princeton Instruments, Model PIMAX2) was used to capture the dynamics of the discharges in the U-shaped tube for *d* = 30 mm and *d* = 1 mm, respectively.

Figure 4.112 shows the dynamics of the discharge in the U-shaped tube for *d* = 30 mm. It can be seen that the plasma bullets compose a bright head and a relatively weak long tail up to several centimeters long, which is different from the plasma bullets in the open air. This observation is in agreement with both the experimental and simulated results. The plasma bullet reaches the beginning point of the curve at 260 ns and passes through the curve at 610 ns. Then it keeps propagating inside the bottom part of the tube along the opposite direction. It is worth emphasizing that the plasma bullet keeps propagating in the bottom part of the tube even after the secondary discharge is ignited, which is different from what was observed when straight tubes were used. When a straight tube is used, the plasma immediately stops propagating as soon as the secondary plasma is ignited.

To understand why the length of the plasma stream decreases with the decrease of *d* and why the propagation of the plasma bullet speeds up as soon as the secondary discharge occurs, the schematic of the charges and the electric field distribution for the propagation of the plasma streams are shown in **Figure 4.113**.

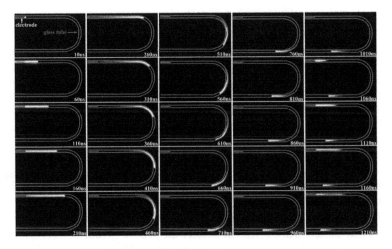

FIGURE 4.112 High-speed images of the discharge in the U-shaped tube for $d = 30$ mm. The exposure time is fixed at 5 ns. Each image is an integrated picture of 20 shots with the same delay time. The time labeled on each image corresponds to the DC pulse rise time in **Figure 4.109b**. The discharge parameters are the same as those in **Figure 4.110**. (From Wu, S. et al., *AIP Adv.*, 5, 027110, 2015.)

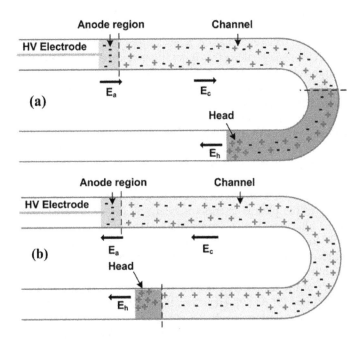

FIGURE 4.113 The schematic of the charges and the electric field distribution for the propagation of the plasma stream. The plasma bullet head inside the bottom part of the tube for (a) the primary discharge and (b) the secondary discharge. (From Wu, S. et al., *AIP Adv.*, 5, 027110, 2015.)

It's commonly agreed that the dominant charges in the plasma bullet head are positive charges. In addition, the electric field E_c is defined as the portion of the whole electric field associated with wall charging, which is produced by the positive surface charges (from the voltage rising time until the voltage starts to fall) and/or the negative surface charge (from the voltage falling phase) in the upper part of the tube. The electric field E_c can be decomposed into a horizontal component E_{ch} and a vertical component E_{cv} [67].

First, the inner surface of the tube is positively charged during the primary discharge, as shown in **Figure 4.113a**. When the plasma stream propagates in the bottom part of the tube, the electric field E_h in front of the plasma bullet head is weakened by the horizontal component E_{ch} produced by the positive charges in the upper part of the tube, which prevents the propagation of the the plasma bullet and shortens the stream length. With the decrease of d, the effects of the field weakening by E_{ch} become more significant because E_{ch} becomes stronger. That's why the smaller the d is, the shorter the plasma stream is.

Second, regarding the acceleration of the plasma bullet, the secondary discharge is ignited at the falling edge of the pulse voltage, which is similar to a negative discharge. During the secondary discharge, the inner surface of the upper part of the tube is negatively charged. Therefore, when the secondary discharge occurs, the direction of E_c is reversed, as shown in **Figure 4.113b**. The horizontal component E_{ch} strengthens the electric field E_h in front of the plasma bullet head, which facilitates the propagation of the plasma stream. That's why the plasma bullet starts to speed up as soon as the secondary discharge is ignited. This explanation is also supported by the result that the plasma bullet propagates along the upper inner surface of the lower part of the tube due to the reversion of the vertical component E_{cv}.

Besides the electric field in the head (E_h), the whole electric field also includes the laplacian field imposed through the power electrode and the plasma column. In both cases (straight or not), the plasma tail impedance should be about the same conductivity, but in the U-shaped tube, the plasma moves in the direction of the higher laplacian field after the U turn, which is not the case during straight propagation. Therefore, this could be another possible reason for the relatively shorter length of a plasma stream in a U-shaped tube.

Finally, according to the observation presented above, we believe that the conductivity of the plasma channel is relatively low (compared to a perfect conductor). Otherwise, if it was more like a perfect conductor, then we would not be able to observe the acceleration behavior at the falling edge of the voltage pulse. On the contrary, we would observe behavior similar to the case for a straight tube [67].

The effect of the dielectric constant on plasma propagation was studied by several groups [51,68,69]. However, their experimental results are quite different. The simulation result [68,69] reveals that, for a constant tube radius of 100 μm, an increase in ε_r by a factor of 10 leads to a decrease in the discharge velocity by about a factor of 2. However, another group reported that, for borosilicate ($\varepsilon_r \approx 4.6$), Teflon, Rilsan, and alumina dielectric materials, the propagation velocities of the plasma bullet are nearly the same [51]. The reasons for the different observations may be related to the different charge accumulation conditions on the tube surfaces with different dielectric constants in different experiments. Further studies are required to clarify this issue.

4.2.6 Gas Flow Rate

For the plasma jet, the gas flow rate of the working gas is the key parameter that affects the mixing of the gas inside the jet with the surrounding gas. The plasma jet length increases almost linearly with the He flow rate when the He flow is laminar. It was pointed out that the minimum He mole fraction necessary to sustain the plasma propagation is almost constant (between 0.45 and 0.5 for different He outlets). When the gas flow rate was increased to 20 SLM (standard liters per minute), the He flow transitions to the turbulent regime. In this case, the plasma plume is quite short. The simulation results of the He mole fraction distribution in the axial plane indicates that the perturbation along the He jet core is very strong, which results in a high concentration of air in the He gas stream and leads to the relatively short plasma plume [70].

The increase of the gas flow velocity can result in an increase of the distance from the nozzle to the position where there is a 50% mixing rate of the working gas inside the gas stream with the surrounding air (when the gas flow is laminar). This effect has been validated by numerical models [7] and observed in experiments [71]. Further experiments suggest that the length increases with V_{flow} in the laminar flow regime and decreases upon transition to the turbulent flow. An increase of V_{flow} also leads to the larger distance from the nozzle where the transformation from an annular to a spherical streamer structure occurs [72].

A detailed study has been carried out to observe the effect of the gas flow rate on the behavior of two plasma plumes that propagate face to face in opposite directions within a tube [73]. The schematic of the experiment setup is shown in **Figure 4.114a**. The high-voltage (HV) electrodes are made of two stainless-steel needles, which are inserted into the quartz tube. The distance between the tips of the two needles of the same size and shape is 77 mm. The radius of the needles is about 100 μm. The inner diameter of the quartz tube is

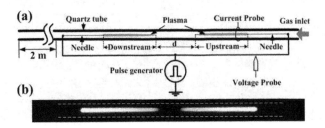

FIGURE 4.114 (a) Schematic of the device and (b) spectrally broadband photograph of the discharge. The exposure time is approximately 100 ms. (From Wu, S. et al., *Phys. Plasmas*, 20, 023503, 2013.)

about 0.8 mm. The working gas flows into the quartz tube from the right end. Research-grade Ar gas (99.999% purity) is used in the experiment. To keep the ambient air contamination in the discharge region as low as possible, another quartz tube with a length of 2 m is connected to the left end of the quartz tube, as shown in **Figure 4.114a**. The two needles connected to the high-voltage pulsed DC power supply are parallel (amplitude up to 10 kV, repetition rates up to 10 kHz, and pulse width variable from 200 ns to DC). The applied voltage of 8 kV, pulse frequency of 9 kHz, and pulse width of 400 ns were fixed in this series of experiments. When HV pulsed DC voltage is applied to the two needles and the working gas is injected into the quartz tube with a gas flow rate of 2 L/min, plasma plumes are generated inside the quartz tube, as shown in **Figure 4.114b**.

When a high voltage is turned on and Ar flows through the quartz tube, two plasma plumes are generated in front of the two needles. **Figure 4.115** shows photographs of the plasma plumes taken with a Sony digital camera (DSC-HX1) at an exposure time of about 100 ms for different Ar flow rates. When the Ar flow rate increases from 0.2 L/min. to 1 L/min, both plasma plumes become longer. The difference between the lengths of the two plasma plumes is very small when the Ar flow rate is lower than 0.5 L/min. When the Ar flow rate is increased to 2 L/min, the plasma plumes become shorter. Moreover, the length of the upstream plasma plume L_{up} becomes even shorter than the length of the downstream plasma plume L_{down}. **Figure 4.116** shows the difference between lengths L_{up}

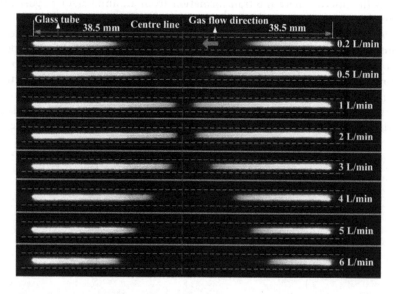

FIGURE 4.115 Photographs of the plasma plumes at different gas flow rates in Ar working gas. The photographs are spectrally broadband. The exposure time is approximately 100 ms. (From Wu, S. et al., *Phys. Plasmas*, 20, 023503, 2013.)

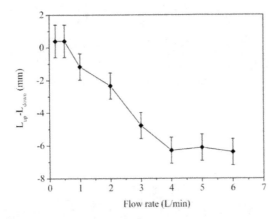

FIGURE 4.116 The length difference of the downstream and the upstream plasma plumes versus the gas flow rate. (From Wu, S. et al., *Phys. Plasmas*, 20, 023503, 2013.)

and L_{down}. Next, a detailed analysis of the effect of the Ar flow rate on the length of the plasma plumes is presented.

To explain why both the plasma plumes get shorter when the flow rate is further increased, the Reynolds numbers of the Ar flow inside the quartz tube were calculated for different gas flow rates. The calculated Reynolds number for a flow rate of 2 L/min is about 4568. Since the critical Reynolds number is 2320 to distinguish between the laminar and transition flows in the tube, the flow becomes hydrodynamically unstable when the flow rate is 2 L/min or higher. Hence, the higher the flow rate is, the stronger the hydrodynamic instability is, which in turn shortens the plasma plume.

To elucidate why the upstream plume becomes shorter than the downstream one with the increase of the gas flow rate, first, we switched the positions of the two needles. The result was exactly the same, which means that any small difference in the sizes and shapes of the two needles had no influence on the observed phenomenon.

Second, for the upstream plasma plume, the gas flows from the tube with the needle in the center to the free space where there is no needle. On the contrary, for the downstream plasma plume, the gas flows from the free space to the place where there is a needle. This could cause the flow mode difference and might result in the observed phenomenon. To verify if this is the actual reason why L_{up} is shorter than L_{down}, another experiment was carried out. As shown in **Figure 4.117**, when a single needle is used as an electrode, the gas flow direction can either be the same as the plasma plume propagation direction or the opposite. **Figure 4.117** shows that when the plasma plume is in the same direction as the gas flow, which is the same as the plasma plume propagation direction, it is longer than the plasma plume when the plasma plume propagation direction and the gas flow direction are the opposite. Due to the very low presence of air in the discharge tube, excited N_2^* species are also not expected

FIGURE 4.117 Photographs of the discharges produced with a single needle electrode in two gas flows with opposite directions. The Ar flow rate is 4 L/min. The photographs are spectrally broadband. The exposure time is approximately 100 ms. (From Wu, S. et al., *Phys. Plasmas*, 20, 023503, 2013.)

FIGURE 4.118 Photographs of the plasma plumes with the addition of nitrogen to the Ar flow. The Ar flow rate is fixed at 4 L/min. The photographs are spectrally broadband. The exposure time is approximately 100 ms. (From Wu, S. et al., *Phys. Plasmas*, 20, 023503, 2013.)

to affect the interaction between the two plasma plumes. To confirm this, 1% of N_2 was added to the Ar gas flow. **Figure 4.118** depicts the propagation of the two plasma plumes with and without the N_2 addition. Detailed analysis shows that the difference between L_{up} and L_{down} remains the same for the two experimental conditions. Therefore, nitrogen-related species also have no effect on the length difference between the upstream and downstream plumes [73].

4.2.7 Applied Voltage Amplitude

Numerous experimental [28,48,53,58] and numerical [5,7,29] studies have been carried out on the effect of the applied voltage U on the plasma jet. A measured dependence of the jet length on the applied voltage at various flow rates indicates that the length increases with U at low voltages and saturates at larger U. This saturation is quite typical for guided streamers that propagate in mixtures of a noble gas with surrounding air. However, this does not happen for streamers in uniform media. When the applied voltage becomes sufficiently high, the streamer penetrates into a region with a high content of air, where the streamer propagation is inhibited. On the other hand, the increasing flow velocity leads to the maximum jet length increases since considerable gas mixing occurs at larger distances from the nozzle exit. In uniform gas flows, the velocity and propagation length are monotonously increasing functions of the applied voltage for the plasma plumes propagating inside the tube [54].

4.2.8 Plasma-Forming Gas

He (pure or with a small amount of various additives) is used as the working gas for most of the plasma jet because discharge ignitions in He require lower applied voltages than in other gases, and the use of He allows one to obtain longer jets. Nevertheless, in several works, plasma jets composed of other noble gases (Ne, Ar, Kr, He–Ar mixtures), pure and with small amounts of molecular additives (O_2, H_2O), were studied [18,64]. It has been shown that plasma jet dynamics and structure can vary substantially depending on the composition of the plasma-forming gas [74].

The effects of adding N_2, O_2, air, and H_2O into the He working gas have been studied experimentally and numerically. The ionization rate shown in **Figure 4.119a** suggests that the plasma jet propagated in the bullet mode. Since the plasma bullet ran out of the tube nozzle at 82 ns, the propagation speed of the plasma bullet is 6.8×10^4 m/s. The ionization of He, N_2, and O_2 in the plasma jet is predominantly due to energetic electrons. Moreover, this ionization takes place in the helium-air mixture layer, with a ring-shaped ionization domain. The dimensions of the plasma channel deduced through the analysis of the electron density distribution (**Figure 4.119b**) are consistent with the plasma channel shown in the long exposure time image. The shrinking plasma channel in the radial direction inside the tube is due to the plasma sheath adjacent to the inner surface of the tube. Although ionization of He, N_2, and O_2 by energetic electrons is likely to be the main mechanism in the bullet, the Penning ionizations of N_2 and O_2 by He* dominate the ionization in the plasma channel and produce enough electrons to maintain its conductivity, so a strong electric field arises in front of the plasma channel (**Figure 4.119d**).

The plasma sheath inside the tube not only compresses electrons into a slim channel along the tube axis but also makes concentrated OH in this channel (**Figure 4.119c**). Although the H_2O concentration inside the tube is only 100 ppm, the OH density is 1.0×10^{18} m^{-3}, which is half the peak OH density value outside the tube. Besides the long lifetime of OH radicals, the comparison between the electron (**Figure 4.119b**) and OH (**Figure 4.119c**)

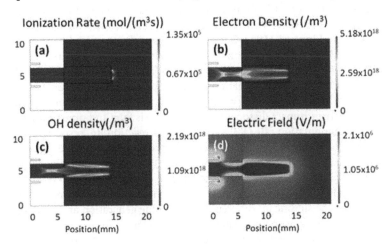

FIGURE 4.119 (a) The ionization rate (including ionization of He, N_2, O_2 and H_2O), (b) electron density, (c) OH density, and (d) electric field of the plasma jet at 176 ns. (From Liu, X.Y. et al., *Phys. Plasmas*, 21, 093513, 2014.)

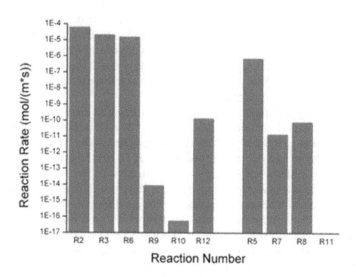

FIGURE 4.120 Spatially averaged reaction rates of reactions related to the production and consumption loss of OH at 176 ns. Reaction number is the same as in Table 4.1. Reactions R2–R12 are related to the OH species production, while reactions R5–R11 are related to the OH consumption loss. (From Liu, X.Y. et al., *Phys. Plasmas*, 21, 093513, 2014.)

density distributions suggests that the energetic electrons produced by the plasma sheath may result in higher OH production rates inside the tube. For the section of the plasma jet outside the tube, the OH channel is wider than the electron channel (compare **Figure 4.119b** and **c**), which is attributed to the higher concentration of H_2O in the gas mixing layer. The OH channel in the ring shape is consistent with the results of the LIF measurements. The peak OH density in **Figure 4.119c** is $2.19 \times 10^{18} \, m^{-3}$, which is reasonably close to the value $1.8 \times 10^{18} \, m^{-3}$ deduced from the experimental measurements.

Figure 4.120 shows that the dissociation of H_2O by electrons (R2, $e + H_2O \Rightarrow e + H + OH$), the recombination of H_2O^+ by electrons (R3, $e + H_2O^+ \Rightarrow H + OH$), and R6 ($O(1D) + H_2O \Rightarrow 2OH$) dominate the production of OH species. The electron-impact dissociation rate of H_2O is four and nine times higher than the recombination rate (R3) and the rate of reaction R6, respectively. This is also the reason for the higher OH production rates inside the tube. The contributions of reactions R9 ($O + H + M \Rightarrow OH + M$), R10 ($H + O_2 \Rightarrow OH + O$), and R12 ($H_2O + O \Rightarrow 2OH$) to OH production appear to be negligible. On the other hand, reaction R5 ($OH + H + M \Rightarrow H_2O + M$) turns out to be the main OH loss mechanism. The significant difference between the generation and loss rates of OH radicals is the reason for the long lifetime of OH.

The effect of different working gases on the OH density distribution has also been studied. **Figure 4.121** shows the variation of the OH species distribution with different gas mixture ratios. The OH radicals mostly reside inside the tube for the He + 0.5% O_2 gas mixture (**Figure 4.121b**). The peak density

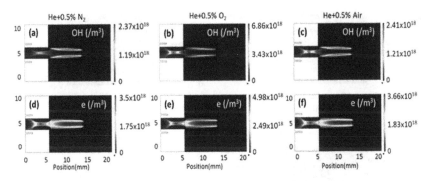

FIGURE 4.121 OH density and electron density distribution at 170 ns with the working gas of He+0.5% N_2 (panels (a) and (d)), He+0.5% O_2 (panels (b) and (e)), and He+0.5% air (panels (c) and (f)) at 176 ns. (From Liu, X.Y. et al., *Phys. Plasmas*, 21, 093513, 2014.)

of OH inside the tube is $6.86 \times 10^{18}\,\text{m}^{-3}$, which is more than three times higher than outside the tube. The addition of O_2 increases the density of O(1D); therefore, reaction R6 (O(1D)+H_2O => 2OH) dominates the OH production and increases the OH density inside the tube. When the working gas was changed to He+0.5% air and He+0.5% N_2 gas mixtures, the peak OH density inside the tube decreased to $1.97 \times 10^{18}\,\text{m}^{-3}$ and $0.7 \times 10^{18}\,\text{m}^{-3}$, respectively (**Figure 4.121a and c**). This confirms the contribution of O(1D) species to OH production in the high O_2 content region [74].

On the other hand, the addition of N_2, O_2, and air do not significantly affect the ring shape of the plasma channel. Indeed, the peak OH density outside the tube increased from $2.19 \times 10^{18}\,\text{m}^{-3}$ (helium only, **Figure 4.119c**) to $2.29 \times 10^{18}\,\text{m}^{-3}$ (He+0.5% N_2, **Figure 4.121a**), $2.4 \times 10^{18}\,\text{m}^{-3}$ (He+0.5% air, **Figure 4.121c**), and $2.43 \times 10^{18}\,\text{m}^{-3}$ (He+0.5% O_2, **Figure 4.121b**). Higher OH densities with the addition of O_2 into the working gas have also been observed in microplasma jets. The addition of N_2, O_2, and air increase the Penning ionization along the plasma channel. Therefore, the peak electron density outside the tube increased from $1.78 \times 10^{18}\,\text{m}^{-3}$ (helium only, **Figure 4.119b**) to $2.09 \times 10^{18}\,\text{m}^{-3}$ (He+0.5% N_2, **Figure 4.121d**), $2.1 \times 10^{18}\,\text{m}^{-3}$ (He+0.5% air, **Figure 4.121f**), and $2.13 \times 10^{18}\,\text{m}^{-3}$ (He+0.5% O_2, **Figure 4.121e**), which eventually enhanced OH production through the electron-impact dissociation of H_2O (reaction R2, e+H_2O => e+H+OH), as well as the neutralization of H_2O+ by electrons (R3, e + H_2O + => H + OH) [74].

The H_2O concentration also affects the OH density distribution. As the H_2O concentration increases from 0.01% to 1% (**Figure 4.122a–c**), not only does the high OH density region move to a position inside the tube, but the propagation speed of the plasma jet also increases. The results in **Figure 4.122d** show that the OH species density inside the tube increases from $0.99 \times 10^{18}\,\text{m}^{-3}$ (0.01% H_2O) to $2.53 \times 10^{19}\,\text{m}^{-3}$ (1% H_2O), which is attributed to the 200-fold increase in the H_2O concentration inside the tube. During the pulse off period, the OH in the high-density region will eventually run out of the tube with the gas flow [74].

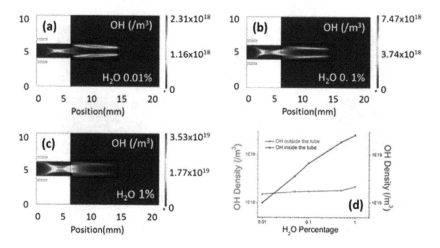

FIGURE 4.122 The OH radical density distribution for H_2O concentrations of (a) 0.01%, (b) 0.1%, and (c) 1%. The peak OH density values inside the tube and outside the tube for different H_2O concentrations (d). (From Liu, X.Y. et al., *Phys. Plasmas*, 21, 093513, 2014.)

Besides the difference n the kinetic and transport coefficients for electrons, the differences in the APPJ parameters for various plasma-forming gases can be caused by such factors as the different role of the Penning ionization, the different mixing rates of the jets with the surrounding air, etc.

4.2.9 Surrounding Gas

For atmospheric pressure plasma jets, the plasma is generated inside a dielectric tube first, then the plasma propagates into a surrounding gas. The diffusion of the surrounding gas affects the propagation of the plasma bullet. The effect of the composition of the surrounding gas on plasma bullet behavior was investigated [8,72,75–77].

Two different discharge configurations were employed to study the effect of the surrounding gas. The first one is the traditional discharge, where the plasma plume is generated in the surrounding air, while the second one generates the plasma inside a bottle. For the second one, the jet was put into a glass bottle, which was first pumped to 10^{-2} Torr and then filled with research-grade He gas [51]. The photographs of the plasmas show clearly that the plasma jet is arrow-like in ambient air and has a diffuse shape in ambient He.

A glass container is employed to study the effect of the surrounding gas on the plasma jet [76]. **Figure 4.123** is the schematic of the experiment setup. The plasma plume is generated by a single electrode plasma jet device. A high-voltage wire electrode, which is made of a copper wire, is inserted into a 4 cm long quartz tube with the right end closed. However, in this report, the plasma plume is generated in a cylindrical glass container rather than in the

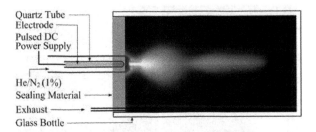

Quartz Tube
Electrode
Pulsed DC
Power Supply

He/N₂ (1%)
Sealing Material
Exhaust
Glass Bottle

FIGURE 4.123 Schematic of the experiment setup. (From Xian, Y. et al., *Plasma Sources Sci. Technol.*, 21, 034013, 2012.)

surrounding air. The inner diameter of the container is 16 cm, and the depth of the container is about 25 cm. There is an exhaust tube with a diameter of 2 mm at the lower part of the stopper. The pressure in the container is at atmospheric pressure. Because the glass container is not vacuum sealed, there is always a trace amount of air inside the glass bottle. A He/N₂ (1%) mixture is used as the working gas with a total flow rate of 2 L/min. N₂ (1%) is added to He to make the plasma bullet brighter so that much clearer plasma bullet photographs can be captured. A high-voltage (HV) pulsed direct current (DC) power supply (amplitude up to ±10 kV, pulse width variable from 200 ns to DC, pulse repetition rate up to 10 kHz) is used to drive the plasma jet device. Before turning on the power supply, we start the He/N₂ (1%) flow for about 10 minutes and maintain the gas flow during the experiment. It is expected that there is only a trace amount of air left in the container. When the HV pulsed DC voltage is turned on, a plasma plume is generated in the glass container, as shown in **Figure 4.123**.

As shown in **Figure 4.124c**, the second plasma bullet starts appearing in the nozzle while the first plasma bullet propagates in the surrounding gas. The second plasma bullet exits the nozzle at about 1.2 μs, as shown in **Figure 4.124e**. Then both bullets propagate in the surrounding gas with approximately the same velocity, as shown in **Figure 4.124e–n**. It is worth mentioning that, when the first bullet exits the nozzle, both the dimension and brightness of the bullet increase dramatically. On the other hand, at the moment when the second plasma bullet exits the nozzle, the dimension and brightness of the bullet increase only slightly, but they keep increasing. As the brightness of the first bullet decreases, both of them reach a similar dimension. Moreover, **Figure 4.124o–q** show that there is another discharge. But the plasma does not propagate further in the surrounding gas. This discharge appears at the falling edge of the applied voltage [76].

To further understand whether the multiple bullet behavior is polarity-related, a negative HV pulsed DC voltage was used to drive the plasma device. The amplitude and repetition frequency of the pulsed DC voltage were the same. The pulse width was decreased from 200 μs to 125 μs. It is found that only one plasma bullet is captured for each voltage pulse when the pulse width is shorter than 80 μs or longer than 105 μs. On the other hand, when the pulse width is increased from 80 μs to 105 μs, two plasma bullets are observed per

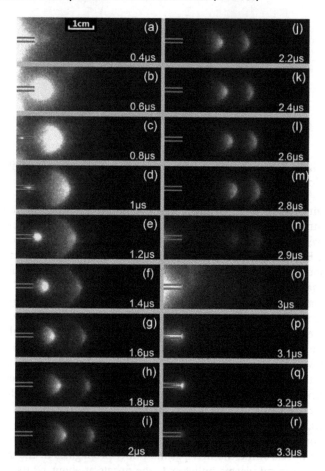

FIGURE 4.124 High-speed photograph of the plasma plume for the positive pulse. (From Xian, Y. et al., *Plasma Sources Sci. Technol.*, 21, 034013, 2012.)

voltage pulse. When the pulse width is in the range from about 98 μs to 102 μs, three plasma bullets may be captured per voltage pulse. Similarly to the positive case, the interval between the bullets first increases, then decreases when the pulse width is increased from 80 μs to 105 μs. **Figure 4.125** shows the plasma bullet behavior for a pulse width of 100 μs. The first, second, and third bullets exit the nozzle at 101.73 μs, 108.9 μs and 115.35 μs, respectively, as shown in **Figure 4.125a, f** and **j**, which corresponds to the rise in voltage from −8 kV to 0 V, as shown in **Figure 4.124b**. According to **Figure 4.125**, the distances between the bullets remain almost the same. It is also interesting to point out that the brightness of the three bullets decreases at the moment they exit the nozzle. The plasma is also generated at the falling edge of the voltage pulse (from 0 V to −8 kV), as shown in **Figure 4.125q–u** [76].

Other studies have been carried out on the effect of He and Ar working gas, which are used separately and in various mixing ratios, to generate a gas flow in

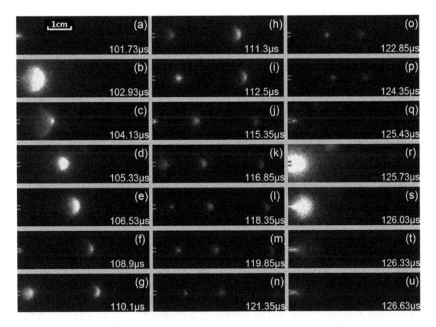

1cm (a) 101.73μs	(h) 111.3μs	(o) 122.85μs
(b) 102.93μs	(i) 112.5μs	(p) 124.35μs
(c) 104.13μs	(j) 115.35μs	(q) 125.43μs
(d) 105.33μs	(k) 116.85μs	(r) 125.73μs
(e) 106.53μs	(l) 118.35μs	(s) 126.03μs
(f) 108.9μs	(m) 119.85μs	(t) 126.33μs
(g) 110.1μs	(n) 121.35μs	(u) 126.63μs

FIGURE 4.125 High-speed photograph of the plasma plume for the negative pulse. (From Xian, Y. et al., *Plasma Sources Sci. Technol.*, 21, 034013, 2012.)

homogenous gas [78]. A number of discharge structures, including stable splitting, random branching, etc., were produced depending on the gas composition.

The effects of the N_2 and O_2 surrounding gases on the propagation of the plasma plume was also studied. It is found that an interesting and unusual feather-like plasma plume can be generated when N_2 is used as the working gas. A combination of nanosecond-precision high-speed photography, electromagnetic measurements, and numerical modeling was used to relate the quantized plasma bullet propagation and the He density distribution to the formation of the unusual feather-like diffuse plasma plume structures. **Figure 4.126** is the schematic of the experiment setup. The high-voltage (HV) wire electrode, which consists of a copper wire with a diameter of 2 mm, is inserted into a 4 cm long quartz tube with one end closed. The inner and outer diameters of the quartz tube are 2 mm and 4 mm, respectively. The quartz tube along with

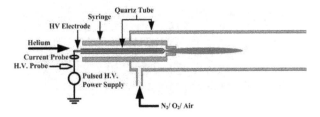

FIGURE 4.126 Schematic of the experiment setup. (From Xian, Y.B. et al., *Appl. Phys. Lett.*, 103, 094103, 2013.)

the HV electrode is inserted into a hollow barrel of a syringe. The diameter of the hollow barrel is about 6 mm and the diameter of the syringe nozzle is 1.2 mm. The distance between the tip of the HV electrode and the nozzle is 1 cm. The syringe is placed into another quartz tube, which can let N_2, O_2, or air flow through. The inner diameter of this quartz tube is 17 mm [79]. When He with a flow rate of 1.0 L/min. is injected into the syringe and when ambient gas (N_2, O_2, or air) with a flow rate of 1.5 L/min is injected into the outer quartz tube, and then the HV pulsed DC voltage is turned on, a homogeneous plasma is generated in front of the end of the inner quartz tube, along the nozzle, and in the ambient gas (**Figure 4.127**). The feather-like plasma plume can only be observed when nitrogen is used as the surrounding gas. It is very likely that the diffusion of He and nitrogen-based species, i.e., the density distribution of these species, plays a major role in the generation of the feather-like plasma plume. On the other hand, the flow rate of the He will affect the concentration distribution of N_2 and He.

The photographs of the plasma plume for different He flow rates in **Figure 4.128** show clearly that the plasma plume becomes shorter and the feather-like plume appears closer to the nozzle at lower He flow rates. To obtain more detailed information about the nitrogen and helium concentration at the starting point of the feather-like plume, the N_2 and He distributions were modeled by using the commercial software Fluent for the flow rates of 0.2 L/min, 0.6 L/min, and 1.0 L/min. As shown in **Figure 4.129**, it is interesting to note that the He concentration on the axis at the starting point of the feather-like plasma plume is about 85% (with respect to its maximum value) for all three He flow rates. Beyond its starting point, the feather-like plume can even expand in radical direction into the region where the He concentration is only around 10%. Importantly, the plasma plume does not start expanding immediately upon exiting the nozzle. **Figure 4.130** shows the magnetic field signals along with the applied voltage waveform for all three cases considered. It shows clearly that, when the plasma plume propagates in

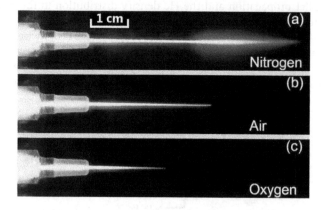

FIGURE 4.127 Photographs of the He plasma jet propagating in ambient (a) N_2, (b) air, and (c) O_2. The flow of He is 1.0 L/min., and the flows of N_2, air, and O_2 are 1.5 L/min. (From Xian, Y.B. et al., *Appl. Phys. Lett.*, 103, 094103, 2013.)

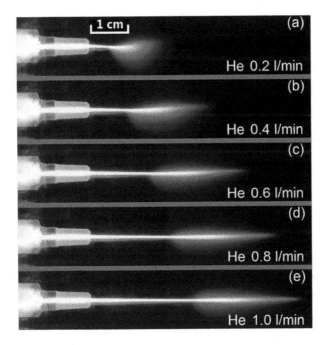

FIGURE 4.128 Photographs of the plasma plumes for the He flow rates of (a) 0.2 L/min., (b) 0.4 L/min., (c) 0.6 L/min., (d) 0.8 L/min., and (e) 1.0 L/min. The flow rate of the surrounding N_2 is 1.5 L/min. (From Xian, Y.B. et al., *Appl. Phys. Lett.*, 103, 094103, 2013.)

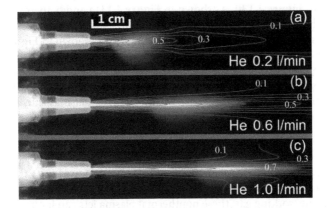

FIGURE 4.129 Numerical modeling of the He concentration distribution for the He flow rates of (a) 0.2 L/min., (b) 0.6 L/min., and (c) 1.0 L/min. The N_2 flow rate with the big quartz tube is fixed at 1.5 L/min. The labels on the contour line denote the relative He concentration with respect to the maximum value. (From Xian, Y.B. et al., *Appl. Phys. Lett.*, 103, 094103, 2013.)

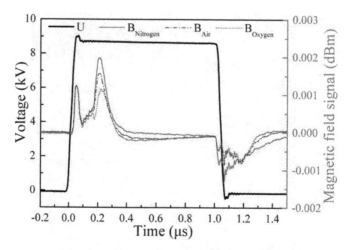

FIGURE 4.130 Applied voltage and magnetic field signals generated by the three plasma plumes when they are propagating in the surrounding N_2, air, and O_2. The flow rates are the same as in **Figure 4.127**. (From Xian, Y.B. et al., *Appl. Phys. Lett.*, 103, 094103, 2013.)

surrounding the N_2, the magnetic field signal generated by the plasma plume is much higher than that in surrounding air or oxygen.

To further understand the feather-like plasma plume, a high-speed ICCD camera (PIMAX2, Princeton Instruments, exposure time is fixed at 10 ns) was used to capture the dynamics of the plasma plumes. **Figure 4.131a–c** show the high-speed photographs of the plasma plumes propagating in ambient N_2, air, and O_2, respectively. The time labeled on each photograph corresponds to the times in **Figure 4.130**. As shown in **Figure 4.131a** and **b**, in ambient N_2 or air, the volume of the plasma bullet increases first and then decreases when it propagates away from the nozzle. On the contrary, as shown in **Figure 4.131c**, when O_2 is used as the surrounding gas, the volume of the plasma bullet keeps decreasing upon propagation. It can be observed that when the plasma bullet propagates in ambient N_2, dim diffuse plasmas appear just behind the bright head of the plasma bullet at 605 ns. The dim diffuse plasmas thus induce the observed feather-like shape of the propagating plasma plume.

4.2.10 Operation Under Variable Pressure

Studies have been carried out on the effects of He and Ar working gas, which are used separately and at various pressures. Recently, it was found that, by using helium as the working gas, these nonthermal plasma jets can ignite a large-volume nonthermal plasma inside a vacuum chamber, and, depending on the pressure, two distinct modes of operation could be sustained: jet mode and diffuse mode (**Figure 4.132**). In the jet mode, the plasma plume's maximum length was measured at 28 cm at 75 Torr inside a 40 cm × 50 cm cross-shaped glass vacuum chamber with a volume of 11 L. At lower pressures, the length contracted, and

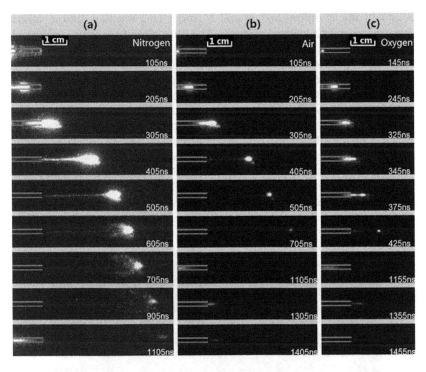

FIGURE 4.131 High-speed optical imaging of the plasma jets propagating in ambient (a) N_2, (b) air, and (c) O_2. The exposure time is fixed at 10 ns. The time labeled on each photograph corresponds to the time in **Figure 4.128**. The flow rates are the same as in **Figure 4.127**. (From Xian, Y.B. et al., *Appl. Phys. Lett.*, 103, 094103, 2013.)

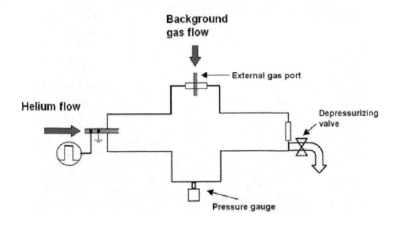

FIGURE 4.132 Schematic of the vacuum chamber, tube reactor, and background gas injection port. (From Akman, M.A., and Laroussi, M., *IEEE Trans. Plasma Sci.*, 41, 839–842, 2013.)

eventually the jet diminished around 20 Torr giving way to a diffuse plasma that expanded with lowered pressure. At 3 Torr, almost the entire volume of the chamber was filled with this diffuse plasma and no jet was observed. Our initial high-speed imaging characterization of the plasma bullets in this setup provided supporting experimental evidence for the photoinization model for propagation proposed. This model predicts the expansion of the plasma in all directions inside the chamber under low pressure conditions as was observed in the diffuse mode of operation in our experiments [77]. The experiment setup is shown in **Figure 4.133**. These parameters were a 7-kV applied voltage pulse with a 5-kHz frequency and a 2-μs pulse width. The voltage pulses were applied to two ring electrodes wrapped around a dielectric tube. Helium was used as the working gas and was fed by a mass flow controller (**Figure 4.134**).

The plume/jet is injected into a chamber where the background pressure can be controlled. Thus, by lowering the pressure, less air molecules will be

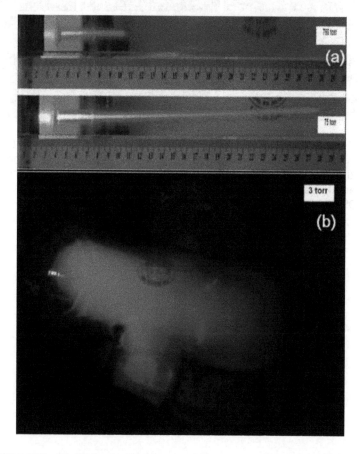

FIGURE 4.133 Nonthermal plasma generated by a tube reactor operated either in (a) jet mode or (b) diffuse mode depending on the pressure alone. (From Akman, M.A., and Laroussi, M., *IEEE Trans. Plasma Sci.*, 41, 839–842, 2013.)

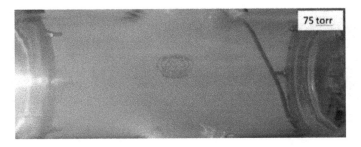

FIGURE 4.134 Diffuse plasma covering the volume of the chamber at 75 Torr when only helium is flowing into the chamber as the working gas and no other background gas is injected. (From Akman, M.A., and Laroussi, M., *IEEE Trans. Plasma Sci.*, 41, 839–842, 2013.)

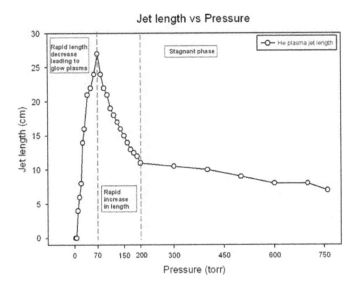

FIGURE 4.135 Plasma plume length versus pressure inside the chamber. (From Laroussi, M., and Akman, M.A., *AIP Adv.*, 1, 032138, 2011.)

interacting with the helium flow and a longer helium channel can be achieved. However, **Figure 4.135** suggests that the behavior of the plume is much more complicated as the pressure changes. In fact, from atmospheric pressure to about 200 Torr, the plume increases relatively slightly in length. Below a pressure of 200 Torr and down to 70 Torr, we observed a rapid increase in the length of the plasma plume. However, below 70 Torr, the length of the plume starts decreasing rather quickly while, at the same time, the plasma starts expanding in all directions, starting at the tip of the acrylic tube inside the chamber. **Figure 4.136**, which shows photographs of the plasma plume/jet inside the chamber at three different pressures—760 Torr, 180 Torr, and

FIGURE 4.136 Plasma plume inside the Pyrex chamber at three pressures: 760 Torr, 180 Torr, and 75 Torr. The voltage pulse amplitude and width, for this figure and the remaining figures are 8 kV and 2 μs, respectively. (From Laroussi, M., and Akman, M.A., *AIP Adv.*, 1, 032138, 2011.)

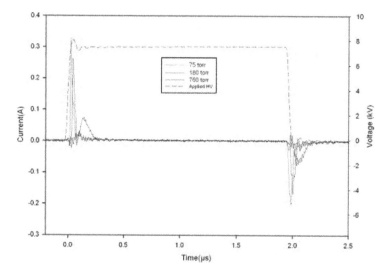

FIGURE 4.137 Current voltage characteristics under three pressure conditions: 760 Torr, 180 Torr, and 75 Torr. (From Laroussi, M., and Akman, M.A., *AIP Adv.*, 1, 032138, 2011.)

75 Torr—illustrates well the dramatic increase of the plume length for pressures below 200 Torr but above 70 Torr. (The photo of the plume at 760 Torr is used as a reference.) The plasma plume/jet length reaches up to 25 cm at a pressure of 75 Torr. This is due to the fact that the ratio of helium mole fraction to that of air stays above the quenching threshold for longer distances.

Figure 4.137 shows the discharge current at the three indicated pressures in the chamber along with the applied high-voltage pulse. The peak amplitudes of the current increase as the pressure decreases, indicating plasmas with higher densities of electrons and ions. Another phenomenon is observed in this figure as the current peaks' temporal position shifts for both the primary and secondary discharges. Primary and secondary discharges are characteristics of pulsed plasma sources of the dielectric barrier type. The primary discharge ignites at the rising edge of the voltage pulse, and the secondary discharge occurs at the falling edge. **Figure 4.137** shows that the primary and secondary discharges take place earlier with decreased pressure. One possible explanation for this can be attributed to the fact that the breakdown voltage of the working gas (helium in this case) decreases with reduced pressure (decreasing pd from about 1000 Torr-cm to about 100 Torr-cm), in accordance with Paschen's Law. Due to the reduced breakdown strength of the gas and an applied pulse with a finite rise time (~30 ns), electrical breakdown can happen earlier at lower electric fields [80].

4.3 Other Physical Effects

4.3.1 Plasma Plume Propagation

Recently, the working gases of most plasma jet devices are noble gases or mixtures of noble gases with small amounts of oxygen or air. An operation with air is more cost effective; however, for air plasma jets, several difficulties are encountered in the plasma generation process. Among these are high plasma or gas temperatures (i.e., heavy particle temperatures) and instabilities that encourage the transition of glow-type discharges into an arc. Nevertheless, several APNP-Js using N_2 or air as the working gas have been developed in recent years, which managed to overcome these problems. Plasmas operated at atmospheric pressure have further provided new fundamental insights. Studies on the dynamics of plasma jets have shown that although they look like a continuous jet to the naked eye, the plasma is actually expelled in bullets with a propagation velocity of several $km \cdot s^{-1}$ or even higher. Further investigations found that the plasma plumes are driven by the electric field rather than by gas flow. It needs to be noted that all these studies on the dynamics of the jet were conducted for an operation with noble gases, in particular argon and helium. It is therefore not clear if the dynamics of air plasma jets are the same as those of jets driven by noble gases.

Pertinent results on the dynamics of an APNP-J operated with air were studied. The gas temperature of the plasma plume decreases rapidly with increasing distance from the nozzle. At a distance of 2 mm, the gas temperature drops to less than 50°C. The propagation velocity of the plasma plume is close to the gas flow velocity and therefore it is much lower than that of a noble gas plasma jet. A plasma jet can be generated by applying a direct current (DC) voltage, which results in either a self-pulsing or a true DC operation of the jet, depending on the applied voltage and gas flow rate. Optical emission spectra show that the atomic oxygen emission intensity from the plasma plume is higher for the self-pulsing mode [81].

Figure 4.138 shows a schematic of the setup of the plasma jet together with a photograph of the air plasma plume that is generated in this configuration. The device consists of three parts: a quartz tube with an inner diameter of about 0.5 mm, a stainless-steel needle with a diameter of 0.2 mm placed in the center of the quartz tube, and a copper ring with a center hole of 0.5 mm diameter

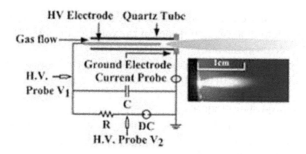

FIGURE 4.138 Schematic of the plasma jet device and the electric circuit together with a photograph of the generated air plasma plume. The capacitance, C, can be varied while the resistance, R, is fixed at 1 MΩ. The operating parameters for the plasma jet shown in the insert are $C = 200$ pF, $V_2 = 7$ kV, and an air flow rate of 2 L·min⁻¹. (From Xian, Y. et al., *Plasma Process. Polym.*, 10, 372–378, 2013.)

attached to the front end of the quartz tube. The stainless-steel needle serves as the high-voltage (HV) electrode, and the copper ring is connected to the ground. The distance between the HV electrode and the ground electrode is about 1.5 mm. The HV electrode is connected in parallel with a capacitor to the DC power supply through a 1-MΩ resistor. The DC voltage provided by the power supply is adjustable up to 20 kV. When air is flowing through the quartz tube and a sufficiently HV is applied, a discharge is ignited in the gap between the electrodes and a plasma plume is launched through the hole of the copper ring into the surrounding air. As shown in **Figure 4.138**, the plasma plume is more than 1 cm long when the voltage provided by the power supply is adjusted to 7 kV and the air flow rate to 2 L · min⁻¹. The voltages, V_1, applied to the needle, and V_2, provided from the power supply, are measured by HV probes and the current through the needle—and therefore the plasma—by a current monitor. **Figure 4.139** shows the typical voltage and current waveforms of the discharge for an applied voltage, V_2, of 6.5 kV and an air flow rate of 2 L · min⁻¹. The voltage recorded in **Figure 4.139** is the voltage, V_1, measured at the needle. Characteristic for these operating parameters are periodic breakdowns of the voltage at the needle, which correlate to a simultaneous increase in current. Therefore, each of these breakdowns corresponds to a discharge and plasma formation. The peak voltage at the needle and the peak discharge current are about 4 kV and 20 A, respectively. Each discharge pulse of this self-pulsing mode lasts for about 50 ns.

Figure 4.140 shows high-speed photographs of the plasma plume that is expelled for the operating parameters as described by the measurements in **Figure 4.139**. Images are taken for different time delays with respect to a voltage breakdown. The time labels for each photograph correspond to the time points indicated in **Figure 4.139a**. Most remarkable, the images reveal that in the self-pulsing mode the plasma jet is expelled in bullets (**Figure 4.141**).

For a more detailed discussion on the dynamics of the plasma plume, the propagation velocities of the plasma plume are plotted versus position in

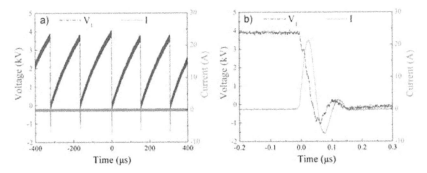

FIGURE 4.139 Voltage, V_1, applied to the HV electrode (needle electrode) and current, I (Circuit and operating parameters, according to **Figure 4.138**: $R = 1$ MΩ, $C = 200$ pF, $V_2 = 6.5$ kV, air flow rate of 2 L·min⁻¹) (a) Subsequent pulses, (b) Time expanded voltage–current characteristics for a single pulse. (From Xian, Y. et al., *Plasma Process. Polym.*, 10, 372–378, 2013.)

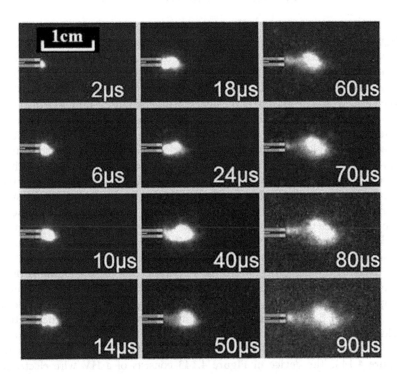

FIGURE 4.140 High-speed photographs of the dynamics of the plasma plume. The exposure time is 2 μs. The discharge parameters are the same as described by the voltage–current characteristics shown in **Figure 4.139** ($R = 1$ MΩ, $C = 200$ pF, $V_2 = 6.5$ kV, air flow rate of 2 L·min⁻¹). Concurrently, the time labels of each photograph correspond to the timescale of the voltage–current characteristics shown in **Figure 4.139a**. (From Xian, Y. et al., *Plasma Process. Polym.*, 10, 372–378, 2013.)

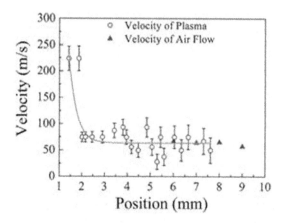

FIGURE 4.141 Propagation velocities of the plasma bullet at different distances from the nozzle along with the gas flow velocities measured by a Pitot tube. (From Xian, Y. et al., *Plasma Process. Polym.*, 10, 372–378, 2013.)

Figure 4.140 from the series of photographs that were taken. Propagation velocities are determined from the location of the head of the plasma plume. In addition, **Figure 4.140** also shows the gas velocities as measured by a Pitot tube. The Pitot tube disturbs and, in fact, determines the flow characteristics when placed directly in front of the nozzle orifice of only 0.5 mm. At a distance of 6–9 mm from the nozzle, where the measurements are taken, no obvious differences in the flow characteristics of the jet can be observed. **Figure 4.140** shows that the plasma plume travels at a velocity of 230 m · s^{-1} as soon as it leaves the nozzle. Then the velocity drops rapidly to about 60 m · s^{-1}, which is close to the air flow velocity of 64 m · s^{-1} measured at 6–9 mm away from the nozzle. When a grounded steel cone is placed in direct contact with the plasma plume, as shown in **Figure 4.142**, the propagation of the air plasma plume is also not affected at all. A different behavior is observed for a plasma jet operated with noble gases, as reported earlier. When a grounded steel cone is placed close to a plasma plume that is generated from noble gas, the jet will follow the electric field and bend toward the steel cone. The plasma plume is then stopped at the steel cone even if it is grounded only through a resistor of several megaohms. In comparison, when the plasma jet is generated from an air flow, the jet neither bends nor stops. The result strongly suggests that the electric field plays much less a role in the propagation of a plasma jet operated with air.

The observation is different from that reported before and shown in **Figure 4.143**. The device in **Figure 4.143** consists of a HV wire electrode inserted into a quartz tube with one closed end. The quartz tube along with the HV electrode is inserted into the hollow barrel of a syringe. When He is injected into the hollow barrel at a flow rate of 2 L/min. and at a high pulsed DC voltage of 6 kV, repetition rate of 10 kHz, and pulse width of 500 ns, a 3 cm-long, cold He plasma plume is generated, as shown in **Figure 4.143a**. When the grounded steel cone is placed close to the plasma plume

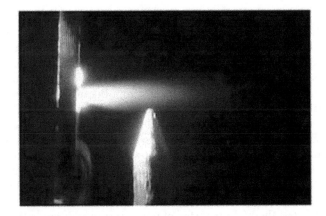

FIGURE 4.142 Photograph of the air plasma plume expelled across a grounded stainless-steel electrode. The trajectory of the plasma jet is not affected by the additional electrode. (From Xian, Y. et al., *Plasma Process. Polym.*, 10, 372–378, 2013.)

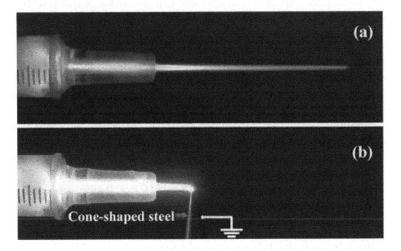

FIGURE 4.143 Photographs of a He plasma plume: (a) The plasma plume propagates in ambient air freely. (b) A steel cone close to the plasma plume stops the propagation. (From Xian, Y. et al., *Appl. Phys. Lett.*, 100, 123702, 2012.)

(but not in direct contact), the plasma plume bends toward the steel cone and stops propagating, as shown in **Figure 4.143b**. This is because the plasma plume in **Figure 4.143** is electrically driven. The grounded steel cone close to the plasma plume affects the electric field distribution and consequently the propagation of the plasma plume. The plasma plume also stops at the steel cone even if it is grounded via a resistor of several MΩ. On the other hand, with regard to the air plasma plume generated by the device, as shown in **Figure 4.142**, when a grounded steel cone is placed in contact with the plasma plume similar to what is shown in **Figure 4.143b**, propagation of the air plasma plume is not affected at all. Therefore, the propagation mechanism must be different and is perhaps not electrically driven [82].

To better understand why the propagation velocity of the plasma plume next to the nozzle is higher than the gas flow velocity, gas temperatures of the plasma inside the discharge gap are obtained from the rotational temperature of the nitrogen according to the emissions from the second positive system. To determine the gas temperature of the plasma, the measured optical emission spectra are compared with simulated spectra. The temperature at which the simulated spectrum gives the best fit to the measured spectrum is the gas temperature of the plasma. The results show that the gas temperature of the plasma in the discharge gap is about 1,200 K for the same experimental conditions as presented in **Figures 4.139** and **4.140**, respectively. In conclusion, the relatively high propagation velocity of the plasma next to the nozzle could be due to both a fast thermal expansion of the plasma in addition to the gas flow itself. With the temperature decreasing rapidly after exiting the nozzle, the propagation velocity of the jet is then also dropping fast with distance.

For an applied voltage, V_2, of 15 kV no intermittent breakdowns of the voltage, V_1, at the needle are observed anymore and the discharge is characterized by an apparent true DC mode with a steady voltage drop, V_1, of 1 kV across the discharge. A transient regime between this true DC mode and the self-pulsing mode is observed for lower voltages, V_2, between 7 kV and 15 kV. For a fixed of the air flow rate of 2 L·min^{-1}, an exclusively self-pulsing operation is then established for even lower voltages, V_2, between 4 kV and 7 kV. Typical time-integrated images of the plasma jet that is expelled under different operating conditions are presented in **Figure 4.144a** and **b**, showing an increase in the length of the jet that is primarily related to the power dissipated in the discharge. An additional effect of the different discharge modes on the jet is not apparent in these images. A transition from the self-pulsing mode to the true DC discharge mode can also be initiated at a higher provided voltage, V_2, when reducing the gas flow rate. In **Figure 4.144c**, a jet is shown accordingly, generated in a true DC discharge mode at a voltage, V_2, of 7 kV and a flow rate of 0.5 L/min. In this case, the appearance of the plasma plume is different since, for the lower flow rate of 0.5 L/min., the gas flow is actually laminar, resulting in a plume that is less disturbed and dispersed by eddies.

It has been established that the plasma bullets that are expelled from a plasma jet that is operated with noble gases are driven by the electric field. The bullets travel at a speed of several km·s^{-1} or even higher, i.e., at velocities that are several orders of magnitude higher than the gas flow velocities. Conversely, the propagation velocities of the air plasma jet presented here are close to the gas flow velocities. This demonstrates that gas flow plays a significant role for jets operated with air. Interestingly, and concurrent with this observation, no plasma plume could be generated in the configuration presented here when it was attempted to operate the jet with helium.

Another remarkable characteristic of the air-operated jet are the different DC discharge modes that are observed. For a fixed flow rate, more and more power is dissipated in the plasma when the voltage is increased. Accordingly, the gas temperature in the plasma increases. As a result, the residual ionization

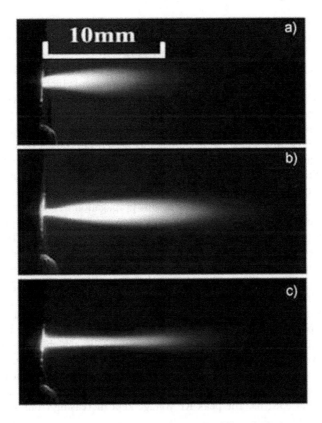

FIGURE 4.144 Photograph of the plasma plumes for different discharge conditions. (a) Self-pulsing mode for $V_2 = 7$ kV, air flow rate 2 L · min^{-1}; (b) True DC mode for $V_2 = 15$ kV, air flow rate: 2 L · min^{-1}; (c) True DC mode for $V_2 = 7$ kV, air flow rate 0.5 L · min^{-1}. (From Xian, Y. et al., *Plasma Process. Polym.*, 10, 372–378, 2013.)

between subsequent breakdowns in the voltage remains at a higher level. Conceivably, the degree of ionization between discharges and during an actual discharge become very similar for even higher voltages, driving higher currents, and eventually a true DC discharge mode—without intermittent voltage breakdowns—is established. A continuous DC discharge mode can also be achieved by lowering the gas flow rate at a given lower voltage. The high plasma temperature that is maintained for lower flow rates again prevents a fast decay of the ionization in the discharge gap [81].

Figure 4.139 Voltage, V_1, applied to the HV electrode (needle electrode) and current, I (Circuit and operating parameter, according to **Figure 4.138**: $R = 1$ MΩ, $C = 200$ pF, $V_2 = 6.5$ kV, air flow rate of 2 L · min^{-1}) (a) Subsequent pulses, (b) Time expanded voltage–current characteristic for a single pulse [81].

Figure 4.140 High-speed photographs of the dynamics of the plasma plume. The exposure time is 2 μs. The discharge parameters are the same as described by the voltage–current characteristic shown in **Figure 4.139**

($R = 1$ MΩ, $C = 200$ pF, $V_2 = 6.5$ kV, air flow rate of 2 L · min⁻¹). Concurrently, the time labels of each photograph correspond to the timescale of the voltage–current characteristic shown in **Figure 4.139a** [81].

Plasma jets generated at atmospheric pressure and flushed in to the open air can be used for direct treatments including biomolecule inactivation, wound healing, and nanostructure fabrication. Different kinds of plasma jets driven by kilohertz high voltage, kilohertz DC pulse, and dual frequency power (kilohertz and megahertz, separating electrodes) have been developed. Recently, dual frequency power on a single electrode, such as pulsed radio frequency and dual radio frequency, have been used to get over the high gas temperature that restricts their use in biomedical applications or extend the plasma plume length. In this section, a 4 cm-long plasma plume produced by a pulsed radio frequency plasma jet and the physics governing plasma propagation were analyzed [83].

The plasma generator used in this work consists of an 8 cm-long glass tube and a copper electrode of the same length along the tube center. The diameter of the copper electrode is 2 mm. The inner and outer diameter of the quartz tube is 5.3 mm and 8 mm, respectively. Both ends of the glass tube are opened so that the plasma's working gas (helium) flows into the glass tube from the injection port at 5 SLM, and the plasma flushes out of the generator from the outlet on the other end. Pulsed modulation divides the RF excitation into power on and power off phases. The pulse frequency is 4 kHz, and the RF is 12.8 MHz. The power on duty cycle is 3.2%, which corresponds to 100 RF cycles. **Figure 4.145** shows the RF voltage curve in one pulse cycle. The trigger signal started at 0 ns. Due to the time delay between the external trigger and the function generator, one hundred RF cycles started at 381 ns, and ended at 8381 ns. The peak RF voltage kept increasing from 300 V to a

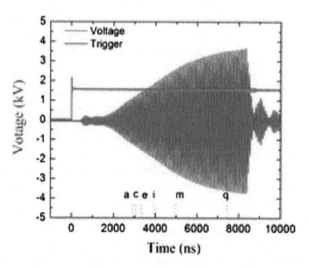

FIGURE 4.145 Applied voltage during one pulse period. The dashed lines with letter a, c, e, i, m, and q correspond to the time 0 of **Figure 4.147** (a), (c), (e), (i), (m), and (q) in sequence. (From Liu, J.H. et al., *Appl. Phys. Lett.*, 98, 151502, 2011.)

maximum of 3.7 kV. Afterward, the fast decay period represented a plasma-quenching process forced by the reduced applied power. The remaining oscillation of the voltage was due to the remaining charges in the jet structure and the electronics matching between the power supply and the jet. In **Figure 4.146** four plasma images with an exposure time of 5 ns were taken at different times during the pulse on period. These images were not normalized to highlight their pattern instead of their relative intensity. The plasma plume length was 2 mm and the bright core was near the electrode when the positive peak RF voltage was 1460 V (**Figure 4.146a**). As the peak RF voltage increased to 3420 V, the plasma length increased to 2.3 cm at 7440 ns (**Figure 4.146d**). This is a 11.5-fold increase over a 2.3-fold increase in the applied RF voltage. On the other hand, a small dark region between the bright core and the plasma tail (see **Figure 4.146a–c**) faded away, as shown in **Figure 4.146d**.

To help with the interpretation of the faster increase in the plasma plume length over the increasing applied RF voltage, the spatial-temporal evolution of the plasma emission at all wavelengths (spectrum-integrated) as well

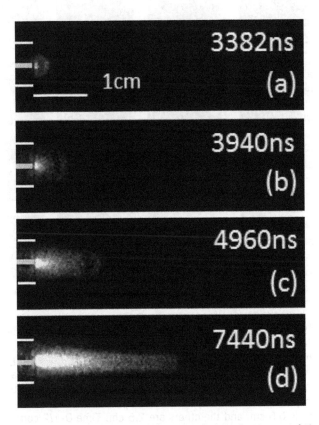

FIGURE 4.146 5 ns exposure time charge-coupled device image of the plasma plume at the RF voltage positive peak. (a–d) are in time sequence. (From Liu, J.H. et al., *Appl. Phys. Lett.*, 98, 151502, 2011.)

as at 391 nm [$N^+_2(B^2\Sigma^+_g) \to N^+_2(X^2\Sigma^+_g) + h\lambda$], 706 nm [$He(^3S_1) \to He(^3P_2) + h\lambda$]}, and 777 nm [$O(^3P_1) \to O(^3S_2) + h\lambda$] are shown in **Figure 4.147**. The plasma first appeared as a positive corona in the positive half RF cycle that starts at 2885 ns, which was indicated by the emission confined in the region less than 0.6 mm from the electrode in the first positive half cycle (**Figure 4.147a**). The formation of the positive glow corona is attributed to a laterally extended space-charge sheath, which leads to the field concentration around the anode and anchors the ionization region to the anode surface. Because the

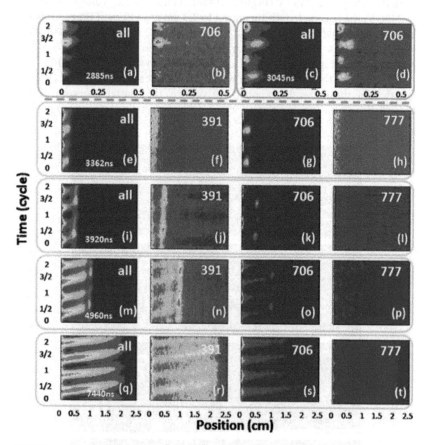

FIGURE 4.147 Spatiotemporal evolution of the optical emission of the plasma plume with different length. The starting time of [(a)–(b)], [(c)–(d)], [(e)–(h)], [(i)–(l)], [(m)–(p)], and [(q)–(t)] are shown in the first figure of each group and correspond to the dashed line a, c, e, i, m, and q of **Figure 4.145**. [(a), (c), (e), (i), (m), and (q)] wavelength integrated, [(f), (j), (n), and (r)] 391 nm, [(b), (d), (g), (k), (o), and (s)] 706 nm and [(h), (l), (p), and (q)] 777 nm. The width of [(a), (b), (c), and (d)] is 0.5 cm, and the others are 2.5 cm. Time 0–1/2 corresponds to the positive half cycle. The position of 0 mm is the copper electrode. The region of no optical emission is in the black area. (From Liu, J.H. et al., *Appl. Phys. Lett.*, 98, 151502, 2011.)

706 nm emission corresponds to high energy electrons of 22 eV, which can be only produced in the strong electric field. The 706 nm emission shown in **Figure 4.147b** confirms this field concentration around the power electrode. The emission intensities of 391 nm and 777 nm were so low that they were not shown for comparison. During the negative period, the negative glow was also confined near the electrode with less emission intensity.

After the full extension of the secondary streamer, the emission intensity near the electrode decreased with the decreasing applied voltage, but the emission in the channel tail was strengthened. This is attributed to the high field concentration captured by the emission at 706 nm (**Figure 4.147k**). In the negative half cycle, the emission pattern is similar to that of the RF discharge between two parallel plates (Liu et al. 2011). The 706 nm emission showed the sheath region close to the electrode, and the 391 nm emission indicated the low electric field in the streamer tail. The plasma streamer length increased to 2.3 cm (**Figure 4.147q**) when the voltage peak increased to 3.5 kV. The 706 nm emission was throughout the whole streamer channel (**Figure 4.147s**), which suggests that the channel was filled with energetic electrons. These energetic electrons caused the dark region between the streamer tail and the bright core close to the electrode to get smaller and disappear (**Figures 4.146, 4.147i, m, and q**). On the other hand, because $N^+_2(B^2\Sigma^+_g)$ is mainly excited by low-energy electrons (excitation $N^+_2(X^2\Sigma^+_g)$ of ~3eV), the 391 nm emission shows a different pattern than the all wavelength and 706 nm emission pattern. The difference between the 391 nm emission and the 706 nm emission (**Figure 4.147r and s**) was also captured by the computer simulation done by Basien and Marode, which shows that the high-field, low-density region was near the electrode and a low-field, high-density region was near the channel tail. The weak temporal dependence of the 777 nm emission (**Figure 4.147h, l, p, and t**) was the same as the 777 nm emission of the RF discharge between two parallel plates. The maximum peak-peak RF voltage increased to 10 kV and the plasma plume length increased to 4 cm (**Figure 4.148**). The bright column in the axis is attributed to the accumulation of the emission along the streamer channel during RF cycles. The rotational temperature was obtained by comparing the simulated spectra

FIGURE 4.148 Pulsed RF plasma plume with applied voltage of 10 kV (max pk-pk). (From Liu, J.H. et al., *Appl. Phys. Lett.*, 98, 151502, 2011.)

of the $C^3\Pi u - B^3\Pi g (\Delta v = -2)$ band transition of nitrogen with the experiment recorded spectra. The simulated spectrum at $V_{rot} = 310 \pm 10$ K was a good fit (not shown) to the experimental spectrum. In order to confirm the same physics between the 7 kV peak-peak case and the 0 kV peak-peak case, a ground electrode was placed at 1.5 cm and 3 cm from the power electrode, respectively. Only the return stroke was observed once the streamer made contact with the ground electrode for both cases. No spark arose during the positive streamer-secondary streamer-spark sequence in the point-plate discharge even when the maximum peak-peak voltage was increased to 11 kV [83].

An atmospheric-pressure argon plasma plume propagated in a branching behavior was also studied based on high-speed ICCD imaging. The plasma plume setup is composed of a medical needle and powered by a high-voltage pulse generator. The applied pulsed voltage was set at 20 kV at a repetition frequency of 2 kHz. Argon gas with 100 sccm was used as the working gas. Due to the nonreproducibility of discharge structures, all the time-integrated images presented here were captured in only one single shot. In addition, for each time delay, four typical images were presented in a row to show the different structures of the discharges, as shown in **Figure 4.149** [84].

The time delays marked in the top left of every image row were the corresponding imaging moments relative to the onset of voltage pulses. The corresponding applied voltage for each time delay was also shown therein. As can been seen, the plasma first propagated in a single primary streamer channel. As the applied voltage increased, small side streamer branches started to appear. The number of side branches was increased with the applied voltage. The figure shows that there was always a primary streamer channel that originated from the needle tip, and the side branches were surrounding this primary channel. The side streamer branches were not extending from the primary streamer but were and propagating around it. This was much clearer when the imaging exposure time increased, as shown in the last image row (exposure time 100 ns). These propagation processes are much different than that of typical plasma jets; in the latter case, the plasma propagated in the form of bulletlike plasma volumes without any side branching. It is also different compared to typical positive streamers generated in point–wire or point–plane gaps. For typical positive streamers, streamer branching is generated around the anode tip. In our case, it was mostly observed surrounding the primary streamer channel, precisely speaking, the argon gas stream [84].

It can be shown that the high-speed images represent a mapping of the spatial distribution of the relevant excited species in the plasma. From optical emission spectroscopy, it was shown that the emissions were mainly from excited Ar* and N_2*. The primary streamer channel was supposed to be formed by the excited argon species, which can be generated under the high external electric field. This is also why the primary streamer channel always propagated from the needle exit and followed the gas stream. The generated side branches are probably related to the diffused air surrounding the argon gas stream. It is possible because the air concentration increased in the space away from the gas stream axis and nitrogen molecules can be easily excited by electrons,

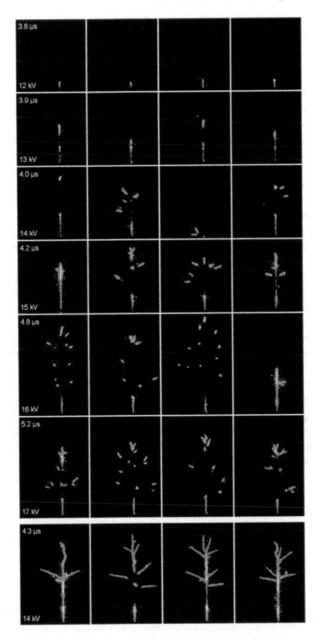

FIGURE 4.149 High-speed image sequences of the argon plasma plume in open air. The time delay and voltage values were marked in each first sub-image of every image row, but the values apply to the entire row. For all images, except for the last image row (100-ns exposure time), the exposure time was set at 20 ns. The nozzle exit (not marked) is located at the bottom of each image. (From Xiong, Q. et al., *IEEE Trans. Plasma Sci.*, 39, 2094–2095, 2011.)

which can be generated due to several effects, such as a high external electric field and photoionization processes.

4.3.2 Plasma Plume Interaction

Investigating plasma plume interaction can give us some insight into the physics of propagating streamers. The property of guided streamers to propagate along a predetermined path allows one to set up a collision of two counter-propagating streamers of the same polarity. The study of the interaction of two positive plasma bullets [85] shows that as soon as the two plasma bullets leave their respective dielectric tubes and approach each other, their interaction leads to dramatically reduced velocities. Simulation of the interaction of two counter-propagating guided streamers [86], in the framework of a 1.5D streamer model, assumed that the streamers propagate toward each other in pure He. The result shows the calculated axial profiles of the electron number density at various time moments after the simultaneous start of two identical streamers moving toward each other with equal velocities. When the distance between the interacting streamers becomes comparable to the streamer radius, the streamers stop, precisely as observed experimentally [85].

Recently, a new experiment was carried out to study two counter-propagating He plasma plumes and the ignition of a third plasma plume without external applied voltage. As shown in **Figure 4.150**, the two high-voltage (HV) electrodes are made of two stainless-steel needles of the same geometry. (The radius of the needle tip is about 100 μm.) The needles are inserted into the quartz tube. The distance between the tips of the two needles is 74 mm. The inner diameter of the quartz tube is about 1 mm. The working gas flows into the quartz tube from the right end of the tube. Research-grade He gas (99.999% purity) is used in the experiment. To keep the ambient air in the quartz tube as low as possible, a glass tube with a length of 2 m is connected to the left end of the quartz tube, as shown in **Figure 4.150**. Thus, the two plasma plumes can be treated as if they were generated in the same gas composition. In other words, the difference between the gas compositions in the downstream and upstream regions caused by the air diffusion from the left end of the tube can be neglected. The two needles are connected to the high-voltage pulsed DC power supply in parallel (amplitude up to 10 kV, repetition rates up to 10 kHz, and pulse length variable

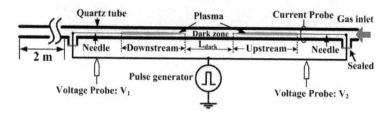

FIGURE 4.150 The schematic of the experiment setup. (From Wu, S. and Lu, X., *Phys. Plasmas*, 21, 023501, 2014.)

FIGURE 4.151 Photos of the plasmas for applied voltages (a) $V_1 = V_2 = 6$ kV and (b) $V_1 = 0$ kV and $V_2 = 6$ kV. Pulse frequency: 5 kHz, pulse width: 800 ns, and He flow rate: 1 L/min. The photos were taken by a digital camera with an exposure time of 1 s. (From Wu, S., and Lu, X., *Phys. Plasmas*, 21, 023501, 2014.)

from 200 ns to DC). When HV pulsed DC voltage is applied to the two needles, and He gas is injected into the quartz tube with a gas flow rate of 1 L/min, two plasma plumes are generated inside the quartz tube [87].

As shown in **Figure 4.151**, for applied voltages of $V_1 = V_2 = 6$ kV, two plasma plumes facing each other are generated inside the tube. The lengths of the upstream and downstream plasma plumes are almost the same, which confirms the gas composition in the upstream and downstream regions can be treated as the same. It's interesting to observe that a dark zone exists in the middle of the two electrodes. On the other hand, for applied voltages of $V_1 = 0$ kV and $V_2 = 6$ kV, a plasma plume is generated between the two tips of the needles. It suggests that the forming of the dark zone is related to electrostatic repulsion induced by the symmetric voltages. To know more about the optical emission characteristics of the plasma plumes, the spatial distribution of the intensities of three optical emission lines are measured by a half-meter spectrometer with slits of 100 μm and grating of 1200 grooves/mm. **Figure 4.153** shows the spatial distribution of optical emission intensities with lines of 391.4 nm [$N_2^+(B^2\Sigma_g^+) \rightarrow N_2^+(X^2\Sigma_g^+) + h\lambda$], 706.5 nm [$He(^3S_1) \rightarrow He(^3P_2) + h\lambda$], and 309.1 nm [$OH(A^2\Sigma^+) \rightarrow OH(X^2\Pi) + h\lambda$]. It clearly indicates that the intensity of the three emission lines approaches zero in the middle of the discharge gap, which corresponds to the dark zone, as mentioned above.

The length of the dark zone is about 4 mm. Moreover, the spatial distribution of the emission intensities is symmetrical. To understand more about the formation of the dark zone in the symmetrical discharges, a fast intensified charge-coupled device camera was used to capture the dynamics of the discharges. **Figure 4.152a** shows the high-speed images of two plasma plumes. It is observed that the two guided streamers propagate toward each other simultaneously, with the velocities decreasing linearly and stopping at the edge of the dark zone. **Figure 4.152b** shows that the velocities of the two counter-propagating streamers are almost the same, which is on the order of 10^4 m/s. To understand the generation mechanism of the dim plasma plume, the dynamics of the plasmas are investigated. The first photo shows that the downstream plasma plume ($V_1 = 6$ kV) is ignited earlier than the upstream plasma plume ($V_2 = 5.5$ kV). This observation is confirmed by repeating the experiment many times. After ignition, the downstream and upstream plasma bullets propagate toward each other and stop at the falling edge of the applied voltage while the second discharge is ignited, as shown in **Figure 4.153** at 768 ns. It is worth pointing out that the structures of the plasma bullets of the

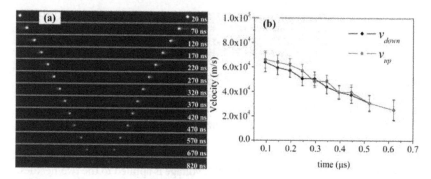

FIGURE 4.152 (a) High-speed photographs for the applied voltages $V_1 = V_2 = 6$ kV. The exposure time is fixed at 5 ns. Each image is an integrated picture of 50 shots with the same delay time. (b) The velocities of the two counter-propagating streamers. (From Wu, S., and Lu, X., *Phys. Plasmas,* 21, 023501, 2014.)

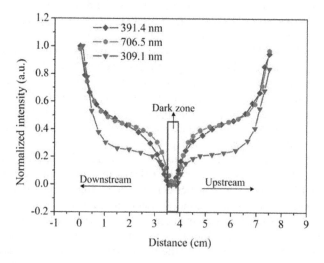

FIGURE 4.153 The spatial distribution of the intensities of three emission lines (N_2^+: 391.4 nm, He: 706.5 nm, and OH: 309.1 nm) along the jet axis. The position of the tip of the left needle is defined as 0 cm. (From Wu, S., and Lu, X., *Phys. Plasmas,* 21, 023501, 2014.)

first discharge and the second discharge are different. For the first discharge, the plasma bullets have a spherical head. On the other hand, the plasma bullets of the second discharge have a sword-like head. Thus, the first discharge is more like a cathode-direct streamer while the second discharge is more like an anode-direct streamer, which is similar to previous reports. It should be emphasized that the two counter-propagating plasmas are not merged till the end of the second discharge [87] (**Figure 4.154**).

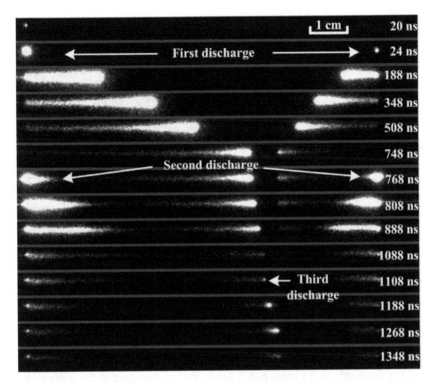

FIGURE 4.154 High-speed images of the plasma for the applied voltages $V_1 = 6$ kV (left side, downstream) and $V_2 = 5.5$ kV (right side, upstream). The exposure time is fixed at 4 ns. Each image is an integrated picture of 800 shots with the same delay time. (From Wu, S., and Lu, X., *Phys. Plasmas*, 21, 023501, 2014.)

4.3.3 Plasma Transfer through a Dielectric

The interactions between the majority of the plasma jets reviewed and solid objects are very mild. It is known that APPJs can be directly touched by human hands without any harm. But it was found that these "gentle" plasma plumes can ignite a secondary plasma discharge inside a dielectric tube, as shown in **Figure 4.155** [88]. The schematic of the experiment setup is shown in **Figure 4.155a**. The high-voltage (HV) wire electrode, which is made of a copper wire, is inserted into a quartz tube with one end closed. The quartz tube along with the HV electrode is inserted into a hollow barrel of a syringe. The distance between the tip of the HV electrode and the nozzle is 1 cm. A glass tube is placed in front of the syringe nozzle at distance d, variable from a few millimeters to several centimeters. When HV pulsed DC voltage (amplitudes up to 10 kV, repetition rate up to 10 kHz, and pulse width variable from 200 ns to DC) is applied to the HV electrode and helium gas is injected

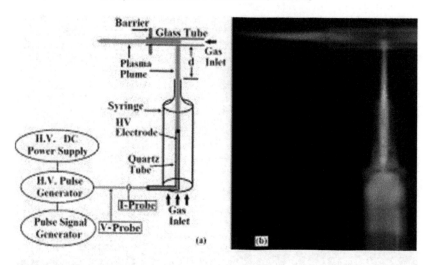

FIGURE 4.155 (a) Schematic of the experiment setup. The inner and outer diameters of the quartz tube are 2 mm and 4 mm, respectively. The inner diameter of the hollow barrel of the syringe is 6 mm and the diameter of the syringe nozzle is about 1.2 mm. The distance between the tip of the quartz tube and the nozzle is 1 cm. The distance d from the nozzle to the glass tube is 1.5 cm. The inner diameter of the glass tube is 2 mm. (b) Photograph of the plasma plumes with the applied voltage V_a of 9 kV, frequency of 4 kHz, pulse width of 800 ns, and helium flow rates of 2 L/min. and 3 L/min. for the syringe and the glass tube, respectively. (From Lu, X. et al., *J. Appl. Phys.*, 105, 043304, 2009.)

into both the hollow barrel and the glass tube with flow rates of 2 L/min. and 3 L/min., respectively, two homogeneous helium plasma plumes are generated in the surrounding air, as shown in **Figure 4.155b**.

In order to avoid direct interaction between the primary and secondary gas flows, a ring-shaped barrier is attached to the outer surface of the glass tube, as shown in **Figure 4.156**. It is found that the barrier does not affect the ignition of the secondary plasma plume. Therefore, the ignition of the secondary plasma plume is not due to the interaction of the two gas flows [88]. To investigate how the secondary plasma plume is ignited, an ICCD camera was used to capture the dynamics of the discharge. **Figure 4.156** shows the photographs of the discharge taken at different delay times. Each picture is an integrated image of over ten shots with the same delay time. According to **Figure 4.156b–f**, the primary discharge resembles a cathode-directed streamerlike discharge, which propagates from the syringe nozzle to the glass tube. The luminance of the primary plasma decreases quickly upon approaching the wall of the glass tube. At about 440 ns, a weak luminous plasma is ignited inside the glass tube, as shown in **Figure 4.156l**. The plasma propagates, reaches the open end of the glass tube, and becomes brighter at 520 ns, as shown in **Figure 4.156n**. Then the plasma continues propagating for about 200 ns before it dies out. The details of the propagation velocities of the primary and secondary plasmas are

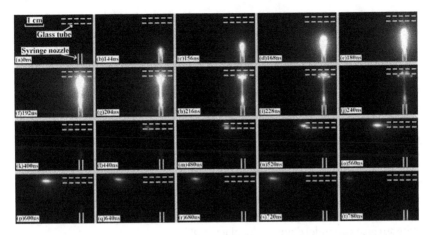

FIGURE 4.156 High-speed photographs of the plasma plume. The exposure time is fixed at 1 ns. (From Lu, X. et al., *J. Appl. Phys.*, 105, 043304, 2009.)

plotted in **Figure 4.157**. Due to the small fluctuation of the plasma bullet location, we repeated the experiment six times. The error bar of the bullet velocity is calculated based on the six experiments. It clearly shows that the velocity of the primary plasma is several times higher than that of the secondary plasma. It should be emphasized that the secondary plasma is ignited at about 200 ns after the primary plasma is extinguished.

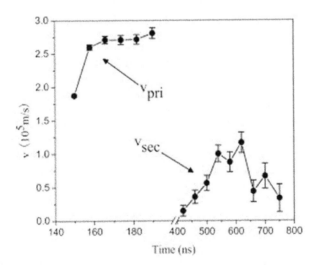

FIGURE 4.157 The propagation velocities of both the primary v_{pri} and the secondary v_{sec} plasma plumes vs. time. (From Lu, X. et al., *J. Appl. Phys.*, 105, 043304, 2009.)

When the glass tube is replaced by an alumina tube or a quartz tube, we are also able to generate the secondary plume with no obvious difference. Therefore, the photons emitted by the primary plume play no direct role in the ignition of the secondary plume. The other possible source is the local electric field created by the charges deposited on the glass tube surface. The charges Q deposited on the glass tube surface can be obtained by integrating the plume current, which has a peak current of about 300 mA. By integrating the plume current, the charges deposited on the glass tube surface are obtained to reach a peak value of about 10^{-8}C. Assuming that the charges are equally distributed along an ideal ring with infinitesimal width on the glass tube surface, the electric field along the glass tube can be calculated according to simple electrostatics. The maximum electric field inside the glass tube can reach more than 20 kV/cm, which is higher than the breakdown voltage of helium at atmospheric pressure. Therefore, the charges deposited on the glass tube surface can indeed ignite the plasma [88].

To understand this phenomenon, numerical simulations of the dynamics of the primary and secondary plasmas were carried out [89]. In these studies, a 25 kV positive voltage pulse with rise and fall times of 25 ns is used to drive the plasma. After rising to its peak of 25 kV, the voltage is held constant for 50 ns. In this simulation, the initial electron density inside the source and transfer channels is zero except for one small electrically neutral plasma cloud in each channel. The plasma transfer through a dielectric barrier is captured by the simulation of the ionization front dynamics, represented by the rate of electron impact ionization, S_e (cm^{3}s^{-1}), and the total electric field E at different time moments after the initiation of the discharge. As a result of the impact of the primary plasma on the top of the dielectric surface and spreading over the surface, the top surface is charged positively. This occurs at a time when a conductive plasma channel is completed from the anode to the top of the dielectric surface. This effectively transfers the anode potential to the surface of the dielectric. About 2 kV is dropped across the top wall of the dielectric. This electric field combined with a small residual background electron density produces a downward directed electron avalanche inside the dielectric tube. By 62 ns, this avalanche has reached the lower inner surface of the dielectric tube which fills the channel with conductive plasma and charges the inner surface of the dielectric. Because the dielectric used for the secondary plasma can be opaque, either electron diffusion or photoionization plays an essential role in the generation of the secondary plasma. The high conductivity of the primary plasma channel, which can effectively transfer the anode potential to the surface of the dielectric, is responsible for the generation of the secondary plasma.

4.3.4 Manipulation of Plasma Jets

Characteristics of electrically driven plasma bullets can be affected by applying, in the direction of their propagation, additional external electric fields [1,23,90]. Typical electrode configurations used in experiments are shown in **Figure 4.158**. Depending on the amplitude, polarity, and duration of

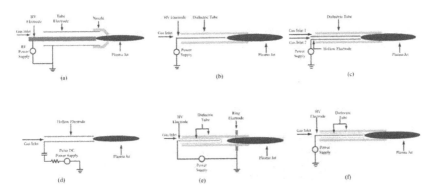

FIGURE 4.158 The schematic of a noble gas plasma jet with different setups of power and ground electrodes.

the applied voltage pulses, it is possible to change the streamer propagation length, as well as to control the streamer velocity and the plasma parameters.

The effect of an applied electric field on a pulsed RF plasma jet array and a pulsed DC plasma jet array were compared [91]. The plasma jet array used for this study employed three dielectric tubes of 5 cm length, one end of which was wrapped with a concentric tin belt 2 cm wide. This tin belt served as the powered electrode. As shown in **Figure 4.159**, both ends of the glass tubes were opened so that the plasma's working gas (helium) flew into these three glass tubes from the injection ports at 5 SLM (standard liters per minute), and the plasma flushed out of the generator from the outlet on the other end. The pulse frequency of the pulsed DC plasma and the pulsed RF plasma both were 4 kHz. The RF of the pulsed RF plasma was 12.5 MHz, and its power on duty cycle was 1.28% (40 RF cycles), which was 3.2 µs and equal to the power on period of the pulsed DC plasma. The maximum peak-to-peak RF voltage was 9 kV (**Figure 4.160a**),

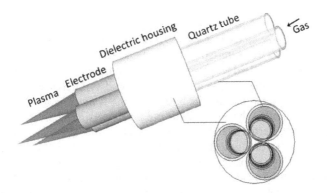

FIGURE 4.159 Schematic of the 3-tube plasma jet array. (From Hu, J.T. et al., *Phys. Plasmas*, 19, 063505, 2012.)

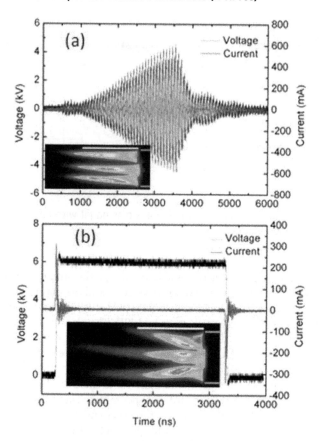

FIGURE 4.160 (a) Current and voltage characteristics (CVC) of a pulsed RF plasma jet array in one pulse cycle. The inserted picture is taken with 10 ms exposure time, and the brown line shows the tube exit position; the white line in the top shows the length of 1 cm. (b) CVC of a pulsed DC jet array. The inserted picture has the same conditions as the top one. (From Hu, J.T. et al., *Phys. Plasmas* 19, 063505, 2012.)

and the pulsed DC voltage was 6.4 kV (RMS value of the max pk-pk RF voltage, **Figure 4.160b**). The inserted picture in **Figure 4.160a** shows that the plasma plume length of the pulsed RF plasma jet array approaches 1.3 cm. Because of the different stray capacitance of the array structure and its different effective impedance in the external structure, the bottom plume is more active. **Figure 4.161** shows the propagation dynamics of the pulsed RF plasma jet array. In the positive half RF cycle started at 2658 ns, the plasma first appears as a positive corona with a length <0.1 cm (not shown, peak RF voltage = 3 kV). Four RF cycles later, the length of the plasma plume was 0.25 cm (peak RF voltage = 3.7 kV, **Figure 4.161a**). Afterward, the length of plasma plume increased to 1.3 cm at the maximum peak voltage of 4.4 kV (**Figure 4.161c**). This is a five-fold increase in length over a 1.2-times voltage increase. The average plasma plume propagation velocity is 2.73×10^6 cm/s [91].

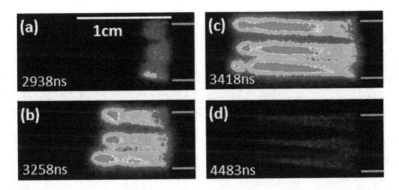

FIGURE 4.161 The pulsed RF plasma jet array development in the pulse on and pulse off periods. Each image was taken with 5 ns exposure time. (From Hu, J.T. et al., *Phys. Plasmas*, 19, 063505, 2012.)

The plasma plume structure changed during its propagation. Although the structure similar to the plasma bullet arose at the negative voltage peak of −4 kV (**Figure 4.161b**), the plasma image (**Figure 4.161c**) taken at the following negative RF voltage peak suggests the plasma plume was dominated by the positive column. In order to study the development of the pulsed RF plasma plume, the propagation dynamics during one RF cycle was shown in **Figure 4.162**. The applied RF voltage amplitude decreased to 0 at a quarter cycle after **Figure 4.161c**; therefore, only accumulated charges on the dielectric tube surface can maintain the electric field and the ionization and excitation in the plume, which led to the weaker optical emission intensity of the plume (**Figure 4.162a**). Because of its long lifetime (compared to an RF cycle in ns scale), the ions and metastables accumulated in the channel made the whole plasma plume respond quickly to the applied RF voltage. **Figure 4.162b** shows the structure of the negative glow, the Faraday dark space, and the positive column from the plasma plume tip to the electrodes at the positive RF voltage peak. Different from secondary electrons produced by ion bombardment on an electrode, the photoionization near the plasma tip can only provide much fewer seed electrons; therefore, the negative glow in the plume head was weaker than that near the ground electrode. These seed electrons were then accelerated toward the tube, which intensified the discharge in the main body of the plasma plume. Finally, this energetic electron stream reached the tube, which worked as a hollow anode. The higher helium concentration intensified the discharge inside the tube further. Consequently, more applied power was consumed by the discharge inside the tube in the positive half cycle, and comparing **Figure 4.162b** and **c** also indicates that the emission intensity of the plasma plume outside the tube was weaker at the positive voltage peak than at the negative voltage peak [91].

The inserted picture in **Figure 4.160b** shows the pulsed DC plasma jet array with a length approaching 2 cm, whose bottom plume was also stronger. **Figure 4.163a–d** shows the plasma plume propagation dynamics of the

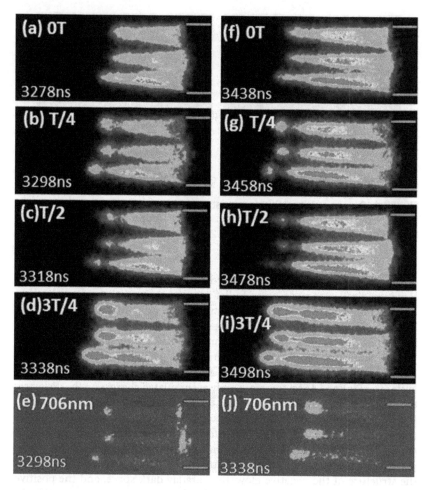

FIGURE 4.162 The pulsed RF plasma jet array dynamics in two RF cycles during the pulse on period. Each image was taken with 5 ns exposure time. [(a)–(d) and (f)–(i)] wavelength-integrated, (e) and (j) 706 nm. (From Hu, J.T. et al., *Phys. Plasmas*, 19, 063505, 2012.)

pulsed DC plasma jet array. The plasma bullet reached the maximum velocity of 7.5×10^6 cm/s (**Figure 4.163c**) in the launching period; afterward, the bullet turned weaker in the propagating period (**Figure 4.163d**). Although Dawson's theorem based on zero applied electric field is highly plausible to explain the formation of the plasma bullet, the plasma bullet propagation dynamics shown in **Figure 4.163a** suggest the applied electric field played a key role in the plasma propagation. During the launching period, not only did the radius of the plasma bullet increase from 0.045 cm (**Figure 4.163b**) to 0.07 cm (**Figure 4.163c**), but also the bullet was brighter because the all wavelength emission has the same pattern as the 391 nm emission. Additionally, this growing plasma bullet (**Figure 4.163c**) and the shrinking plasma bullet

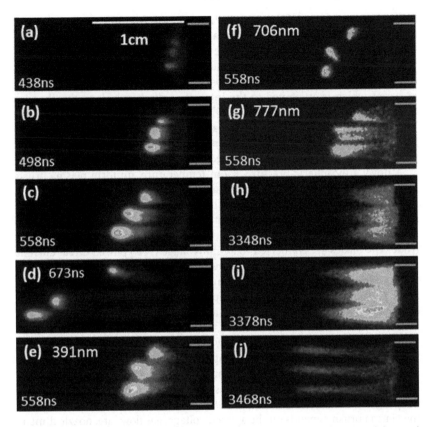

FIGURE 4.163 The pulsed DC plasma jet array dynamics in the pulse rising and falling phase. Each image was taken with 5 ns exposure time. [(a)–(d) and (h)–(j)] wavelength-integrated, (e) 391 nm, (f) 706 nm, and (g) 777 nm. (From Hu, J.T. et al., *Phys. Plasmas*, 19, 063505, 2012.)

in the propagating period (**Figure 4.163d**) support the theoretical prediction that the ion density in the streamer tip increased in the high field region and then dropped until the streamer propagation stopped. According to the photon-ionization theory, the photo electrons created at a distance from the bullet tip were accelerated toward the bullet filled with positive ions. Besides the electric field of the bullet on its own, the highly stressed applied electric field and the lower air mole fraction in the vicinity of the tube not only facilitated the energy coupled to this electron stream $\varepsilon = E \times j_e$ (E is the applied electric field, j_e is the electron flux) but it also intensified the excitation and ionization; therefore, the growing plasma bullet in the launching period was only near the tube exit. After that, as the bullet moved further in the propagating period, the energy coupled from the weaker applied electric field decreased, but the energy consumed more quickly because of the higher air mole fraction, so the bullet turned weaker until it disappeared [91].

By putting the optic fiber at a distance of 2 cm from the center tube and moving along the plasma plume propagating direction, the optical emission intensity as a function of plasma plume position is obtained. The higher emission intensity at 309 nm and 777 nm indicates the higher concentration of OH radicals and atomic oxygen in the active plume zone of the pulsed RF plasma jet array, which is attributed to the intensified discharge along the channel under the fast oscillating RF electric field and longer plasma working time. The possibility of controlling the streamer plasma parameters can be helpful in numerous emerging applications in plasma chemistry, health care and medicine, food processing, and nanotechnology [92–96].

A novel atmospheric-pressure, room-temperature, uniform plasma brush is discussed in this section. The plasma brush can be operated in several different discharge modes depending on the distance between the nozzle and the sample. A schematic of the experiment setup is shown in **Figure 4.177a**. The device comprises two blades to guide the gas flow and also serve as the electrode. The blades are connected to a high-voltage (HV) pulsed direct current (DC) power supply (up to 10 kV with repetition rates up to 10 kHz and pulse widths variable from 200 ns to DC) through a 60 kΩ ballast resistor R and a 36 pF capacitor C. The resistor and capacitor are used to limit the discharge current. The radius of the blade edge is about 50 μm and the dimensions of the plasma brush nozzle are about 25 mm × 1 mm. Helium, argon, or their mixtures with O_2 can be used. The gas flow rate is controlled by a mass-flow controller. When the HV pulsed DC voltages are applied to the blades and He is supplied at a flow rate of 1 L/min., a homogeneous plasma brush is generated. It can be touched by a human finger, as shown in **Figure 4.177b** taken with a Sony digital camera (DSC-HX1) at an exposure time of 100 ms. The length of the plasma brush depends on the applied voltage, gas flow rate, nozzle dimensions, and the distance between the nozzle and the treated object. The other advantage is that both conducting and insulating materials can be treated by this plasma brush uniformly [97].

Five widely used N-APPJs (**Figure 4.164**) are selected for comparing the absolute concentrations of hydroxyl and atomic oxygen [98]. For geometry configuration 1 [2], two flat ring-electrodes are attached to the surface of two centrally perforated quartz disks with a diameter of 15 mm and thickness of 1 mm. The holes in the center of the quartz disk are 2 mm in diameter. The distance between the two dielectric disks is 10 mm. More details about the device can be found in Ref. [2]. This device is able to generate a plasma plume up to 5 cm long, which was the longest room-temperature N-APPJ ever reported at the time. For geometry configuration 2, a naked stainless-steel pin is placed in the center of the dielectric tube with an tip 10 mm away from the nozzle end. The radius of curvature at the tip of the pin is 0.1 mm. A cylindrical grounded ring-electrode is placed outside at the exit of the tube with its inner diameter of 2 mm. When no object is placed in front of the plasma plume, the discharge with this configuration is more or less like a DBD. However, when an object is placed in front of the plasma plume, the main discharge current could flow from the pin electrode to the object to be treated. It is no longer operating as a DBD under such conditions, and there is

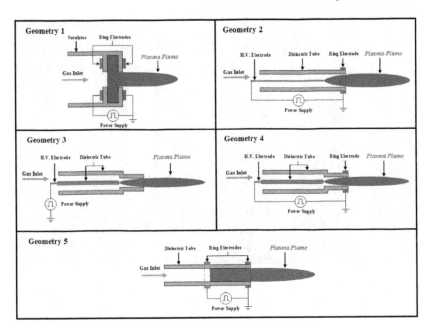

FIGURE 4.164 Schematic of the five N-APPJs.

potential for transition to arc discharge, which must be avoided when it is used for applications such as plasma medicine. Configuration 3 consists of a dielectric tube guiding the working gas and a quartz tube–covered pin electrode. The quartz tube for covering the electrode has a thickness of 0.5 mm and an outer diameter of 3 mm with its tip 10 mm from the nozzle end. The dielectric tube for guiding the working gas has an inner diameter of 2 mm at the tube end. For this device, whether an object is placed far away from or close to the nozzle end, there is no risk of arcing [99]. To enhance the discharge inside the dielectric tube, a ground electrode is added, as shown in geometry configuration 4. A ground ring-electrode is placed outside of the dielectric tube, which can enhance the discharge inside the dielectric tube and generate more reactive species inside the tube [100]. For those reactive species with a relatively long lifetime, they might be able to flow to the downstream and reach the object to be treated. However, the electric field along the plasma plume in the open space is weakened, which could result in a lower outcome of reactive species generated in the downstream. Configuration 5 consists of a quartz tube with two conductive ring-electrodes on the outside of the tube, as drawn in **Figure 4.164** [101]. The one close to the nozzle end is connected to the ground. The distance between the two ring electrodes is 10 mm. The inner diameter of the quartz tube is 2 mm.

Figure 4.165 shows the two-dimensional OH LIF intensity distribution of the five N-APPJs shown in **Figure 4.164**. It shows clearly that the OH-LIF intensity for devices 2 and 5 are higher than the others. Besides, the LIF intensity of all the N-APPJs has high intensity along the center axis, and it decreases in the

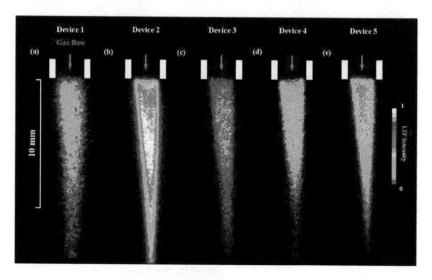

FIGURE 4.165 Two-dimension OH-LIF intensity distribution of the five N-APPJs shown in **Figure 4.164**. (a–e) corresponds to geometry 1–5 in **Figure 4.164**. (From Yue, Y.F. et al., *IEEE Trans. Radiat. Plasma. Med. Sci.* 1, 541–549, 2017.)

radial direction. **Figure 4.6** shows the absolute OH concentration of the five N-APPJs along the axial direction calibrated by the model described above. The OH concentration generated by device 2 is the highest; it reaches 10^{14} cm^{-3}. The second highest OH concentration is generated by device 5; it has a maximum of 6×10^{13} cm^{-3}. Because of the small integral interval, **Figure 4.166** is presented by fitting the absolute densities of integration windows in axial positions. The OH concentration generated by device 4 is higher than that of device 3. The difference between devices 3 and 4 is the ground electrode. The ground electrode added to device 4 can enhance the discharge inside the tube but not the outside, which means more OH generated inside the tube can flow to the downstream and contribute to the total OH concentration in the downstream near the nozzle end.

As we know, the working gas composition, the gas flow rate, and the parameters of the applied voltage all have effects on the OH concentration. Next, their effects on the OH concentration were investigated. **Figure 4.167** shows the OH concentration at 5 mm away from the nozzle in effluent for different applied voltages. For all the N-APPJs, the OH concentrations increase approximately linearly with the increase of the applied voltages. **Figure 4.168** shows the OH concentration versus the frequencies of the voltage pulse. The OH concentrations of all five N-APPJs increase with the frequency for all the devices too, but it is not linear. This might be due to the limited lifetime of the OH. Pei et al. show that OH concentration decreases up to 25% at 1 ms in the afterglow because of diffusions and termination reactions, which is consistent with our observation [102]. The observations of the dependence of the OH concentration on the applied voltage and pulse frequency are also consistent with the work reported in Ref. [102].

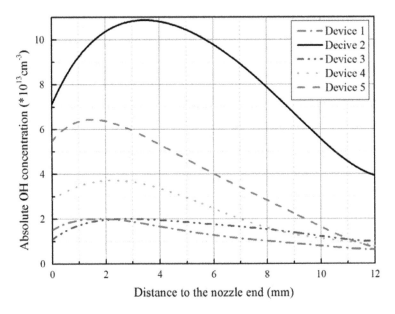

FIGURE 4.166 Absolute OH concentration vs. the distance from the nozzle. (From Yue, Y.F. et al., *IEEE Trans. Radiat. Plasma. Med. Sci.*, 1, 541–549, 2017.)

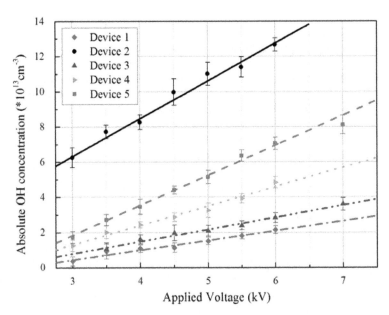

FIGURE 4.167 OH concentration of the five N-APPJs at 5 mm away from the nozzle vs. the applied voltage. (From Yue, Y.F. et al., *IEEE Trans. Radiat. Plasma. Med. Sci.*, 1, 541–549, 2017.)

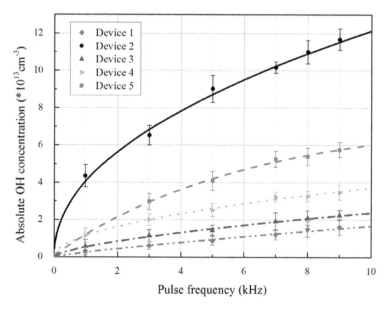

FIGURE 4.168 OH concentration of the five N-APPJs at 5 mm away from the nozzle vs. the frequency of the voltage pulse. (From Yue, Y.F. et al., *IEEE Trans. Radiat. Plasma. Med. Sci.*, 1, 541–549, 2017.)

It is widely accepted that the OH radicals are mainly generated by the dissociation of H_2O through the collision dissociation reactions between H_2O and electrons and metastable state He [103]. Thus, the H_2O concentration in the working gas could be crucial for OH concentration. The OH lifetime is at millisecond range in helium atmospheric plasma jets; thus, the gas flow rate could also contribute to the OH distribution [102]. Next, the effects of the H_2O concentration on the working gas and the gas flow rate on the OH concentration were investigated. **Figure 4.169** shows the OH concentration at 3 mm away from the nozzle. The OH concentration increases dramatically with the increase of the H_2O concentration at first. Then it decreases slowly with the further increase of the H_2O concentration. For all five N-APPJs, the OH concentrations reach their peaks at about 70 ppm of H_2O. Device 2 generates the highest OH concentration. It is worth pointing out that, according to different reports, the optimal value of H_2O for generating OH may be different with respect to different experimental conditions (ambient air, power supply, etc.). In the reports of Yonemori et al. [104] and Voráč et al. [105], the highest densities of OH are obtained at 200–300 ppm and at 650 ppm (using RF power supply), respectively. In reference [105], especially, the peak OH density is observed at the plasma edge, which seems inconsistent with our result (showing the peak at in the center of the plasma plume). However, the contradiction can be explained by the transportation of OH by gas flow from upstream to downstream and the overlapping donut-shape distribution of OH. More details about the distribution issue of OH can be found in our report Ref. [42].

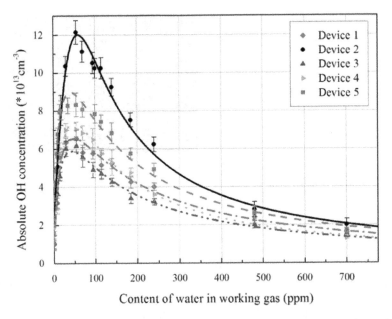

FIGURE 4.169 OH concentration of the five N-APPJs at 3 mm away from the nozzle vs. H_2O concentration in the working gas. (From Yue, Y.F. et al., *IEEE Trans. Radiat. Plasma. Med. Sci.*, 1, 541–549, 2017.)

Figure 4.170a presents the OH concentration of the five N-APPJs at 5 mm away from the nozzle for different gas flow rates. It is found that the OH concentration increases slightly with the increase of the gas flow rate. To better understand the effect of the gas flow on OH concentration, two-dimensional LIF measurements were conducted. It is found that the gas flow rate has a similar effect on the five N-APPJs. **Figure 4.170b** shows the two-dimensional LIF intensity distribution of device 5 for different gas flow rates. With the increase of the gas flow rate, a great increase in LIF intensity can be observed further downstream from the nozzle end.

As we know, gas flow can contribute to the OH concentration in the following three aspects. First, higher gas flow results in faster gas flow rate, thus there is a shorter time delay for the OH generated in the tube to reach further downstream, and a higher OH concentration downstream. Second, higher gas flow will minimize the diffusion of the surrounding air into the plasma plume. Because of the vibrational and rotational states of N_2 and O_2, and also the electric negative characteristics of O_2, less air will result in a higher electron temperature and electron density, and thus a higher OH concentration. Third, due to the humidity of air, some of the H_2O will also diffuse into to the plasma plume; if the H_2O in the working gas stream is low, the H_2O from the surrounding air could enhance the generation of OH, and thus the higher gas flow could lower OH concentration. For the results presented in **Figure 4.170**, the H_2O in the working gas is 115 ppm, which is higher than what is needed when the OH concentration reaches its peak, as shown in **Figure 4.169**. That could be the reason why the OH concentration always increases with the increase of the gas flow rate.

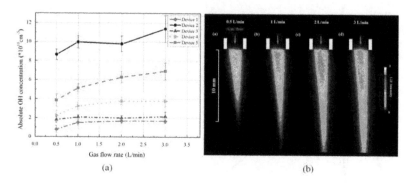

(a)   (b)

FIGURE 4.170 (a) OH concentration of the five N-APPJs at 5 mm away from the nozzle vs. the gas flow rate and (b) two-dimensional LIF intensity distribution of device 5 for different gas rates. (From Yue, Y.F. et al., *IEEE Trans. Radiat. Plasma. Med. Sci.*, 1, 541–549, 2017.)

Figure 4.171 shows the O-TALIF intensity distribution of the five N-APPJs under the experimental conditions described above. For each device, the TALIF measurement is conducted at a distance of 1 mm, 3 mm, 5 mm, 7 mm, and 9 mm from the nozzle. It shows that device 2 gives the strongest O TALIF signals. To further understand the O concentration of the plasma plumes, the TALIF intensity was calibrated. **Figure 4.172** shows the calibrated O concentration along the axial direction. Device 2 has the O concentration up to 10^{16} cm^{-3}, which is about two orders of magnitude higher than the OH concentration. Similarly, for the other four devices, the O concentrations are on the order of 10^{15} cm^{-3}, and they are also about two orders of magnitude higher than their OH concentrations.

Next, several parameters were also varied to investigate their effects on the O concentration, including the applied voltage, the pulse frequency, the percentage of O_2 in the working gas, and the gas flow rate. **Figures 4.173** and **4.174** show

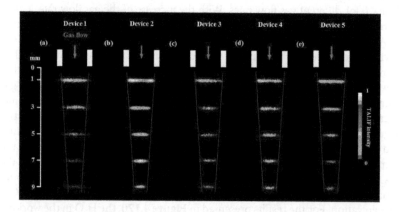

FIGURE 4.171 O-TALIF intensity distribution of the five N-APPJs (a–e) shown in **Figure 4.164** geometry 1–5. (From Yue, Y.F. et al., *IEEE Trans. Radiat. Plasma. Med. Sci.*, 1, 541–549, 2017.)

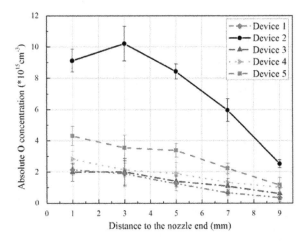

FIGURE 4.172 Absolute O concentration of the five N-APPJs vs. the distance to the nozzle. (From Yue, Y.F. et al., *IEEE Trans. Radiat. Plasma. Med. Sci.*, 1, 541–549, 2017.)

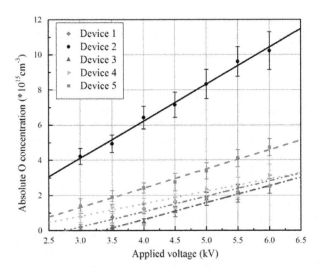

FIGURE 4.173 O concentration at 5 mm away from the nozzle of the five N-APPJs vs. the applied voltage. (From Yue, Y.F. et al., *IEEE Trans. Radiat. Plasma. Med. Sci.*, 1, 541–549, 2017.)

the effect of the applied voltage and the pulse frequency on the O concentration at 5 mm away from the nozzle. The O concentration increases linearly with the increase of the applied voltage for all five N-APPJs. Such behavior is similar to that of the OH discussed before. The relationship between the O concentration and the pulse frequency is shown in **Figure 4.174**. The O concentrations for all

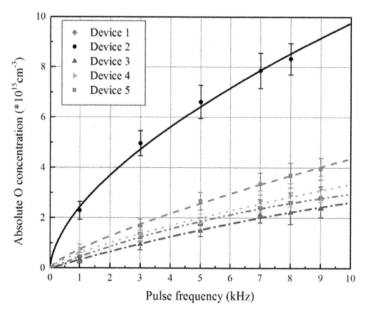

FIGURE 4.174 O concentration at 5 mm away from the nozzle of the five N-APPJs vs. the pulse frequency. (From Yue, Y.F. et al., *IEEE Trans. Radiat. Plasma. Med. Sci.*, 1, 541–549, 2017.)

five N-APPJs increase with the increase of the frequency, but not linearly, which is also similar to that of the OH, as shown in **Figure 4.168**.

For the five N-APPJs discussed in this paper, dissociation of O_2 by collisions with electrons and metastable state He are believed to be the main O atom production paths [106]. Thus, the O_2 concentration in the working gas and the diffusion of air, which is related to the gas flow rate, could affect the O concentration. **Figure 4.175** presents the O concentration at 3 mm away from the nozzle of the five N-APPJs for different O_2 percentages in the working gas. It shows that the O concentration increases dramatically with the increase of O_2 percentage in the working gas at first, and then it decreases slowly with the further increase of the O_2 percentage. The O concentrations reach their maximum when about 0.5%–1% O_2 is added for all five devices. **Figure 4.176** illustrates the O concentration at 5 mm away from the nozzle with different gas flow rates for all five devices. Similarly, it is found that O concentration increases with the increase of the gas flow rate and saturates when the gas flow rate is higher than certain values. As we know, the lifetime of atomic O in a helium atmospheric plasma jet could be up to the millisecond range; thus, O atoms could be delivered from the upstream in the tube to the downstream in the effluent by gas flow [107]. Therefore, the O concentration generated inside the tube could be delivered to the downstream (**Figure 4.177**).

The OH and O concentrations generated by the five typical N-APPJs are measured and the effects of the applied voltage, the pulse frequency, the gas composition, and the gas flow rate on their concentrations were investigated.

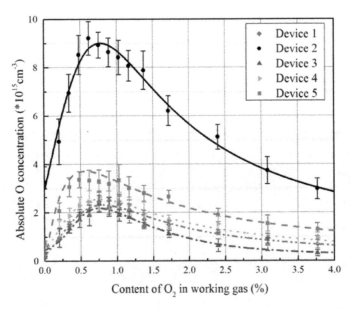

FIGURE 4.175 O concentration at 3 mm away from the nozzle of the five N-APPJs vs. O_2 percentage in the working gas. (From Yue, Y.F. et al., *IEEE Trans. Radiat. Plasma. Med. Sci.*, 1, 541–549, 2017.)

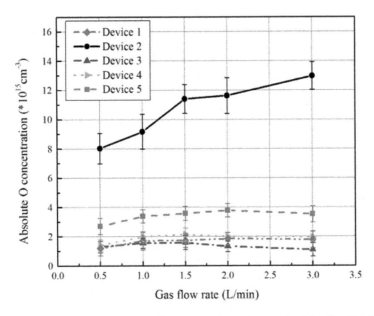

FIGURE 4.176 O concentration at 5 mm away from the nozzle of the five N-APPJs vs. the gas flow rate. (From Yue, Y.F. et al., *IEEE Trans. Radiat. Plasma. Med. Sci.*, 1, 541–549, 2017.)

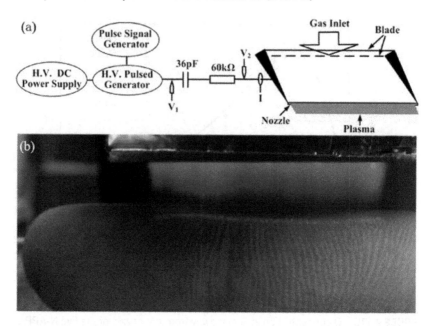

FIGURE 4.177 (a) Schematic of the device. (b) Photograph of the plasma brush touched by a human finger (the distance between the blade and the finger is about 5 mm). (From Lu, X. et al., *Plasma Sources Sci. Technol.,* 20, 065009, 2011.)

It is found that the jet with a naked HV pin electrode (device 2) generates the highest concentrations of OH and O, which are 10^{14}/cm^3 for OH and 10^{16}/cm^3 for O. The average power dissipated to the plasma is 0.88 W, 0.44 W, 0.09 W, 0.53 W, and 0.22 W for devices 1–5, respectively. Thus, according to the measured OH and O concentrations, if we divide OH or O concentration by the average power dissipated to each of the devices, i.e., the OH or O production efficiency from the energy point of view, then devices 2, 3, and 5 have a similar efficiency. However, as mentioned in Section 4.3, there is a risk of transition to arc discharge for device 2, so this device is thus not the best choice for medical applications. For all five N-APPJs investigated in the paper, it is worth pointing out that the OH concentration reaches its peak when about 70 ppm H_2O is added to the working gas while O concentration reaches its peak when about 0.75% of O_2 is added to the working gas, which corresponds to about 7500 ppm of O_2. So, the required optimized O_2 concentration is almost two orders of magnitude higher than the required optimized H_2O concentration. According to OH and O concentration measurements, the peak O concentrations for all the devices are also about two orders of magnitude higher than the peak OH concentrations. On the other hand, as we know, the dissociation energy of H_2O to O+OH is about 4.3 eV and O_2 to 2O is about 5.1 eV. So, the dissociation energy for O_2 is actually higher than H_2O. But the measured OH maximum concentration is much lower than O. This could be related to the driven voltage. The OH concentrations of the plasma jets reach up to the order of 10^{13}/cm^3 for the jet driven by a 20 kHz power supply, 10^{14}/cm^3 for the jet driven by a an RF power supply, and 10^{15}/cm^3 for the jet driven by a microwave power source. Similarly, the O concentrations of the plasma jets reach up to the order of 10^{14}/cm^3 for the jet driven by an RF power supply, and 10^{16}/cm^3 for the jet driven by a microwave power source. Therefore, since only pulsed DC voltage is applied to all the devices, when different power supplies, such as kHz AC power supply or RF power supply, are used, they must also affect the reactive species concentrations. In order to have a clear picture of the reactive species concentrations, the reactive species concentration measurements of the N-APPJs driven by different power sources are urgently needed. It should be emphasized that, for the studies presented in this paper, no object is placed in front of the plasma plume, which is different from the reality when a plasma jet is used for applications. The OH and O concentration could be significantly different, and their concentrations also depend on the properties of the objects, such as the wetness, roughness, the conductivity of the surface of the object, and the distance between the nozzle and the object [108]. This will be the focus of our future studies. Finally, in this paper, we only focus on the OH and O concentrations. Other reactive species, such as NO, NO_2, O_3, etc. could also play some role in various applications. The best ways of generating these reactive species could belong to other devices rather than device 2 and the density distribution variation with other the parameters could also be different from that of the OH and O.

Conducting materials to be treated are electrically connected to the ground. As the distance between the brush nozzle (anode) and the object (cathode) decreases from 10 mm to 2 mm, the discharge mode of the plasma brush changes from a corona discharge to a glow discharge, as illustrated in **Figure 4.178**. A corona discharge occurs when the distance between the nozzle and the object is larger than 10 mm. The plasma brush extends into the surrounding air for a few millimeters, as shown in **Figure 4.178a**. **Figure 4.178b** shows that when the distance between the brush nozzle and the sample is reduced to 6 mm, a glow discharge emerges. Negative glow, Faraday dark space, and positive column are observed from bottom to top. If the distance is further reduced to 4 mm, the plasma resembles a glow discharge, as shown in **Figure 4.178c**. The width of the Faraday dark space decreases compared to that in **Figure 4.178b**, but the length of the positive column is constant. This is different from a traditional glow discharge at a low pressure, in which the length of the positive column decreases with the decreasing discharge gap. **Figure 4.178d** reveals that the Faraday dark space disappears when the distance is reduced to 2 mm.

Figure 4.179a and **b** depict the voltage–current waveforms when the object is 6 mm and 2 mm away from the nozzle, respectively. As shown in **Figure 4.179a** for a distance of 6 mm, before the discharge current reaches the main peak (labeled 4), there is a small current peak of about 10 mA (labeled 2). The main current pulse has an amplitude of about 70 mA and a width of about 300 ns. As shown in **Figure 4.179b** for a distance of 2 mm, only one discharge current peak with an amplitude of about 400 mA can be observed. The rise time of the discharge current is about 10 ns. The discharge current diminishes from 400 mA to 100 mA quickly, in about 30 ns,

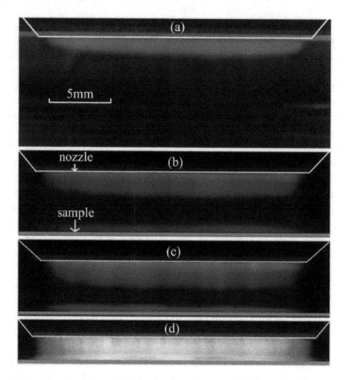

FIGURE 4.178 An HV electrode made of two blades, with the radius of the blade edge being about 50 μm and the distance between the two blades about 1 mm: (a)–(d) photos of the plasma brush with the distance between the blade and the object being 10 mm, 6 mm, 4 mm, and 2 mm, respectively (exposure time = 100 ms). (From Lu, X. et al., *Plasma Sources Sci. Technol.*, 20, 065009, 2011.)

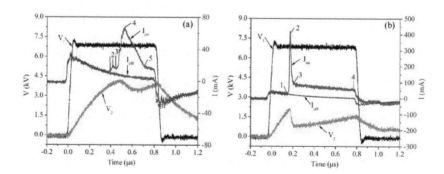

FIGURE 4.179 *I–V* characteristics of the plasma brush with V_1 = applied voltage, V_2 = voltage at the nozzle, I_{on} = total current (plasma on), I_{off} = displacement current (plasma off): V_1 = 7 kV, pulse width = 800 ns, pulsing frequency = 5 kHz, working gas = He, He flow rate = 1.5 L/min, discharge gap distance: (a) 6 mm and (b) 2 mm. (From Lu, X. et al., *Plasma Sources Sci. Technol.*, 20, 065009, 2011.)

but drops slowly afterward. It should be mentioned that the stray capacitor plays an important role during the discharge, and the main peak of the discharge current (from 1 to 3 as labeled in **Figure 4.179b**) possibly arises from the discharge of the stray capacitance, whereas the slowly decreasing part of the discharge current (from 3 to 4 as labeled in **Figure 4.179b**) is due to the discharge of the main circuit. Therefore, by integrating the main peak of the discharge current in **Figure 4.179b**, the total charges can be calculated to be about 6×10^{-9} C. The voltage drop during the main discharge is about 1 kV, and according to $C = Q/\Delta V$, the stray capacitance is estimated to be about 6 pF. To further investigate the mechanism of the plasma brush, a fast ICCD camera was used to capture the dynamics of the discharge. **Figures 4.180 and 4.181**

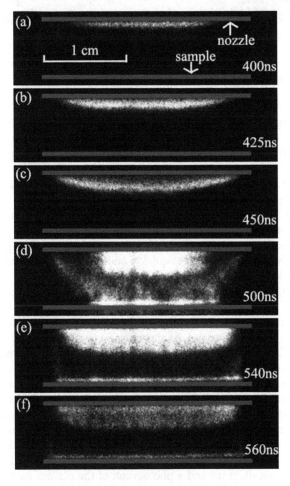

FIGURE 4.180 (a–f) High-speed photographs of the plasma brush for an exposure time of 2.5 ns. The time labels in each photograph correspond to the times shown in **Figure 4.179a**. The discharge gap distance is 6 mm. (From Lu, X. et al., *Plasma Sources Sci. Technol.*, 20, 065009, 2011.)

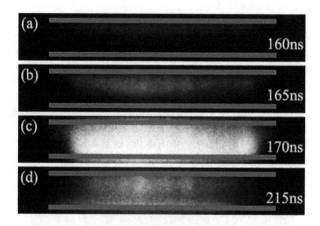

FIGURE 4.181 (a–d) High-speed photographs of the plasma brush. The exposure time is fixed at 2.5 ns. The time labels in each photograph correspond to the times shown in **Figure 4.178b**. The discharge gap distance is 2 mm. (From Lu, X. et al., *Plasma Sources Sci. Technol.*, 20, 065009, 2011.)

depict photographs taken at different delay times for distances of 6 mm and 2 mm, respectively. The time labeled on each photograph in **Figures 4.180** and **4.181** corresponds to the time shown in **Figure 4.179a** and **b**, respectively. For a distance of 6 mm, as shown in **Figure 4.180a**, the plasma is initiated from the nozzle at 400 ns corresponding to 1 as labeled in **Figure 4.179a**. Then it becomes brighter, as shown in **Figure 4.180b**. This corresponds to the small current peak 2 as labeled in **Figure 4.179a**. Afterward, the plasma propagates toward the sample (cathode) leaving a dark zone near the nozzle (anode). The luminance of the plasma decreases corresponding to 3 in **Figure 4.179a**. Afterward, the plasma propagates quickly and reaches the cathode, as shown in **Figure 4.180d**. Hence, it takes about 100 ns for the plasma to cross the 6 mm gap. **Figure 4.180e** illustrates a stable glow discharge. The negative glow, Faraday dark space, and positive column can be readily discerned from bottom to top, and finally, the glow discharge reaches a steady state and the luminance of the plasma starts to decrease. As shown in **Figure 4.181a** and **c** corresponding to 1 and 2 in **Figure 4.179b** for a distance of 2 mm, the plasma resembles a cathode-directed streamer-like discharge propagating from the nozzle to the sample in about 10 ns. The light intensity of the plasma increases dramatically as soon as it reaches the sample, and afterward the intensity decreases.

The generation of stable, room-temperature, atmospheric-pressure plasma jet discharges operated in open air without any external power or gas supply is reported in this section. This portable plasma source is handheld (named "plasma flashlight" here) and is powered by a 12 V DC battery. **Figure 4.182** shows the circuit diagram and a photograph of the plasma flashlight device and also a sketch of the biofilm treatment. This handheld plasma jet is driven by a 12 V battery and does not require any external generator or wall power; neither does it require any external gas feed or handling (e.g., valves, mass flow controllers, etc.) system. With the input voltage of 12 V, the output voltage of

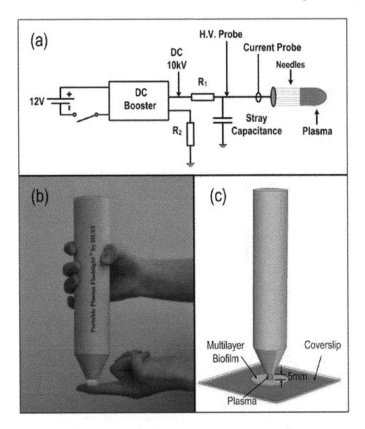

FIGURE 4.182 (a) Schematic of the plasma jet setup, (b) photograph of the portable handheld plasma flashlight device, and (c) schematic of the biofilm treatment. (From Pei, X. et al., *J. Phys. Appl. Phys.*, 45, 165205, 2012.)

the DC booster reaches 10 kV (**Figure 4.182a**). An array of 12 stainless-steel needles is used as an electrode. The radius of the needle tips is ~50 µm. The ballast resistors R_1 and R_2 (both 50 MΩ) shown in **Figure 4.182a** are used to limit the discharge current. This is made to minimize the plasma heating and electric shock effects on the human body and to make it safe to touch, as pictured in **Figure 4.182b**. The gas temperature of the plasma plume is measured within the 20°C–28°C range, which is very close to room temperature. The current and voltage waveforms are recorded by a Tektronix DPO7104 wideband digital oscilloscope and are shown in **Figure 4.183**. From **Figure 4.183** one can clearly see that the discharge appears as a periodic sequence of nanosecond repetitive pulses with a pulse repetition rate of approximately 20 kHz. **Figure 4.183b** shows the current and voltage waveforms of a typical single pulse. The current pulse has a full-width at half-maximum of ~100 ns and a peak value of ~6 mA. Using these current–voltage waveforms, the power dissipated into the plasma was estimated to be ~60 mW. The reactive plasma species

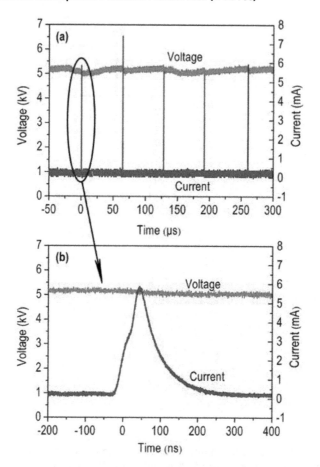

FIGURE 4.183 Current–voltage waveforms of the plasma discharge: (a) multiple repetitive pulses and (b) a single pulse. (From Pei, X. et al., *J. Phys. Appl. Phys.*, 45, 165205, 2012.)

produced by this flashlight can penetrate to the bottom layer of a 25.5 μm-thick *Enterococcus faecalis* biofilm and produce a strong bactericidal effect. This is the thickest reported biofilm inactivated using room-temperature air plasmas.

4.3.5 Media Activated by N-APPJs: Reactive Species and Biological Effects

Plasma biomedicine has attracted ever increasing interest from researchers because of its great potential in healthcare fields such as sterilization, dentistry, wound treatment, blood coagulation, and cancer therapy [110–112]. Nonequilibrium atmospheric pressure plasmas (N-APP) have been widely used in plasma biomedicine because they can generate abundant reactive oxygen and nitrogen species (RONS) while keeping a low temperature, generally below 40°C. It has been confirmed that N-APP can efficiently kill various microbes, even highly resistant forms, such as bacterial spores and biofilm [109]. Though the mechanisms of plasma–cell interaction are not fully understood, RONS generated by the plasma in the gas and/or liquid phases is believed to play key roles in the treatments. Recently, plasmas generated in air and in contact with water are of great interest in the plasma medicine community [113–117]. On one hand, this is because most biological organisms live in a wet or aqueous liquid environment and plasma is the first contact with the liquid layer covering the living systems in most cases. On the other hand, air plasma can produce large quantities of RONS compared to rare gas plasma. These species generated in the gas and at the gas–liquid interface can dissolve into water and initiate complex chemical and biocidal processes in water [118,119]. This kind of water, which was exposed to plasma, is the plasma-activated water mentioned in previous papers, and it can also inactivate bacteria suspended in it [120–124].

The main reactive species produced by plasma, such OH· radicals, atomic oxygen, ozone, hydrogen peroxide, and nitric oxide, are thought to have significant influences on biological systems [113]. Short-lived aqueous active species have been reported to be the dominant inactivation agents when the bacteria suspension is treated by plasma directly or by plasma afterglow. For example, Xiong et al. reported that aqueous OH·contributes dominantly in the *Candida glabrata* inactivation process [125]. Ikawa et al. reported that aqueous O_2^- is the cause of bactericidal effects when the pH of a bacteria suspension is below 4.7 [126]. But for the antimicrobial actions of PAW, conditions are different. The solutions are usually acidified by the dissolution of NOx species formed by air plasma in water, and short-lived species, such as OH, NO_2, and NO radicals are unstable and rapidly convert to stable species, like H_2O_2, NO_3^- and NO_2^- through chemical reactions. It has been demonstrated previously that acidic pH plays a significant role in the bactericidal effects of PAW, but the acidic environment itself was not sufficient for bacteria inactivation. At present, the antimicrobial properties of PAW are tentatively attributed to the synergetic effect of acidic pH, H_2O_2, and nitrite/nitrate in the solution while the exact inactivation mechanism of these species is still not very clear [123,126–128]. In this section, a DC-driven atmospheric pressure air plasma array source was used to generate plasma in air and in contact with water surface to produce PAW. The key properties of air plasma were investigated by optical diagnosis methods. The pH value and the concentrations of RONS (H_2O_2, NO_2^-, NO_3^-, OH, and $ONOO^-$) in PAW were measured by chemical and optical methods. The bacteria *E. coli* was used as a microbial model to study the biocidal effects of PAW. The property of PAPBS was also studied in some cases to better understand the effect of pH on the bactericidal process of plasma-treated solutions. We found that peroxynitrite produced in acidic conditions plays a key role in the antibacterial process of PAW, and the bactericidal effect of peroxynitrite decreases significantly but is still detectable in nonacidic environments. Hydrogen peroxide or acidified nitrites alone cannot result in an obvious bactericidal effect, but they are the preconditions to produce peroxynitrite in PAW.

The schematic of the experiment setup is shown in **Figure 4.184a**. The air plasma source is driven by a homemade DC power supply, whose output voltage can be adjusted from 12 kV to 20 kV. The high-voltage output of the power supply

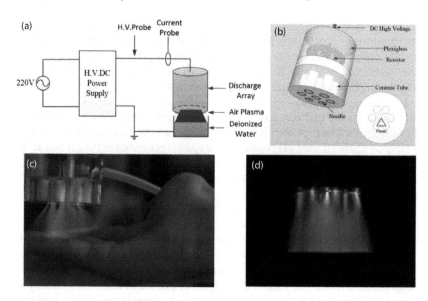

FIGURE 4.184 (a) Schematic of the discharge system to produce plasma-activated water. (b) The structure of the discharge array. (c) Discharge to human palm. (d) Discharge to water surface. (From Ma, M. et al., *IEEE Trans. Plasmas,* To be published.)

is connected to the discharge array (**Figure 4.184b**). When the power is turned on, a large-area diffuse air plasma is generated between the discharge array and the counter electrode, such as hand and deionized water (**Figure 4.184c** and **d**). The air plasma array contains a cellular structure with one central ceramic tube surrounded by six other tubes, and the diameter of the air plasma array is 55 mm. The tip of the needle has a radius of about 50 μm and the distance from the tip of the needle to the inner surface of tube end is 5 mm. Each needle electrode connects to the DC high voltage through a resistor of 100 MΩ to prevent the undesired glow-to-arc transition. Plasma activated water (PAW) and plasma activated PBS (PAPBS) were created by this air plasma source, as shown in **Figure 4.184d**.

The current and voltage waveforms are shown in **Figure 4.185**. The voltage of the DC power supply was 18 kV and the distance from the ceramic tube nozzle to the water surface was fixed at 15 mm. Throughout the paper, these parameters were kept constant unless indicated otherwise.

Figure 4.185a shows that the discharge appears as a periodic sequence of nanosecond repetitive pulses with a pulse repetition rate of approximately 5 kHz. The peak current can reach 60 mA. **Figure 4.185b** zooms a typical single pulse. The discharge current lasted for about 150 ns. The mean power dissipated into the plasma was estimated to be 14.5 W through the following equation:

$$P = \frac{1}{T}\int_0^T u(t)\cdot i(t)\,dt$$

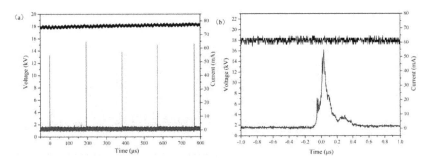

FIGURE 4.185 (a) Typical current-voltage waveforms of the plasma (b) close look at a single pulse. (From Ma, M. et al., *IEEE Trans. Plasmas*, To be published.)

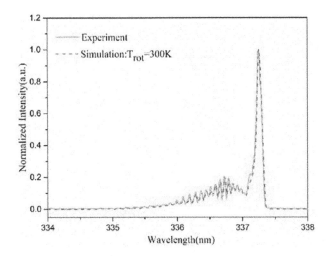

FIGURE 4.186 Experimental and simulated emission spectra of N_2 second positive system 0–0 transition. (From Ma, M. et al., *IEEE Trans. Plasmas*, To be published.)

Figure 4.186 shows that the simulated spectra at $T_{rot} = 300$ K gives the best fit to the experimental spectra; therefore, the estimated gas temperature of the air plasma would be 300 K during the experiments (maximum discharge time is less than 5 min.). **Figure 4.187a** and **b** show the emission spectra of the plasma from 200 nm to 500 nm and 500 nm to 800 nm, respectively. The excited N_2 and N_2^+ are the dominant reactive species in the air plasma; in addition, an atomic oxygen emission at 777 nm and an excited OH ($A^2\Sigma^+ \rightarrow X^2\Pi, \Delta v = 0$) at 309 nm are also found. The O_3 and NO_2 concentration at different radial positions are shown in **Figure 4.188**. The maximum concentration of O_3 and NO_2 are 85 ppm and 130 ppm in the discharge center respectively. The O_3 concentration decreased dramatically when the radius was larger than 2 cm,

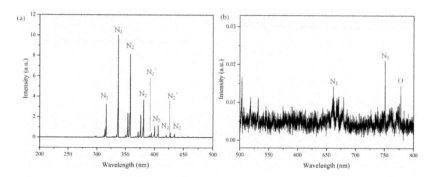

FIGURE 4.187 Optical emission spectra of the plasma: (a) 250–500 nm and (b) 500–800 nm. The grating is 1200 g/mm, and the entrance and exit slits of the spectroscope are fixed at 100 μm. (From Ma, M. et al., *IEEE Trans. Plasmas*, To be published.)

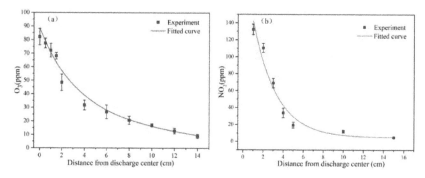

FIGURE 4.188 (a) O_3 and (b) NO_2 concentration of the plasma at different radial positions. (The vertical position is at the center of the plasma.) (From Ma, M. et al., *IEEE Trans. Plasmas*, To be published.)

while the NO_2 concentration decreased monotonously as the radius increased. The NO concentration in the plasma volume was only 1~2 ppm; this might because NO rapidly converts to NO_2 when reacting with O_2.

The bactericidal effect of PAW or PAPBS was quantified by the number of viable *E. coli* (represented in units of CFU/mL) after exposure to PAW or PAPBS. **Figure 4.189** shows the inactivation efficiency of *E. coli* by PAW and PAPBS with different plasma treatment times and pH values. The pellet was added to PAW or PAPBS immediately after plasma treatment. The bactericidal effect of PAW was not obvious within 1 min., then enhanced significantly with longer plasma treatment time. PAW can achieve a complete bacteria inactivation (detection limit is 10 CFU ml^{-1} in this work) when the treatment time is longer than 5 min. As for PAPBS, its bactericidal effect is much weaker than PAW. *E. coli* reduction by 1.8 log was measured in PAPBS treated for 5 min., and the pH value was estimated to be one of the reasons. The pH value of PAW decreased to pH $\approx$ 2.8 for the 5 min. treatment while the pH value of PAPBS

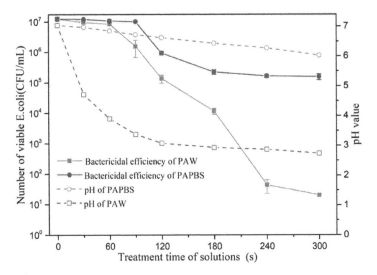

FIGURE 4.189 The bactericidal effect and pH value of PAW and PAPBS with different plasma treatment times. (From Ma, M. et al., *IEEE Trans. Plasmas*, To be published.)

decreased very little (pH ≈ 6 for a 5 min. treatment) due to its sufficient buffer capacity. Although acidification alone was not sufficient for bacteria inactivation, it significantly affected the aqueous-phase chemistry of reactive oxygen and nitrogen species in the liquid. The effects of pH and the specific contributions of each aqueous RONS will be discussed next.

The OH$^{\bullet}_{aq}$ concentration dissolved in the PAW was measured immediately after plasma treatment, and its concentration is shown in **Figure 4.190**. Although the OH$^{\bullet}_{aq}$ concentration increased with longer plasma treatment time, the OH$^{\bullet}_{aq}$ in the PAW was still below 0.1 μM at the end of treatment. This result is much lower compared to around 10 μM by the direct plasma treatment of TA solution. Such a low concentration is due to the short lifetime of OH$^{\bullet}_{aq}$ especially in the liquid phase. OH$^{\bullet}_{aq}$ can rapidly convert to stable H_2O_2 via reaction (4.6) [127]

$$OH^{\bullet} + OH^{\bullet} \rightarrow H_2O_2 \qquad (4.6)$$

The small amount of OH$^{\bullet}$ detected in the PAW may be attributed to the decomposition of peroxynitrite via reaction (4.7) at pH < 6.8.

$$O = NOOH \leftrightarrow OH^{\bullet} + NO_2^{\bullet} \qquad (4.7)$$

Therefore, the contribution of OH$^{\bullet}$ radicals in the PAW inactivation process can be ignored due to its low concentration.

Nitrites and nitrates in the PAW are produced by the dissolution of nitrogen oxides in the gas phase. The pH of the PAW decreased as NO_2^- and NO_3^- formed in the PAW. The process can be described by reactions (4.8) and (4.9) [127,129]

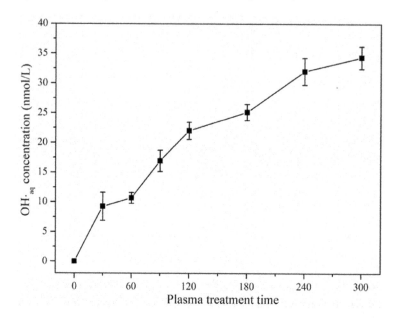

FIGURE 4.190 The OH$^\bullet_{aq}$ concentration in the PAW dependent on plasma treatment time. (From Ma, M. et al., *IEEE Trans. Plasmas*, To be published.)

$$NO_2 + NO_2 + H_2O \rightarrow NO_2^- + NO_3^- + 2H^+ \qquad (4.8)$$

$$NO + NO_2 + H_2O \rightarrow 2NO_2^- + 2H^+ \qquad (4.9)$$

Hydrogen peroxide is produced in the PAW through the recombination reaction of OH$\cdot$ radicals generated by plasma at the gas/liquid interface through reaction (4.6).

The concentrations of hydrogen peroxide and nitrite and nitrate ions in PAW as a function of discharge treatment time are shown in **Figure 4.191**. The concentration of H_2O_2 and NO_3^- kept increasing while the concentration of NO_2^- decreased after its maximum value at 90 s. However, continuously increasing NO_2^- concentration in the plasma-treated solution was reported by P. Lukes et al. [127]. This is because the pH value of the solution was controlled at 6.9, and samples for ion chromatography analysis were fixed by a phosphate buffer immediately to stop acidic decomposition of nitrites via reaction (4.10)

$$3NO_2^- + 3H^+ + H_2O \rightarrow 2NO + NO_3^- + H_3O^+ \qquad (4.10)$$

In this work, the NO_2^- concentration was measured immediately after treatment without adding a phosphate buffer. This is because the bacteria need to be suspended in the PAW for 15 min. in our experiments, which is close to the time needed for ion chromatography analysis of one sample. So, the NO_2^- concentration shown in **Figure 4.190** can reflect the real situation during the PAW inactivation process. In addition, Though the actual production of NO_2^- was

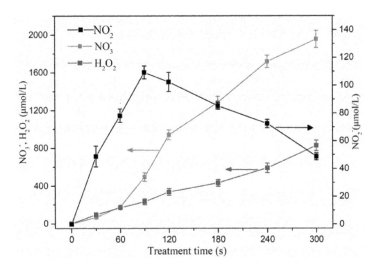

FIGURE 4.191 Nitrite (NO_2^-), nitrate (NO_3^-) and hydrogen peroxide (H_2O_2) concentrations measured in PAW as a function of discharge treatment time. (From Ma, M. et al., *IEEE Trans. Plasmas*, To be published.)

enhanced with increasing the discharge treatment time, it cannot compensate for the decomposition of nitrites when the pH value of PAW decreases to 3.

In order to analyze the effect of hydrogen peroxide and acid environment on the bactericidal activity of PAW, *E. coli* was exposed to various concentrations of H_2O_2 solutions, which were made by diluting 30% H_2O_2 solution (Sinopharm Chemical Reagent Co., Ltd, China) and different pH citrate-Na buffers.

Figure 4.192a indicates that no significant bacteria inactivation occurred after 15 min. exposure to 2 mM H_2O_2 solution. This concentration is much higher than the H_2O_2 concentration in 0.8 mM of PAW. The acid environment only has moderate antibacterial activity when the pH decreases to 2.8, as shown in **Figure 4.192b**. Therefore, the hydrogen peroxide or acid

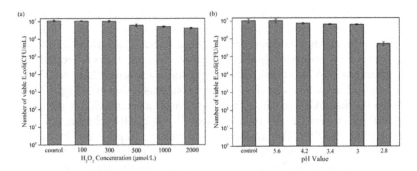

FIGURE 4.192 The bacteria inactivation results of (a) hydrogen peroxide and (b) acid citrate-Na buffers, respectively. (From Ma, M. et al., *IEEE Trans. Plasmas*, To be published.)

environment alone cannot cause obvious inactivation effects alone, which is in agreement with previous investigations [123,126].

As for O_3, the determination of dissolved ozone in the PAW by the DPD (N,N-diethyl-p-phenylenediamine) colorimetric method [130] was negative. The detection limit of the kit used in the experiment (HuanKai Biotechnology, China) is 0.05 mg/L. Considering the very low solubility of ozone at atmospheric pressure [131] and ozone elimination via reaction (4.11) with nitrites in the liquid, it is assumed that ozone did not contribute to the antibacterial process of PAW.

$$NO_2^- + O_3 \rightarrow NO_3^- + O_2 \tag{4.11}$$

The bactericidal effect of peroxynitrous acid (O=NOOH) and its conjugate base peroxynitrite (O=NOO⁻) are known from the literature to be important in direct plasma treatment or the PAW inactivation process [132,133]. It is reported that peroxynitrite can be formed under alkaline conditions through reaction (4.12) of nitric oxide and superoxide anion radicals, or by reaction (4.13) of $NO_2 \cdot$ with an OH• radical [133–135]. However, in acidic PAW, reaction (4.14) should be the main route to form peroxynitrite. The pK_a of ONOOH was reported to be 6.8 [136], so the dominant form in acidic PAW should be ONOOH, and in neutral or alkaline conditions, it should be the anionic form (ONOO⁻).

$$O_2^- \cdot + NO \cdot \rightarrow O = ONOO^- \tag{4.12}$$

$$OH \cdot + NO_2 \cdot \rightarrow ONOO^- + H^+ \tag{4.13}$$

$$NO_2^- + H_2O_2 + H^+ \rightarrow O = NOOH + H_2O \tag{4.14}$$

In previous work, bacteria inactivation was performed in acidic PAW while samples for peroxynitrite measurement were fixed by a phosphate buffer [124]. The results obtained in this way cannot reflect the extensive amount of peroxynitrite in PAW, and whether peroxynitrite can still play a role in neutral environments is not clear. To test this, PAW and PAPBS, both treated for 5 min., were used. The pH value of PAW and PAPBS was 2.7 and 6.2 respectively after treatment. An appropriate volume of NaOH solution was added to one part of PAW immediately after treatment to change its pH to around 7 and the volume of PAW was almost unchanged due to a relatively high NaOH concentration. The bactericidal effect of three solutions is shown in **Figure 4.193**. We can see that PAW can cause a complete bacteria inactivation while its bactericidal effect substantially reduced after adding NaOH solution, but it still stronger than PAPBS. The amount of peroxynitrite (represented by the relative fluorescence intensity of DCFH-DA) and the concentration of nitrite is shown in **Figure 4.194**. The amount of peroxynitrite increased softly in PAW after adding NaOH solution because peroxynitrite decays easier under acidic conditions. The highest concentration of nitrite and the lowest amount of peroxynitrite was detected in PAPBS due to the solution being kept almost

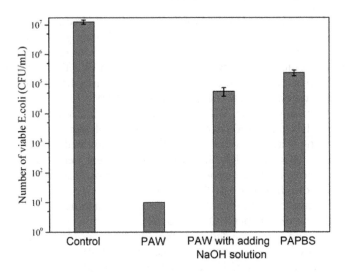

FIGURE 4.193 Bacteria inactivation results for three plasma-activated solutions. (From Ma, M. et al., *IEEE Trans. Plasmas*, To be published.)

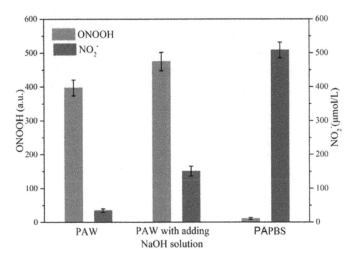

FIGURE 4.194 The relative amount of peroxynitrite (fluorescence intensity with reference subtracted) and NO_2^- concentration in three plasma-activated solutions. (From Ma, M. et al., *IEEE Trans. Plasmas*, To be published.)

neutral during the plasma treatment process. It can be concluded that peroxynitrite detected in PAW correlates directly with the bactericidal efficacy, and the bactericidal effect of peroxynitrite decreases significantly but is still detectable in non-acidic environment even though its concentration becomes a little higher in nonacidic environments. In addition, we can deduce that NO_2^- has no obvious bactericidal effect in neutral solutions.

To exclude the influence of acidified nitrites, the bactericidal effect of hydrochloric acid (pH = 2.8) and hydrochloric acid, adding different concentrations of $NaNO_2$ solutions, were measured. No obvious enhanced bactericidal effect was found even when the concentration of $NaNO_2$ increased to 200 uM. It further confirmed that a nitric acid solution of the same pH did not lead to the same bactericidal effects compared to PAW.

To further study whether ONOOH is the key agent in bacteria inactivation, the bactericidal effect of PAW at different times after plasma treatment and the corresponding relative amount of ONOOH were examined. The time evolution of H_2O_2, NO_2^- and NO_3^- in PAW after treatment were also measured. (The pH value of PAW was kept at 2.8 during the post-discharge time.) The results are shown in **Figures 4.195** and **4.196**.

The inactivation efficacy of PAW decreased significantly after 2 h. Meanwhile, the amount of ONOOH and the concentration of NO_2^- decreased dramatically and cannot be detected after 5 h. Reactions (4.10) and (4.14) are the two routes for the decay of NO_2^- dissolved in PAW. The mechanism of the decomposition of ONOOH is complex. Briefly, ONOOH decomposed into about 70% NO_3^- via reaction (4.15) and 30% $OH·/NO_2·$ via reaction (4.7).

$$O=NOOH \rightarrow NO_3^- + H^+ \quad\quad (4.15)$$

A slight increase in nitrate and a decrease in hydrogen peroxide concentration during the post-discharge time shown in **Figure 4.13** further confirmed the reactions mentioned above. The log reduction of *E. coli* in PAW was almost unchanged (around 1.31) after 5 h, and this result is close to the bactericidal effect of citrate-Na buffers at pH 2.8 shown in **Figure 4.192b**

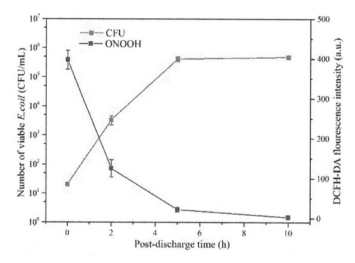

FIGURE 4.195 Time-dependence of inactivation efficacy of PAW and its corresponding relative amount of ONOOH. (From Ma, M. et al., *IEEE Trans. Plasmas*, To be published.)

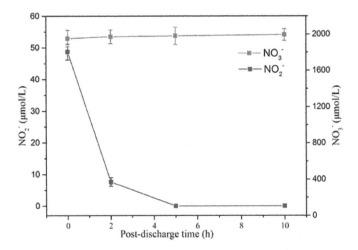

FIGURE 4.196 Time evolution of H_2O_2, NO_2^-, and NO_3^-, in PAW after plasma treatment. (From Ma, M. et al., *IEEE Trans. Plasmas*, To be published.)

(log reduction was 1.28). Considering the pH of PAW was also 2.8, it is reasonable to assume that the bactericidal effect of PAW after 5 h has no difference compared to the acid solution of the same pH value.

The interaction between nonequilibrium atmospheric pressure plasma (N-APP) and skin is a key issue of application of plasma in dermatology [137]. Two kinds of treatments have been carried out by plasma. The first treating object is a wound with a skin lesion. The damaged stratum corneum of the epidermis cannot block the transportation of reactive species of plasma into the deep tissue. The biological gelatin model is used to simulate the skin tissue without the stratum corneum of the epidermis. The measurement suggests that the penetration depth of plasma in the biological gelatin model is 1.5 mm [138,139]. This penetration depth can make reactive species work to their advantage, such as killing bacteria and producing epidermal growth factor; therefore, various plasma sources have been used in the stimulation of wound healing [140,141]. The second treating object is a skin disease of the complete skin structure, such as atopic dermatitis, psoriasis, melanoma, etc. [142,143]. The stratum corneum of the epidermis and the abnormal hyperplasia stratum corneum is a challenge for the penetration of reactive species of plasma into the deep tissue. To master and optimize the penetration depth of reactive species in the skin will improve therapeutic effect of plasma on these diseases.

The preliminary studies on the interaction between APNP and skin indicate that the stratum corneum, which is the outermost layer of epidermis and composed of 15–20 layers of flattened cells with no nuclei and cell organelles, can block the transportation of APNP through the skin [144]. The treatment of human skin by argon plasma jets (Kinpen 09) indicated that the penetration depth of plasma in stratum corneum is only 10 μm [145]. Using solid Colo-357 tumours in the TUM-CAM model TUNEL-staining showed plasma-induced (Kinpen 09) apoptosis up to a depth of tissue penetration (DETiP) of 48.8 ± 12.3 μm [146].

Recently, the experiment treatment of cancer cells by plasma-activated water (PAW) and plasma-activated buffered solutions (PABS) indicated that cancer cells have specific vulnerabilities to the aqueous plasma reactive species [147,148]. These aqueous plasma reactive species can inhibit proliferation of chronic chemo-resistant ovarian cancer cells and selectively kill glioblastoma brain tumor cells by down-regulating a survival signaling molecule [149,150]. The biocompatible PAW and PABS might be a promising tool for skin cancer therapy and skin disease treatment. The actual therapy of dermatology needs the deeper penetration depth of plasma; for example, melanoma is usually in the corium layer [151], and pathogenic fungus can infect the corium layer and subcutaneous tissue [152]; therefore, more efficient ways to optimize the transportation of gaseous and aqueous plasma reactive species in the skin are needed.

In this section, in this work, a DC-driven air plasma source is used to produce the gaseous and aqueous reactive species, and the dorsal skin of lab mice is the object of gaseous and aqueous reactive species. The measurement of reactive species and their derivatives beneath the dorsal skin of lab mice indicate that the aqueous reactive species of air plasma are more efficient in transdermal penetration.

The air plasma source employed in this study was developed on the basis of the "Plasma flashlight" [109]. A higher power DC source was used to drive the air plasma jet array, as shown in **Figure 4.197a**. The applied voltage was 18 kV. The discharge in each tube appears as a periodic sequence of nanosecond repetitive pulses with a pulse repetition rate of approximately 20 kHz. The average power of this source was 3 W. The air plasma source can treat the human hand directly (**Figure 4.197b**) and the gas temperature of plasma is 300 K. The O_3 concentration of the air plasma was measured by UV absorption spectroscopy [153]. The NO and NO_2 concentration of the air plasma was

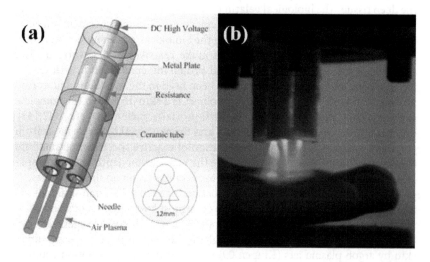

FIGURE 4.197 (a) Schematic structure of the air plasma source. (b) The side view of the air plasma produced by the air plasma source over a human hand. (From Liu, X. et al., *J. Phys. D Appl. Phys.*, 51, 075401, 2018.)

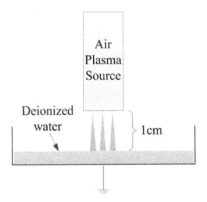

FIGURE 4.198 The discharge system to produce aqueous reactive species in deionized water. (From Liu, X. et al., *J. Phys. D Appl. Phys.*, 51, 075401, 2018.)

measured by a multi-gas detector (RAE Systems, PGM-7800). The plasma between the air plasma source and the deionized water produced the aqueous reactive species in the plasma activated water as shown in **Figure 4.198**. The deionized water was placed in a metal container with ground connection.

The transdermal experiments were carried out on the rat skin obtained from 6- to 8- week-old female animals (Beijing Vital River Laboratory Animal Technology Co., Ltd, China). 3 cm × 3 cm skin was taken from the back of the rat. The thickness of the skin was 0.75 ± 0.2 mm. Medical cotton balls with normal saline were used to remove subcutaneous fat. The improved Franz vertical transdermal diffusion device was used to study the penetration of gaseous and aqueous plasma reactive species. The rat skin was fixed on the top of the accepting pool by pinch cock, with the corneum upward. 360 μL deionized water was fed into the accepting pool precisely to make the corium layer completely attach to the deionized water and prevent bubbles from accumulating below the corium layer.

The air plasma produced by the plasma jet array was applied to the skin of the mouse on the accepting pool (**Figure 4.199**). The O_3 concentration in the middle of gas gap was 85 ppm. The NO concentration was below the lower limit of the detector. The NO_2 concentration was 130 ppm. **Figure 4.200** shows 10 min. treatment did not increase the aqueous H_2O_2, NO_2^-, and NO_3^- concentrations in the deionized water of the accepting pool. The concentration of these aqueous active species was close to 0 with the consideration of measurement error. This indicates that the gaseous plasma active species cannot penetrate 0.75 ± 0.2 mm thick mouse skin.

Figure 4.201 lists four possible ways of transdermal delivery of gaseous plasma active species. Because of the large contact area between air plasma and the skin, the transdermal transport via a tortuous pathway within extracellular lipids (2nd way in **Figure 4.201**) seemed to be the main pathway. However, as the outermost layer of the epidermis, the stratum corneum was made up of 10–20 layers cornocytes, and this compact structure not only blocked most of

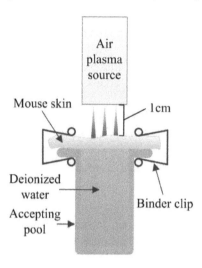

FIGURE 4.199 The schematic of air plasma applied to the skin of a mouse over the accepting pool. (From Liu, X. et al., *J. Phys. D Appl. Phys.*, 51, 075401, 2018.)

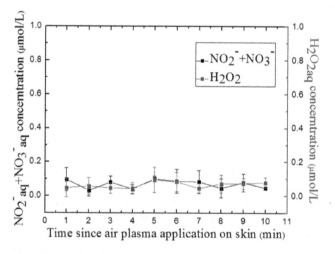

FIGURE 4.200 H_2O_2, NO_2^-, and NO_3^- concentration in the deionized water in the accepting pool as a function of treatment time. (From Liu, X. et al., *J. Phys. D Appl. Phys.*, 51, 075401, 2018.)

the gaseous plasma active species, but also made it difficult to absorb the active species by extracellular lipids [154]. Therefore, the large contact area did not improve the transdermal delivery of gaseous plasma active species. The plasma can propagate along the center hole (diameter of 5 μm) of optic fiber; therefore, the narrow space between the hair and hair follicles provides a possible way of transdermal delivery for the active plasma species (1st way in **Figure 4.201**), which will be discussed in the following paper. Plasma can also transport

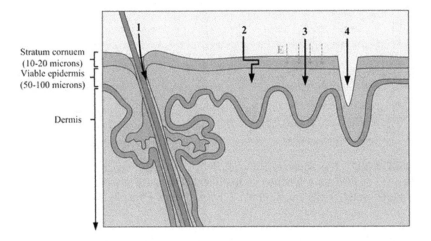

FIGURE 4.201 The possible pathways into the skin of transdermal delivery of gaseous plasma active species. (1) Transport through hair follicles and sweat ducts. (2) Transdermal transport via a tortuous pathway largely within extracellular lipids. (3) Transport directly across the stratum corneum by electroporation. (4) Microneedles remove stratum corneum to make micron-scale (or larger) pathways into the skin. (From Liu, X. et al., *J. Phys. D Appl. Phys.*, 51, 075401, 2018.)

various charge particles to the skin, so the accumulation of charged particles on the skin can produce a strong electric field. The simulation suggested the electric field on the surface of the epidermis of an apple was 8.32×10^6 V/m when the apple was treated by air plasma, which was larger than the electric field of 6×10^6 V/m inducing cell apoptosis. The electric field in the bulk epidermis was still 6×10^5 V/m [11]. This electric field was larger than the electric field of 2×10^4 V/m [162] in the epidermis produced by ~200 V pulse voltage [156] used for electroporation transdermal delivery of drugs. Although the air plasma has the stronger electric field, the result shown in **Figure 4.201** suggests it is still difficult for plasma to enhance the transdermal delivery of gaseous active species. In addition, the strong electric field and high-density active species can change the structure of the stratum corneum, which eventually results in a topically applied substance to better penetrate through the skin barrier following plasma treatment [157]. In view of the blockage effect of the stratum corneum, a microneedle was used to destroy the integrity of the stratum corneum. As the 4th way shown in **Figure 4.201**, the microneedle (AYT 540) was used to punch the skin and generate 8×10 holes with a diameter of 0.25 mm and a depth of 0.3 mm. These holes did not improve the transdermal delivery of gaseous reactive species in the following treatment by air plasma. The concentrations of aqueous H_2O_2, NO_2^-, and NO_3^- in the deionized water of the accepting pool were still close to 0. The consumption of gaseous reactive species by dermis below the stratum corneum and the recovery of the hole because of skin elasticity are the reason for that.

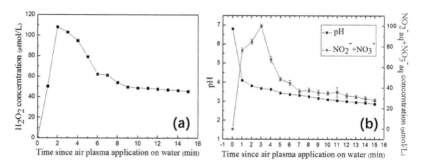

FIGURE 4.202 The aqueous H_2O_2 (a), NO_2^- and NO_3^- concentration and pH value (b) of PAW as a function of treatment time since plasma application on deionized water. (From Liu, X. et al., *J. Phys. D Appl. Phys.*, 51, 075401, 2018.)

The PAW produced by the air plasma source as shown in **Figure 4.198** was used to test the efficiency of transdermal penetration of aqueous plasma active species. A 2 min. treatment of the deionized water by an air plasma source made the aqueous H_2O_2 concentration reach its maximum value of 110 μmol/L. The aqueous H_2O_2 concentration decreased afterward, and its concentration was 46 μmol/L at the end of a 15 min. treatment (**Figure 4.202**). The aqueous NO_2^- and NO_3^- concentration reached the peak value of 100 μmol/L at 3 min., and decreased to 28 μmol/L at the end of treatment. The pH of PAW was 3.2 at the end of treatment. During the air plasma treatment, reactions 1, 2, 3 result in the fast increasing aqueous H^+, H_2O_2, NO_3^-, NO_2^- concentration. The accumulation of H^+ resulted in reaction 4 dominated in the heavily acidic conditions and the decreasing H_2O_2, NO_3^-, NO_2^- concentration after their peak value [127].

1. $OH\bullet + OH\bullet \rightarrow H_2O_2$ Reaction 1

2. $2NO_2(g) + H_2O(l) \rightarrow 2H^+ + NO_3^- + NO_2^-$ Reaction 2

3. $NO(g) + NO_2(g) + H_2O(l) \rightarrow 2H^+ + 2NO_2^-$ Reaction 3

4. $NO_2^- + H_2O_2 + H^+ \leftrightarrow O=NOOH + H_2O$ Reaction 4

In order to study the transdermal penetration effect of aqueous plasma active species, 100 μL PAW was dropped on the skin (**Figure 4.203**). The PAW was the deionized water treated by air plasma for 15 min. The H_2O_{2aq} concentration, $NO_2^-{}_{aq}$ and $NO_3^-{}_{aq}$ concentration in the PAW was 46 μmol/L and 28 μmol/L, respectively. **Figure 4.204** shows the concentration of $H_2O_{2\ aq}$, $NO_2^-{}_{aq}$ and $NO_3^-{}_{aq}$ in the deionized water of the accepting pool. The concentration of $NO_2^-{}_{aq}$ and $NO_3^-{}_{aq}$ in the accepting pool reached the maximum value of 7.3 μmol/L at 1.5 min. after dropping the PAW, and the concentration decreased to 0 at 5 min. This indicates that the aqueous plasma active species can penetrate the skin (**Figure 4.204**).

Stratum corneum is the main barrier for transdermal delivery. The aqueous plasma active species can penetrate through the stratum corneum in

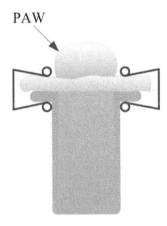

PAW

FIGURE 4.203 The schematic of a drop of PAW applied to the skin of a mouse over the accepting pool. (From Liu, X. et al., *J. Phys. D Appl. Phys.*, 51, 075401, 2018.)

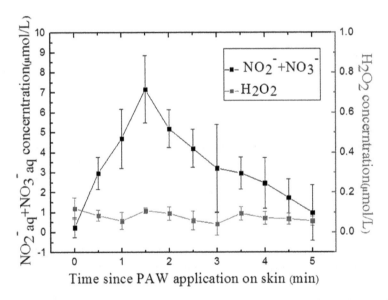

FIGURE 4.204 H_2O_2, NO_2^-, and NO_3^- concentration in the deionized water in the accepting pool as a function of treatment time. (From Liu, X. et al., *J. Phys. D Appl. Phys.*, 51, 075401, 2018.)

three ways (**Figure 4.205**). The first way is the shunt route, which provides a parallel pathway by which aqueous plasma active species can be absorbed by sweat ducts and hair follicles without hindrance by the stratum corneum. But its contribution is limited, because the area of sweat ducts and hair follicles is only 10^{-2}–10^{-3} of total skin area. The second and third ways are the intercellular and transcellular routes. For the transcellular route, aqueous plasma

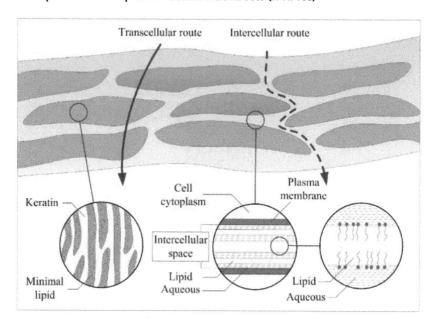

FIGURE 4.205 The possible pathways into the skin of transdermal delivery of aqueous plasma active species. (From Liu, X. et al., *J. Phys. D Appl. Phys.*, 51, 075401, 2018.)

active species pass directly through the corneal cells and intermediary intercellular lipid matrix, while for the intercellular route, aqueous plasma active species diffuse around the corneal cells in a tortuous manner, remaining constantly in the intercellular matrix [158]. Once these solutes penetrate through stratum corneum, the transportation through dermis is more efficient, which eventually results in the transdermal penetration of aqueous plasma active species [158]. In addition, the decreasing $NO_2^-{}_{aq}$ and $NO_3^-{}_{aq}$ in the accepting pool after its peak value is attributed to the fact that the $NO_2^-{}_{aq}$ and $NO_3^-{}_{aq}$ can enter into epidermic cells and turn into nitrite, arginine, and other nitrogenous compounds after prolonged exposure to endosexine of the skin [159]. Although aqueous plasma active species can penetrate through the skin, the concentration of $H_2O_2{}_{aq}$ in the accepting pool was close to 0 with the consideration of measurement error. This is because $H_2O_2{}_{aq}$ is consumed by the antioxidant substances in the skin such as β-carotene, lycopene, vitamin A/C/D, various enzymes and etc. during its penetration [144].

The increasing concentration of subcutaneous NO_2^- and NO_3^- can help treatment of some common diseases. The co-application of a mild acid and nitrite on the skin can liberate NO in a controlled fashion [160], which show therapeutic efficiency in dermatophyte fungal tinea infections, viral infections, molluscum contagiosum, viral warts, and in vitro shows of activity against bacterial pathogens, including the acne bacillus and Staphylococcus aureus [161]. The glyceryl trinitrate and S-nitroso-N-acetylpenicillamine were used for anal Wssures or cutaneous leishmaniasis [162, topical diazeniumdiolates [163]. In addition, the transdermal delivery of NO_2^- and NO_3^- may affect the role of NO-mediated signaling in the skin cancer and the results are complicated. For example, the enhanced release of NO in the basal component of dysplastic nevi and in primary melanomas may be responsible for the observed effects in the papillary dermis of these lesions [164]. NOS_2 immunoreactivity was relatively weak in well-differentiated adenocarcinomas and strong (although heterogeneous) in poorly differentiated ones, suggested that NOS_2 expression may reflect the degree of proliferation of the tumor [165]. Therefore, the transdermal delivery of plasma reactive species can also enhance the treatment of such diseases.

The pH of the PAW used for the transdermal penetration test was 3.2. NaOH was added into the PAW to exclude the influences of heavily acidic conditions on the transdermal penetration. The pH of the PAW was 7 after adding of NaOH. The $NO_2^-{}_{aq}$ and $NO_3^-{}_{aq}$ concentration in the PAW increased to 87.9 μmol/L after adding NaOH, which was attributed to the fact that Reaction 4 is a reversible reaction and O=NOOH decompose to NO_2^- in the weak acidic environment. **Figure 4.206** indicates the peak value of $NO_2^-{}_{aq}$ and $NO_3^-{}_{aq}$ concentration in the accepting pool increased to 17.5 μmol/L compared to 7.3 μmol/L in **Figure 4.204**. The penetration efficiency of $NO_2^-{}_{aq}$ and $NO_3^-{}_{aq}$ were the same for cases without and with the addition of NaOH. **Figure 4.207** shows the ratio of $[NO_x^-$ in the accepting

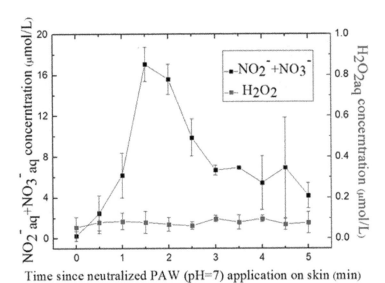

FIGURE 4.206 H_2O_2, NO_2^-, and NO_3^- concentration in the deionized water in the accepting pool as a function of treatment time (The PAW is neutralized by NaOH). (From Liu, X. et al., *J. Phys. D Appl. Phys.*, 51, 075401, 2018.)

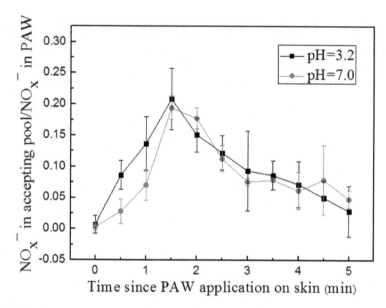

FIGURE 4.207 The ratio of $NO_2^-{}_{aq}$ and $NO_3^-{}_{aq}$ in the accepting pool/$NO_2^-{}_{aq}$ and $NO_3^-{}_{aq}$ in the PAW for the PAW drop on the skin with pH of 3.2 and 7.0. (From Liu, X. et al., *J. Phys. D Appl. Phys.*, 51, 075401, 2018.)

pool/[NO_x^- in PAW] are the same for the PAW dropped on the skin with a pH of 3.2, and 7.0. H_2O_{2aq} in the accepting pool was still below the lower limit of the detector. The comparison between heavily acidic PAW and neutralized PAW suggests that the pH of PAW did not affect the transdermal penetration of aqueous plasma active species.

References for Part II

1. N. Mericam-Bourdet, M. Laroussi, A. Begum and E. Karakas, Experimental investigations of plasma bullets, *J. Phys. Appl. Phys.* 42 (2009) 055207.
2. M. Laroussi and X. Lu, Room-temperature atmospheric pressure plasma plume for biomedical applications, *Appl. Phys. Lett.* 87 (2005) 113902.
3. J.L. Walsh, F. Iza, N.B. Janson, V.J. Law and M.G. Kong, Three distinct modes in a cold atmospheric pressure plasma jet, *J. Phys. Appl. Phys.* 43 (2010) 075201.
4. Y. Sakiyama, D.B. Graves, J. Jarrige and M. Laroussi, Finite element analysis of ring-shaped emission profile in plasma bullet, *Appl. Phys. Lett.* 96 (2010) 041501.
5. G.V. Naidis, Modelling of streamer propagation in atmospheric-pressure helium plasma jets, *J. Phys. Appl. Phys.* 43 (2010) 402001.
6. J.-P. Boeuf, L.L. Yang and L.C. Pitchford, Dynamics of a guided streamer ('plasma bullet') in a helium jet in air at atmospheric pressure, *J. Phys. Appl. Phys.* 46 (2013), 015201.
7. G.V. Naidis, Modelling of plasma bullet propagation along a helium jet in ambient air, *J. Phys. Appl. Phys.* 44 (2011) 215203.
8. D. Breden, K. Miki and L.L. Raja, Self-consistent two-dimensional modeling of cold atmospheric-pressure plasma jets/bullets, *Plasma Sources Sci. Technol.* 21 (2012) 034011.
9. N.Y. Babaeva, W. Tian and M.J. Kushner, The interaction between plasma filaments in dielectric barrier discharges and liquid covered wounds: Electric fields delivered to model platelets and cells, *J. Phys. Appl. Phys.* 47 (2014) 235201.
10. S.A. Norberg, W. Tian, E. Johnsen and M.J. Kushner, Atmospheric pressure plasma jets interacting with liquid covered tissue: Touching and not-touching the liquid, *J. Phys. Appl. Phys.* 47 (2014), 475203.
11. H. Cheng, X. Liu, X. Lu and D. Liu, Active species delivered by dielectric barrier discharge filaments to bacteria biofilms on the surface of apple, *Phys. Plasmas* 23 (2016) 073517.
12. T. Unfer and J.P. Boeuf, Modelling of a nanosecond surface discharge actuator, *J. Phys. Appl. Phys.* 42 (2009) 194017.
13. X.Y. Liu, X.K. Pei, X.P. Lu and D.W. Liu, Numerical and experimental study on a pulsed-dc plasma jet, *Plasma Sources Sci. Technol.* 23 (2014), 035007.
14. D.X. Liu, P. Bruggeman, F. Iza, M.Z. Rong and M.G. Kong, Global model of low-temperature atmospheric-pressure He + H_2O plasmas, *Plasma Sources Sci. Technol.* 19 (2010), 025018.
15. G.J.M. Hagelaar and L.C. Pitchford, Solving the Boltzmann equation to obtain electron transport coefficients and rate coefficients for fluid models, *Plasma Sources Sci. Technol.* 14 (2005) 722–733.
16. T. Murakami, K. Niemi, T. Gans, D. O'Connell and W.G. Graham, Chemical kinetics and reactive species in atmospheric pressure helium–oxygen plasmas with humid-air impurities, *Plasma Sources Sci. Technol.* 22 (2013) 015003.
17. A. Bourdon, V.P. Pasko, N.Y. Liu, S. Célestin, P. Ségur and E. Marode, Efficient models for photoionization produced by non-thermal gas discharges in air based on radiative transfer and the Helmholtz equations, *Plasma Sources Sci. Technol.* 16 (2007) 656.

18. M.B. Zhelezniak, A.K. Mnatsakanian and S.V. Sizykh, Photoionization of nitrogen and oxygen mixtures by radiation from a gas discharge, *Teplofiz. Vysok. Temp.* 20 (1928) 423–428.

19. Q. Xiong, X. Lu, J. Liu, Y. Xian, Z. Xiong, F. Zou, C. Zou et al., Temporal and spatial resolved optical emission behaviors of a cold atmospheric pressure plasma jet, *J. Appl. Phys.* 106 (2009) 083302.

20. E. Karakas and M. Laroussi, Experimental studies on the plasma bullet propagation and its inhibition, *J. Appl. Phys.* 108 (2010) 063305.

21. X.P. Lu, Z.H. Jiang, Q. Xiong, Z.Y. Tang, Z.L. Xiong, J. Hu, X.W. Hu and Y. Pan, Effect of E-field on the length of a plasma jet, *IEEE Trans. Plasma Sci.* 36 (2008) 988–989.

22. S. Spiekermeier, D. Schoerder, V. Schulz-von der Gathen, M. Boeke and J. Winter, Helium metastable density evolution in a self-pulsing mu-APPJ, *J. Phys. Appl. Phys.* 48 (2015) 035203.

23. A. Shashurin, M.N. Shneider and M. Keidar, Measurements of streamer head potential and conductivity of streamer column in cold nonequilibrium atmospheric plasmas, *Plasma Sources Sci. Technol.* 21 (2012) 034006.

24. A. Begum, M. Laroussi and M.R. Pervez, Atmospheric pressure He-air plasma jet: Breakdown process and propagation phenomenon, *AIP Adv.* 3 (2013) 062117.

25. H. Cheng, X. Lu and D. Liu, The effect of tube diameter on an atmospheric-pressure micro-plasma jet, *Plasma Process. Polym.* 12 (2015) 1343–1347.

26. M.A. Lieberman and A. Lichtenberg, *Principles of Plasma Discharges and Materials Processing*, 2nd ed. John Wiley & Sons, New York.

27. D. Breden and L.L. Raja, Computational study of the interaction of cold atmospheric helium plasma jets with surfaces, *Plasma Sources Sci. Technol.* 23 (2014) 065020.

28. Z. Xiong and M.J. Kushner, Atmospheric pressure ionization waves propagating through a flexible high aspect ratio capillary channel and impinging upon a target, *Plasma Sources Sci. Technol.* 21 (2012) 034001.

29. G.V. Naidis, Simulation of streamers propagating along helium jets in ambient air: Polarity-induced effects, *Appl. Phys. Lett.* 98 (2011) 141501.

30. S. Wu, X. Lu, D. Liu, Y. Yang, Y. Pan and K. Ostrikov, Photoionization and residual electron effects in guided streamers, *Phys. Plasmas* 21 (2014) 103508.

31. X. Lu, Z. Jiang, Q. Xiong, Z. Tang and Y. Pan, A single electrode room-temperature plasma jet device for biomedical applications, *Appl. Phys. Lett.* 92 (2008) 151504.

32. S. Wu, X. Lu and Y. Pan, Effects of seed electrons on the plasma bullet propagation, *Curr. Appl. Phys.* 13 (2013) S1–S5.

33. L. Nie, Y. Xian, X. Lu and K. Ostrikov, Visible light effects in plasma plume ignition, *Phys. Plasmas* 24 (2017) 043502.

34. D-X. Liu, M-Z. Rong, X-H. Wang, F. Iza, M.G. Kong and P. Bruggeman, Main species and physicochemical processes in cold atmospheric-pressure He + O2 plasmas, *Plasma Process. Polym.* 7 (2010) 846–865.

35. M.M. Pejovic, G.S. Ristic and J.P. Karamarkovic, Electrical breakdown in low pressure gases, *J. Phys. Appl. Phys.* 35 (2008) R91.

36. Y. Gosho and A. Harada, Role of exo-electrons in the breakdown of a positive point-to-plane air gap, *J. Phys. Appl. Phys.* 16 (1983) 1047.

37. S. Reuter, J. Winter, S. Iseni, S. Peters, A. Schmidt-Bleker, M. Dünnbier, J. Schäfer, R. Foest and K-D. Weltmann, Detection of ozone in a MHz argon plasma bullet jet, *Plasma Sources Sci. Technol.* 21 (2012) 034015.

38. J. Jarrige, M. Laroussi and E. Karakas, Formation and dynamics of plasma bullets in a non-thermal plasma jet: Influence of the high-voltage parameters on the plume characteristics, *Plasma Sources Sci. Technol.* 19 (2010) 065005.

39. K. Urabe, T. Morita, K. Tachibana and B.N. Ganguly, Investigation of discharge mechanisms in helium plasma jet at atmospheric pressure by laser spectroscopic measurements, *J. Phys. Appl. Phys.* 43 (2010) 095201.

40. S. Wu, Q. Huang, Z. Wang and X. Lu, The effect of nitrogen diffusion from surrounding air on plasma bullet behavior, *IEEE Trans. Plasma Sci.* 39 (2011) 2286–2287.

41. Y. Xian, Y. Yue, D. Liu, Y. Yang, X. Lu and Y. Pan, On the mechanism of ring-shape structure of plasma bullet, *Plasma Process. Polym.* 11 (2014) 1169–1174.

42. Y. Yue, F. Wu, H. Cheng, Y. Xian, D. Liu, X. Lu and X. Pei, A donut-shape distribution of OH radicals in atmospheric pressure plasma jets, *J. Appl. Phys.* 121 (2017) 033302.

43. S. Yonemori and R. Ono, Flux of OH and O radicals onto a surface by an atmospheric-pressure helium plasma jet measured by laser-induced fluorescence, *J. Phys. Appl. Phys.* 47 (2014) 125401.

44. C. Zou and X. Pei, OH radicals distribution in a nanosecond pulsed atmospheric pressure plasma jet, *IEEE Trans. Plasma Sci.* 42 (2014) 2484–2485.

45. Z. Xiong, X. Lu, Q. Xiong, Y. Xian, C. Zou, J. Hu, W. Gong et al. Measurements of the propagation velocity of an atmospheric-pressure plasma plume by various methods, *IEEE Trans. Plasma Sci.* 38 (2010) 1001–1007.

46. X. Pei, Y. Lu, S. Wu, Q. Xiong and X. Lu, A study on the temporally and spatially resolved OH radical distribution of a room-temperature atmospheric-pressure plasma jet by laser-induced fluorescence imaging, *Plasma Sources Sci. Technol.* 22 (2013) 025023.

47. K. Ostrikov (Ken), F. Beg and A. Ng, Colloquium, *Rev. Mod. Phys.* 88 (2016) 011001.

48. Y. Xian, X. Lu, Y. Cao, P. Yang, Q. Xiong, Z. Jiang and Y. Pan, On plasma bullet behavior, *IEEE Trans. Plasma Sci.* 37 (2009) 2068–2073.

49. L. Nie, L. Chang, Y. Xian and X. Lu, The effect of seed electrons on the repeatability of atmospheric pressure plasma plume propagation: I. Experiment, *Phys. Plasmas* 23 (2016) 093518.

50. Y.B. Xian, H.T. Xu, X.P. Lu, X.K. Pei, W.W. Gong, Y. Lu, D.W. Liu and Y. Yang, Plasma bullets behavior in a tube covered by a conductor, *Phys. Plasmas* 22 (2015) 063507.

51. E. Robert, V. Sarron, D. Ries, S. Dozias, M. Vandamme and J-M. Pouvesle, Characterization of pulsed atmospheric-pressure plasma streams (PAPS) generated by a plasma gun, *Plasma Sources Sci. Technol.* 21 (2012) 034017.

52. B.L. Sands, S.K. Huang, J.W. Speltz, M.A. Niekamp, J.B. Schmidt and B.N. Ganguly, Dynamic electric potential redistribution and its influence on the development of a dielectric barrier plasma jet, *Plasma Sources Sci. Technol.* 21 (2012) 034009.

53. Q. Xiong, X. Lu, K. Ostrikov, Z. Xiong, Y. Xian, F. Zhou, C. Zou, J. Hu, W. Gong and Z. Jiang, Length control of He atmospheric plasma jet plumes: Effects of discharge parameters and ambient air, *Phys. Plasmas* 16 (2009) 043505.

54. J.L. Walsh, P. Olszewski and J.W. Bradley, The manipulation of atmospheric pressure dielectric barrier plasma jets, *Plasma Sources Sci. Technol.* 21 (2012) 034007.

55. Q. Xiong, X. Lu, Y. Xian, J. Liu, C. Zou, Z. Xiong, W. Gong et al., Experimental investigations on the propagation of the plasma jet in the open air, *J. Appl. Phys.* 107 (2010) 073302.

56. J. He, J. Hu, D. Liu and Y-T. Zhang, Experimental and numerical study on the optimization of pulse-modulated radio-frequency discharges, *Plasma Sources Sci. Technol.* 22 (2013) 035008.

57. S. Wu, H. Xu, X. Lu and Y. Pan, Effect of pulse rising time of pulse dc voltage on atmospheric pressure non-equilibrium plasma, *Plasma Process. Polym.* 10 (2013) 136–140.

58. R.J. Leiweke and B.N. Ganguly, Effects of pulsed-excitation applied voltage rise time on argon metastable production efficiency in a high pressure dielectric barrier discharge, *Appl. Phys. Lett.* 90 (2007) 241501.

59. Y. Xian, P. Zhang, X. Lu, X. Pei, S. Wu, Q. Xiong and K. Ostrikov, From short pulses to short breaks: Exotic plasma bullets via residual electron control, *Sci. Rep.* 3 (2013) 1599.

60. C. Jiang, M.T. Chen and M.A. Gundersen, Polarity-induced asymmetric effects of nanosecond pulsed plasma jets, *J. Phys. Appl. Phys.* 42 (2009) 232002.

61. Z. Xiong, X. Lu, Y. Xian, Z. Jiang and Y. Pan, On the velocity variation in atmospheric pressure plasma plumes driven by positive and negative pulses, *J. Appl. Phys.* 108 (2010) 103303.

62. J-S. Oh, J.L. Walsh and J.W. Bradley, Plasma bullet current measurements in a free-stream helium capillary jet, *Plasma Sources Sci. Technol.* 21 (2012) 034020.

63. G.B. Sretenovic, I.B. Krstic, V.V. Kovacevic, B.M. Obradovic and M.M. Kuraica, Spectroscopic measurement of electric field in atmospheric-pressure plasma jet operating in bullet mode, *Appl. Phys. Lett.* 99 (2011) 161502.

64. X. Lu, M. Laroussi and V. Puech, On atmospheric-pressure non-equilibrium plasma jets and plasma bullets, *Plasma Sources Sci. Technol.* 21 (2012) 034005.

65. J. Gou, Y. Xian and X. Lu, Low-temperature, high-density plasmas in long micro-tubes, *Phys. Plasmas* 23 (2016) 053508.

66. X. Lu, S. Wu, J. Gou and Y. Pan, An atmospheric-pressure, high-aspect-ratio, cold micro-plasma, *Sci. Rep.* 4 (2014) 7488.

67. S. Wu, H. Xu, Y. Xian, Y. Lu and X. Lu, Propagation of plasma bullet in U-shape tubes, *AIP Adv.* 5 (2015) 027110.

68. J. Jansky and A. Bourdon, Simulation of helium discharge ignition and dynamics in thin tubes at atmospheric pressure, *Appl. Phys. Lett.* 99 (2011) 161504.

69. J. Jansky, F. Tholin, Z. Bonaventura and A. Bourdon, Simulation of the discharge propagation in a capillary tube in air at atmospheric pressure, *J. Phys. Appl. Phys.* 43 (2010) 395201.

70. R. Xiong, Q. Xiong, A.Y. Nikiforov, P. Vanraes and C. Leys, Influence of helium mole fraction distribution on the properties of cold atmospheric pressure helium plasma jets, *J. Appl. Phys.* 112 (2012) 033305.

71. E. Karakas, M. Koklu and M. Laroussi, Correlation between helium mole fraction and plasma bullet propagation in low temperature plasma jets, *J. Phys. Appl. Phys.* 43 (2010) 155202.

72. W-C. Zhu, Q. Li, X-M. Zhu and Y-K. Pu, Characteristics of atmospheric pressure plasma jets emerging into ambient air and helium, *J. Phys. Appl. Phys.* 42 (2009) 202002.

73. S. Wu, Z. Wang, Q. Huang, X. Tan, X. Lu and K. Ostrikov, Atmospheric-pressure plasma jets: Effect of gas flow, active species, and snake-like bullet propagation, *Phys. Plasmas* 20 (2013) 023503.

74. X.Y. Liu, X.K. Pei, K. Ostrikov, X.P. Lu and D.W. Liu, The production mechanisms of OH radicals in a pulsed direct current plasma jet, *Phys. Plasmas* 21 (2014) 093513.

75. R.J. Leiweke, B.L. Sands and B.N. Ganguly, Effect of gas mixture on plasma jet discharge morphology, *IEEE Trans. Plasma Sci.* 39 (2011) 2304–2305.

76. Y. Xian, X. Lu, J. Liu, S. Wu, D. Liu and Y. Pan, Multiple plasma bullet behavior of an atmospheric-pressure plasma plume driven by a pulsed dc voltage, *Plasma Sources Sci. Technol.* 21 (2012) 034013.

77. M.A. Akman and M. Laroussi, Insights into sustaining a plasma jet: Boundary layer requirement, *IEEE Trans. Plasma Sci.* 41 (2013) 839–842.

78. S. Hofmann, A. Sobota and P. Bruggeman, Transitions between and control of guided and branching streamers in DC nanosecond pulsed excited plasma jets, *IEEE Trans. Plasma Sci.* 40 (2012) 2888–2899.

79. Y.B. Xian, D.D. Zou, X.P. Lu, Y. Pan and K. Ostrikov, Feather-like He plasma plumes in surrounding N-2 gas, *Appl. Phys. Lett.* 103 (2013) 094103.

80. M. Laroussi and M.A. Akman, Ignition of a large volume plasma with a plasma jet, *AIP Adv.* 1 (2011) 032138.

81. Y. Xian, S. Wu, Z. Wang, Q. Huang, X. Lu and J.F. Kolb, Discharge dynamics and modes of an atmospheric pressure non-equilibrium air plasma jet, *Plasma Process. Polym.* 10 (2013) 372–378.

82. Y. Xian, X. Lu, S. Wu, P.K. Chu and Y. Pan, Are all atmospheric pressure cold plasma jets electrically driven? *Appl. Phys. Lett.* 100 (2012) 123702.

83. J.H. Liu, X.Y. Liu, K. Hu, D.W. Liu, X.P. Lu, F. Iza and M.G. Kong, Plasma plume propagation characteristics of pulsed radio frequency plasma jet, *Appl. Phys. Lett.* 98 (2011) 151502.

84. Q. Xiong, A.Y. Nikiforov, X.P. Lu and C. Leys, A branching streamer propagation argon plasma plume, *IEEE Trans. Plasma Sci.* 39 (2011) 2094–2095.

85. C. Douat, G. Bauville, M. Fleury, M. Laroussi and V. Puech, Dynamics of colliding microplasma jets, *Plasma Sources Sci. Technol.* 21 (2012) 034010.

86. G.V. Naidis, Simulation of interaction between two counter-propagating streamers, *Plasma Sources Sci. Technol.* 21 (2012) 034003.

87. S. Wu and X. Lu, Two counter-propagating He plasma plumes and ignition of a third plasma plume without external applied voltage, *Phys. Plasmas* 21 (2014) 023501.

88. X. Lu, Q. Xiong, Z. Xiong, J. Hu, F. Zhou, W. Gong, Y. Xian et al., Propagation of an atmospheric pressure plasma plume, *J. Appl. Phys.* 105 (2009) 043304.

89. Z. Xiong, E. Robert, V. Sarron, J-M. Pouvesle and M.J. Kushner, Atmospheric-pressure plasma transfer across dielectric channels and tubes, *J. Phys. Appl. Phys.* 46 (2013) 155203.

90. G.V. Naidis and J.L. Walsh, The effects of an external electric field on the dynamics of cold plasma jets-experimental and computational studies, *J. Phys. Appl. Phys.* 46 (2013) 095203.

91. J.T. Hu, X.Y. Liu, J.H. Liu, Z.L. Xiong, D.W. Liu, X.P. Lu, F. Iza and M. Kong, The effect of applied electric field on pulsed radio frequency and pulsed direct current plasma jet array, *Phys. Plasmas* 19 (2012) 063505.

92. X. Yan, Z. Xiong, F. Zou, S. Zhao, X. Lu, G. Yang, G. He and K. Ostrikov, Plasma-induced death of HepG2 cancer cells: Intracellular effects of reactive species, *Plasma Process. Polym.* 9 (2012) 59–66.

93. M. Ishaq, M. Evans and K. Ostrikov, Effect of atmospheric gas plasmas on cancer cell signaling, *Int. J. Cancer* 134 (2014) 1517–1528.

94. Z. Xiong, S. Zhao, X. Mao, X. Lu, G. He, G. Yang, M. Chen, M. Ishaq and K. Ostrikov, Selective neuronal differentiation of neural stem cells induced by nanosecond microplasma agitation, *Stem Cell Res.* 12 (2014) 387–399.

95. Z. Chen, X. Cheng, L. Lin and M. Keidar, Cold atmospheric plasma discharged in water and its potential use in cancer therapy, *J. Phys. Appl. Phys.* 50 (2017) 015208.

96. W. Yan, Z.J. Han, W.Z. Liu, X.P. Lu, B.T. Phung and K. Ostrikov, Designing atmospheric-pressure plasma sources for surface engineering of nanomaterials, *Plasma Chem. Plasma Process.* 33 (2013) 479–490.

97. X. Lu, S. Wu, P.K. Chu, D. Liu and Y. Pan, An atmospheric-pressure plasma brush driven by sub-microsecond voltage pulses, *Plasma Sources Sci. Technol.* 20 (2011) 065009.

98. Y. Yue, X. Pei and X. Lu, Comparison on the absolute concentrations of hydroxyl and atomic oxygen generated by five different nonequilibrium atmospheric-pressure plasma jets, *IEEE Trans. Radiat. Plasma Med. Sci.* 1 (2017) 541–549.

99. X. Lu, Z. Jiang, Q. Xiong, Z. Tang, X. Hu and Y. Pan, An 11cm long atmospheric pressure cold plasma plume for applications of plasma medicine, *Appl. Phys. Lett.* 92 (2008) 081502.

100. J.L. Walsh and M.G.Kong, Contrasting characteristics of linear-field and cross-field atmospheric plasma jets, *Appl. Phys. Lett.* 93 (2008) 111501.

101. M. Teschke, J. Kedzierski, E.G. Finantu-Dinu, D. Korzec and J. Engemann, High-speed photographs of a dielectric barrier atmospheric pressure plasma jet, *IEEE Trans. Plasma Sci.* 33 (2005) 310–311.

102. X. Pei, S. Wu, Y. Xian, X. Lu and Y. Pan, On OH density of an atmospheric pressure plasma jet by laser-induced fluorescence, *IEEE Trans. Plasma Sci.* 42 (2014) 1206–1210.

103. P. Bruggeman and D.C. Schram, On OH production in water containing atmospheric pressure plasmas, *Plasma Sources Sci. Technol.* 19 (2010) 045025.

104. S. Yonemori, Y. Nakagawa, R. Ono and T. Oda, Measurement of OH density and air–helium mixture ratio in an atmospheric-pressure helium plasma jet, *J. Phys. Appl. Phys.* 45 (2012) 225202.

105. J. Voráč, A. Obrusník, V. Procházka, P. Dvořák and M. Talába, Spatially resolved measurement of hydroxyl radical (OH) concentration in an argon RF plasma jet by planar laser-induced fluorescence, *Plasma Sources Sci. Technol.* 23 (2014) 025011.

106. T. Murakami, K. Niemi, T. Gans, D. O'Connell and W.G. Graham, Interacting kinetics of neutral and ionic species in an atmospheric-pressure helium–oxygen plasma with humid air impurities, *Plasma Sources Sci. Technol.* 22 (2013) 045010.

107. Y. Yue, Y. Xian, X. Pei and X. Lu, The effect of three different methods of adding O_2 additive on O concentration of atmospheric pressure plasma jets (APPJs), *Phys. Plasmas* 23 (2016) 123503.

108. D. Riès, G. Dilecce, E. Robert, P.F. Ambrico, S. Dozias and J-M. Pouvesle, LIF and fast imaging plasma jet characterization relevant for NTP biomedical applications, *J. Phys. Appl. Phys.* 47 (2014) 275401.

109. X. Pei, X. Lu, J. Liu, D. Liu, Y. Yang, K. Ostrikov, P.K. Chu and Y. Pan, Inactivation of a 25.5 μm Enterococcus faecalis biofilm by a room-temperature, battery-operated, handheld air plasma jet, *J. Phys. Appl. Phys.* 45 (2012) 165205.

110. G. Fridman, M. Peddinghaus, M. Balasubramanian, H. Ayan, A. Fridman, A. Gutsol and A. Brooks, Blood coagulation and living tissue sterilization by floating-electrode dielectric barrier discharge in air, *Plasma Chem. Plasma Process.* 26 (2006) 425–442.

111. H. Hashizume, T. Ohta, J. Fengdong, K. Takeda, K. Ishikawa, M. Hori and M. Ito, Inactivation effects of neutral reactive-oxygen species on *Penicillium digitatum* spores using non-equilibrium atmospheric-pressure oxygen radical source, *Appl. Phys. Lett.* 103 (2013) 153708.

112. S.K. Kang, H.Y. Kim, G.S. Yun and J.K. Lee, Portable microwave air plasma device for wound healing, *Plasma Sources Sci. Technol.* 24 (2015) 035020.

113. X. Lu, G.V. Naidis, M. Laroussi, S. Reuter, D.B. Graves and K. Ostrikov, Reactive species in non-equilibrium atmospheric-pressure plasmas: Generation, transport, and biological effects, *Phys. Rep.* 630 (2016) 1–84.

114. M.J. Traylor, M.J. Pavlovich, S. Karim, P. Hait, Y. Sakiyama, D.S. Clark and D.B. Graves, Long-term antibacterial efficacy of air plasma-activated water, *J. Phys. Appl. Phys.* 44 (2011) 472001.

115. D. Dobrynin, G. Friedman, A. Fridman and A. Starikovskiy, Inactivation of bacteria using dc corona discharge: Role of ions and humidity, *New J. Phys.* 13 (2011) 103033.

116. J.L. Brisset and J. Pawlat, Chemical effects of air plasma species on aqueous solutes in direct and delayed exposure modes: Discharge, post-discharge and plasma activated water, *Plasma Chem. Plasma Process.* 36 (2016) 1–27.

117. S.G. Joshi, M. Cooper, A. Yost, M. Paff, U.K. Ercan, G. Fridman, G. Friedman, A. Fridman and A.D. Brooks, Nonthermal dielectric-barrier discharge plasma-induced inactivation involves oxidative DNA damage and membrane lipid peroxidation in *Escherichia coli Antimicrob, Agents Chemother.* 55 (2011) 1053–1062.

118. P. Lukes, J.-L. Brisset and B.R. Locke, Biological effects of electrical discharge plasma in water and in gas–liquid environments, in: *Plasma Chemistry and Catalysis in Gases and Liquids*, V.I. Parvulescu, M. Monicagureanu and P. Lukes (Eds), Wiley-VCH Verlag GmbH & Co. KGaA, Somerset, NJ, 2012, pp. 309–352.

119. P. Lukes, B.R. Locke and J.-L. Brisset, Aqueous-phase chemistry of electrical discharge plasma in water and in gas–liquid environments, in: *Plasma Chemistry and Catalysis in Gases and Liquids*, V.I. Parvulescu, M. Monicagureanu and P. Lukes (Eds.), Wiley-VCH Verlag GmbH & Co. KGaA, Somerset, NJ, 2012, pp. 243–308.

120. G. Kamgang-Youbi, J-M. Herry, J-L. Brisset, M-N. Bellon-Fontaine, A. Doubla and M. Naïtali, Impact on disinfection efficiency of cell load and of planktonic/adherent/detached state: Case of Hafnia alvei inactivation by plasma activated water, *Appl. Microbiol. Biotechnol.* 81 (2008) 449–457.

121. M. Naïtali, G. Kamgang-Youbi, J-M. Herry, M-N. Bellon-Fontaine and J-L. Brisset, Combined effects of long-living chemical species during microbial inactivation using atmospheric plasma-treated water, *Appl. Environ. Microbiol.* 76 (2010) 7662–7664.

122. G. Kamgang-Youbi, J-M. Herry, T. Meylheuc, J-L. Brisset, M-N. Bellon-Fontaine, A. Doubla and M. Naïtali, Microbial inactivation using plasma-activated water obtained by gliding electric discharges, *Lett. Appl. Microbiol.* 48 (2009) 13–18.

123. K. Oehmigen, J. Winter, M. Hähnel, C. Wilke, R. Brandenburg, K-D. Weltmann and T. von Woedtke, Estimation of possible mechanisms of *Escherichia coli* inactivation by plasma treated sodium chloride solution, *Plasma Process. Polym.* 8 (2011) 904–913.

124. K. Hensel, E. Spetlikova, L. Sikurova and P. Lukes, Formation of ROS and RNS in water electro-sprayed through transient spark discharge in air and their bactericidal effects, *Plasma Process. Polym.* 10 (2013) 649–659.

125. Q. Xiong, H. Liu, W. Lu, Q. Chen, L. Xu, X. Wang, Q. Zhu, X. Zeng and P. Yi, Inactivation of Candida glabrata by a humid DC argon discharge afterglow: Dominant contributions of short-lived aqueous active species, *J. Phys. Appl. Phys.* 50 (2017) 205203.

126. S. Ikawa, K. Kitano and S. Hamaguchi, Effects of pH on bacterial inactivation in aqueous solutions due to low-temperature atmospheric pressure plasma application, *Plasma Process. Polym.* 7 (2010) 33–42.

127. P. Lukes, E. Dolezalova, I. Sisrova and M. Clupek, Aqueous-phase chemistry and bactericidal effects from an air discharge plasma in contact with water: Evidence for the formation of peroxynitrite through a pseudo-second-order post-discharge reaction of H_2O_2 and HNO_2, *Plasma Sources Sci. Technol.* 23 (2014) 015019.

128. K. Oehmigen, M. Hähnel, R. Brandenburg, C. Wilke, K.D. Weltmann and T.V. Woedtke, The role of acidification for antimicrobial activity of atmospheric pressure plasma in liquids, *Plasma Process. Polym.* 7(2010) 250–257.

129. J.Y. Park and Y.N. Lee, Solubility and decomposition kinetics of nitrous acid in aqueous solution, *J. Phys. Chem.* 92 (1988) 6294–6302.

130. K.A.H. Buchan, D.J. Martin-Robichaud and T.J. Benfey, Measurement of dissolved ozone in sea water: A comparison of methods, *Aquac. Eng.* 33 (2005) 225–231.

131. W. Tian and M.J. Kushner, Long-term effects of multiply pulsed dielectric barrier discharges in air on thin water layers over tissue: Stationary and random streamers, *J. Phys. Appl. Phys.* 48 (2015) 494002.

132. L. Zhu, C. Gunn and J.S. Beckman, Bactericidal activity of peroxynitrite, *Arch. Biochem. Biophys.* 298 (1992) 452–457.

133. J.S. Beckman and W.H. Koppenol, Nitric oxide, superoxide, and peroxynitrite: The good, the bad, and ugly, *Am. J. Physiol. Cell Physiol.* 271 (1996) C1424–C1437.

134. J.S. Beckman, T.W. Beckman, J. Chen, P.A. Marshall and B.A. Freeman, Apparent hydroxyl radical production by peroxynitrite: Implications for endothelial injury from nitric oxide and superoxide, *Proc. Natl. Acad. Sci.* 87 (1990) 1620–1624.

135. G.L. Squadrito and W.A. Pryor, Oxidative chemistry of nitric oxide: The roles of superoxide, peroxynitrite, and carbon dioxide, *Free Radic. Biol. Med.* 25 (1998) 392.

136. S. Goldstein, J. Lind and G. Merényi, Chemistry of peroxynitrites as compared to peroxynitrates, *Chem. Rev.* 105 (2005) 2457–2470.

137. G. Lloyd, G.Friedman, S. Jafri, G. Schultz, A. Fridman and K. Harding, Gas plasma: Medical uses and developments in wound care, *Plasma Process. Polym.* 7 (2010) 194–211.

138. E.J. Szili, J.W. Bradley and R.D. Short, A 'tissue model' to study the plasma delivery of reactive oxygen species, *J. Phys. Appl. Phys.* 47 (2014) 152002.

139. T. He, D. Liu, H. Xu, Z. Liu, D. Xu, D. Li, Q. Li, M. Rong and M.G. Kong, A 'tissue model' to study the barrier effects of living tissues on the reactive species generated by surface air discharge, *J. Phys. Appl. Phys.* 49 (2016) 205204.

140. A.V. Nastuta, I. Topala, C. Grigoras, V. Pohoata and G. Popa, Stimulation of wound healing by helium atmospheric pressure plasma treatment, *J. Phys. Appl. Phys.* 44 (2011) 105204.

141. G. Isbary, W. Stolz, T. Shimizu, R. Monetti, W. Bunk, H-U. Schmidt, G.E. Morfill et al., Cold atmospheric argon plasma treatment may accelerate wound healing in chronic wounds: Results of an open retrospective randomized controlled study in vivo, *Clin. Plasma Med.* 1 (2013) 25–30.

142. S. Emmert, F. Brehmer, H. Hänßle, A. Helmke, N. Mertens, R. Ahmed, D. Simon et al., Atmospheric pressure plasma in dermatology: Ulcus treatment and much more, *Clin. Plasma Med.* 1 (2013) 24–29.

143. J. Heinlin, G. Isbary, W. Stolz, G. Morfill, M. Landthaler, T. Shimizu, B. Steffes, T. Nosenko, J. Zimmermann and S. Karrer, Plasma applications in medicine with a special focus on dermatology, *J. Eur. Acad. Dermatol. Venereol.* 25 (2011) 1–11.

144. J. Lademann, C. Ulrich, A. Patzelt, H. Richter, F. Kluschke, M. Klebes, O. Lademann, A. Kramer, K.D. Weltmann and B. Lange-Asschenfeldt, Risk assessment of the application of tissue-tolerable plasma on human skin, *Clin. Plasma Med.* 1 (2013) 5–10.

145. J.W. Fluhr, S. Sassning, O. Lademann, M.E. Darvin, S. Schanzer, A. Kramer, H. Richter, W. Sterry and J. Lademann, In vivo skin treatment with tissue-tolerable plasma influences skin physiology and antioxidant profile in human stratum corneum, *Exp. Dermatol.* 21 (2012) 130–134.

146. L.I. Partecke, K. Evert, J. Haugk, F. Doering, L. Normann, S. Diedrich, F-U. Weiss et al., Tissue tolerable plasma (TTP) induces apoptosis in pancreatic cancer cells in vitro and in vivo, *BMC Cancer* 12 (2012) 473.

147. H. Tanaka, M. Mizuno, K. Ishikawa, K. Nakamura, H. Kajiyama, H. Kano, F. Kikkawa and M. Hori, Plasma-activated medium selectively kills glioblastoma brain tumor cells by down-regulating a survival signaling molecule, AKT kinase, *Plasma Med.* 1 (2011) 265–277.

148. D. Yan, A. Talbot, N. Nourmohammadi, X. Cheng, J. Canady, J. Sherman and M. Keidar, Principles of using cold atmospheric plasma stimulated media for cancer treatment, *Sci. Rep.* 5 (2015) 18339.

149. F. Utsumi, H. Kajiyama, K. Nakamura, H. Tanaka, M. Mizuno, K. Ishikawa, H. Kondo, H. Kano, M. Hori and F. Kikkawa, Effect of indirect nonequilibrium atmospheric pressure plasma on anti-proliferative activity against chronic chemo-resistant ovarian cancer cells in vitro and in vivo, *PLoS One* 8 (2013) e81576.

150. D. Yan, H. Cui, W. Zhu, N. Nourmohammadi, J. Milberg, L.G. Zhang, J.H. Sherman and M. Keidar, The specific vulnerabilities of cancer cells to the cold atmospheric plasma-stimulated solutions, *Sci. Rep.* 7 (2017) 4479.

151. F.S. Hodi, S.J. O'Day, D.F. McDermott, R.W. Weber, J.A. Sosman, J.B. Haanen, R. Gonzalez et al., Improved survival with Ipilimumab in Patients with Metastatic Melanoma, *N. Engl. J. Med.* 363 (2010) 711–723.

152. J.A. Ribes, C.L. Vanover-Sams and D.J. Baker, Zygomycetes in human disease, *Clin. Microbiol. Rev.* 13 (2000) 236–301.

153. X. Pei, J. Liu, Y. Xian and X. Lu, A battery-operated atmospheric-pressure plasma wand for biomedical applications, *J. Phys. Appl. Phys.* 47 (2014) 145204.

154. M.R. Prausnitz, P.M. Elias, T.J. Franz, M. Schmuth and J.C. Tsai, Skin barrier and transdermal drug delivery, *Med. Ther.* 19 (2008) 2065–2073.

155. S. Corovic, I. Lackovic, P. Sustaric, T. Sustar, T. Rodic and D. Miklavcic, Modeling of electric field distribution in tissues during electroporation, *Biomed. Eng. OnLine* 12 (2013) 16.

156. A-R. Denet, R. Vanbever and V. Préat, Skin electroporation for transdermal and topical delivery, *Adv. Drug Deliv. Rev.* 56 (2004) 659–674.

157. O. Lademann, H. Richter, M.C. Meinke, A. Patzelt, A. Kramer, P. Hinz, K-D. Weltmann, B. Hartmann and S. Koch, Drug delivery through the skin barrier enhanced by treatment with tissue-tolerable plasma, *Exp. Dermatol.* 20 (2011) 488–490.

158. M.H. Abraham, H.S. Chadha and R.C. Mitchell, The factors that influence skin penetration of solutes, *J. Pharm. Pharmacol.* 47 (1995) 8–16.

159. M. Kelm, Nitric oxide metabolism and breakdown, *Biochim. Biophys. Acta BBA Bioenerg.* 1411 (1999) 273–289.

160. A.D. Ormerod, M.I. White, S.A.A. Shah and N. Benjamin, Molluscum contagiosum effectively treated with a topical acidified nitrite, nitric oxide liberating cream, *Br. J. Dermatol.* 141 (1999) 1051–1053.

References for Part II

161. R. Weller, A.D. Ormerod, R.P. Hobson and N.J. Benjamin, A randomized trial of acidified nitrite cream in the treatment of tinea pedis, *J. Am. Acad. Dermatol.* 38 (1998) 559–563.
162. G. Dorfman, M. Levitt and C. Platell, Treatment of chronic anal fissure with topical glyceryl trinitrate, *Dis. Colon Rectum* 42 (1999) 1007–1010.
163. D.J. Smith and M.L. Simmons, Transdermal delivery of nitric oxide from diazeniumdiolates, *J. Controlled Release* 51 (1998) 153–159.
164. V. Den Oord, Expression of the neuronal isoform of nitric oxide synthase (nNOS) and its inhibitor, protein inhibitor of nNOS, in pigment cell lesions of the skin, *Br. J. Dermatol.* 141 (1999) 12–19.
165. M. Kagoura, C. Matsui, M. Toyoda and M. Morohashi, Immunohistochemical study of inducible nitric oxide synthase in skin cancers, *J. Cutan. Pathol.* 28 (2001) 476–481.

PART III

Diagnostics of Atmospheric Pressure Plasmas

5

Introduction to Plasma Diagnostics

IN THE PAST DECADE, insight has been gained into many fundamental properties of cold plasmas at atmospheric conditions [1–14]. Established diagnostic techniques had to be adjusted to the special circumstances atmospheric pressure plasmas provide. Typically, atmospheric pressure plasma jets generate small-scale plasmas resulting in high gradients in space and in time. For an analysis of their reactivity, generation processes of reactive species, transport processes, energy dissipation processes, and surface interaction, including particle flow, need to be studied. Nonequilibrium processes governing these plasmas pose a challenge to many diagnostic techniques that are successful in thermal or low-pressure plasmas. The following chapter describes fundamental plasma properties of nonthermal plasmas and introduces suitable diagnostic techniques.

5.1 Fundamental Properties of Plasmas Relevant for Diagnostics

For a complete picture of a plasma, position and velocity would need to be known at all time points. Since this cannot be achieved, a description of plasmas is performed by distribution functions $f(\vec{r}, \vec{v}, t)$. f describes the probability of finding a particle at the position $\vec{r}$ with the velocity $\vec{v}$ at the time t. From the velocity, the kinetic energy can be calculated. A fundamental parameter governing most reaction kinetics in nonequilibrium plasmas is the electrons' distribution function as a function of energy, called the electron energy distribution function (EEDF). Distribution functions characterize the energy content of a plasma. In thermal equilibrium, all particles adopt a Maxwellian distribution of their energies. Kinetic energy of a particle $E_{kin} = 1/2\, mv^2$ allows to move from velocity to an energy distribution function.

The number of electrons in an energy interval dE is related to the number of electrons in a velocity range dv according to:

$$F(E) = 4\pi v^2 f_e(v) \frac{dv}{dE}. \tag{5.1}$$

The energy distribution function is then

$$F(E) = 4\pi \frac{2E}{m} \sqrt{\frac{1}{2mE}} f_e(v). \tag{5.2}$$

The Maxwell distribution is the distribution of maximal entropy. Normalized to 1, the probability of finding a particle within the velocity interval dv is given by:

$$f(v) = n\left(\frac{m}{2\pi k_B T}\right)^{3/2} e^{\frac{-1/2\, mv^2}{k_B T}} \tag{5.3}$$

from the mean energy of n particles in the velocity interval dv, $\langle E \rangle = \int 1/2\, mv^2 d^3v / \int f(v) d^3v = 3/2\, k_B T$, the definition of the electron temperature follows. For electrons of Maxwellian energy distribution, a mean energy of 1 eV corresponds to $T = 11{,}594$ K. The electron temperature is one of the most fundamental plasma properties needed to characterize a plasma. In nonthermal plasma jets, electrons derive their energy from the electric power introduced to the plasma region. While their average energy can reach several eV, heavy particles remain cold with temperatures close to room temperature. Ionization occurs mainly through electron impact. Collisions of molecules and atoms deplete electrons in the high-energy tail of the energy distribution function. Resultantly, nonequilibrium plasmas have EEDFs that do not follow a Maxwellian distribution, and while the term electron temperature is still used to describe the electron energy, it is in these cases not based on the thermodynamic concept of temperature.

Due to the light mass of electrons, the electric excitation in nonthermal plasmas leads to a high electron temperature. In atmospheric pressure plasmas, molecules from ambient gas or from a feed gas admixture are present. Here, additional energy dissipation pathways are present. Molecules have the additional degrees of freedom of electronic excitation leading to vibrational and rotational molecular excitation. Each excitation mechanism can lead to a respective temperature development. The hierarchy in temperature distribution is $T_e > T_v > T_i \approx T_{rot} \approx T_{gas}$. Vibrational-translational energy relaxation often is a slow process so that the vibrational temperature T_v can easily be higher than the ion temperature T_i. Translational-rotational energy relaxation is often a fast process so that T_{rot} is often in the same order as the ion temperature and the total neutral gas temperature T_{gas}, which lies at or near room temperature.

Vibrational excitation of molecules plays an important role in the generation of reactive species in nonthermal atmospheric pressure plasma jets.

If energy states have enough time to reach equilibrium conditions, the ratio of the population density N_i and N_j of two different energy states follows the Boltzmann distribution:

$$\frac{N_i}{N_j} = \frac{g_1}{g_2} e^{-\Delta E/kT} \tag{5.4}$$

where g_1 and g_2 are accounting for degenerate levels, ΔE is the energy difference of the two levels, k is the Boltzmann constant, and T is the temperature.

5.2 Fundamentals of Plasma Spectroscopy

A characteristic of plasmas is the emission and absorption of electromagnetic radiation. A great variety of information can be gained from a spectroscopic analysis of this radiation. Nonthermal plasmas at atmospheric pressure exhibit small dimensions and high gradients in time and space [16]. Typical invasive diagnostics such as probes are in most cases too intrusive to yield reliable results, or their respective physical principle or interaction model used for evaluation of the data cannot be applied, as is the case for Langmuir-probe measurements. Therefore, non-invasive optical diagnostic techniques have been the method of choice for a study of atmospheric pressure nonequilibrium plasmas [16–21]. The following section describes the physics background for spectroscopic analysis of plasma properties. These techniques range from optical emission spectroscopy, laser methods such as laser induced fluorescence techniques, scattering techniques such as Rayleigh-, Raman-, and Thomson scattering, and optical absorption spectroscopy measurements [11,17–19] to techniques known from chemical analysis such as mass spectrometry, gas chromatography, or electron paramagnetic resonance spectroscopy (see **Figure 5.1**).

5.2.1 Interaction of Light and Particles

An incident electromagnetic wave induces a dipole moment by forcing a charge to oscillate. This is the dominant pathway of energy exchange of photons and electrons. A bound electron can change its energy state by interaction with electromagnetic waves. An absorption of photonic energy can lead to a transition from a lower electronic state $|1\rangle$ to a higher electronic state $|2\rangle$ of an atom or a molecule. The probability of an electronic transition from $|1\rangle$ to $|2\rangle$ by photon absorption is given by the Einstein coefficient for absorption B_{12}. If only radiative transitions are given, depopulation of level $|2\rangle$ can occur by spontaneous emission—described by the Einstein coefficient A_{21} for spontaneous emission—or by stimulated emission—described by B_{21}. In a given radiation field with the energy density $\rho(v)$ the absorption rate for a state $|1\rangle$ of population density n_1 is given by

$$n_1 B_{12}\rho(v) = A_{21}n_2 + B_{21}\rho(v)n_2 \tag{5.5}$$

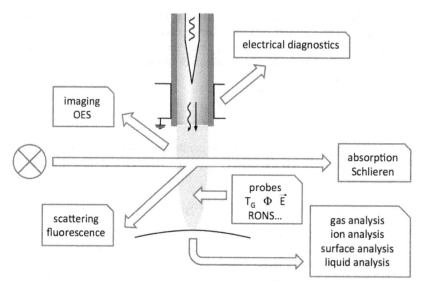

FIGURE 5.1 Diagnostic methods used on atmospheric pressure plasma jets. (From Reuter, S. et al., *J. Phys. D Appl. Phys.*, 51, 2018. Attribution 3.0 Unported. Creative Commons Attribution 3.0 licence in Ref. [67], http://creativecommons.org/licenses/by/3.0/, 2017.)

Spectroscopic notation of electronic quantum states of light atoms for which spin orbit LS coupling is valid, follows:

$$n\ell^q \ {}^{2S+1}L_{L+S} \tag{5.6}$$

The main quantum number is n. The numbers of electrons in the shell are given by q. ℓ is the orbital angular momentum quantum number, the sum of which is L, the total orbital angular momentum quantum number. S is the total spin angular momentum quantum number. LS coupling refers to the fact that in simple atoms, L and S do not affect each other and $J = L + S$ is the total angular momentum quantum number. $2S + 1$ is the degree of degeneracy of the energy state.

Figure 5.2 shows the energy level diagram for the singlet and triplet states of helium. Electronic transitions are only possible between singlet and between triplet states. For diatomic molecules, the following notation is common [22]:

$$^{2S+1}\Lambda_{g,u}^{+/-} \tag{5.7}$$

$\pm$ and g/u denote the symmetry of the eigenfunction of the electron. For molecules, the orbital and angular momentum Λ are given by Greek letters (Σ, Π, Δ, ...) that denote the projection of the respective vectors on the molecular axis. Metastable molecules can be identified through a small letter (a, b, c), however, for the case of metastable nitrogen, capital letters are used for historic reasons. In this notation, the first metastable electronically excited state of molecular oxygen, for example, is $O_2(a^1\Delta_g)$.

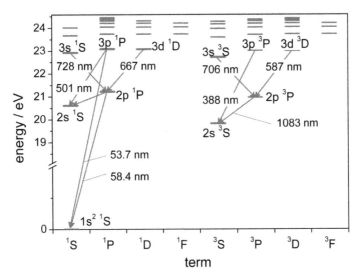

FIGURE 5.2 Energy levels of helium with the wavelengths of allowed radiative transitions. (From Reuter, S., *Formation Mechanisms of Atomic Oxygen in an Atmospheric Pressure Plasma Jet Characterised by Spectroscopic Methods*, Cuvillier Verlag, Göttingen, Germany, 2008. With permission.)

Radiative transitions are allowed only for certain quantum number transitions, listed in **Table 5.1**. If a radiative transition to a lower energy level is forbidden, the respective energy state is metastable. Since a fast radiative transition is not allowed, metastable species have a lifetime that is multiple orders of magnitude higher than energy levels where a radiative transition is allowed. In **Figure 5.2**, the triplet state 2s 3S of helium is a metastable energy state with an energy of 19.82 eV. It carries high energy and leads to Penning ionization in helium plasmas with molecular admixture.

Radiative energy transitions of species in a plasma can be probed by spectroscopy. The observed spectral lines' main characteristics are shown in **Figure 5.3**.

Table 5.1 Allowed Optical Transitions			
Quantum Number	**0**	**± 1**	**g**
ΔJ	x[a]	x	n.a.
$\Delta S \, \& \, \Delta \Sigma$	x		n.a.
ΔL	x[a]	x	n.a.
u	n.a.	n.a.	x

[a] Forbidden, if L_1 and $L_2 = 0$.

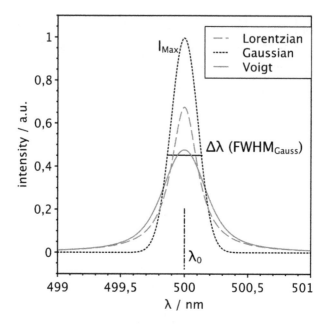

FIGURE 5.3 Line profiles relevant for spectroscopy.

λ_0 denotes the central wavelength of the emission or absorption line. It is determined by the energy gap ΔE between the energy state i and the energy state k of the electronic transition that is probed by the spectroscopic measurement. The wavelength λ_{ik} of a transition from state i to state k is related to the photon energy ΔE by

$$\lambda_{ik} = \frac{hc}{\Delta E} \tag{5.8}$$

hc is the Planck constant times the speed of light c. The region inside the full width at half max is called line core and the region outside the full width at half maximum (FWHM) is called wings of the line. Differences between Gaussian and Lorentzian line profile can be identified best in the line wings. In high-resolution spectroscopy, information about the Lorentz profile can still be gained from the line wings, even in pressure broadened lines, because the Gaussian profile quickly drops to zero in the line wings.

5.2.2 Quenching

Spectral information from species concentration and energy state population through spectroscopy can be gained, if the respective state can be probed with electromagnetic radiation. In case of atmospheric pressure plasmas, processes are dominated by collisions involving atoms and molecules—charged, excited, or neutral. These collisions can lead to an energy exchange that causes a depopulation of the probed state and will thus lead to a loss in spectral signal. These quenching collisions need to be accounted for and taken into consideration for an evaluation of spectroscopic data. In some cases, though, collisions can form part of the spectroscopic technique itself.

5.2.3 Spectral Line Shapes and Their Broadening

Not only position and intensity of a spectral line yields information about the medium it is observed from, but also the line shape and the broadening of the line hold a multitude of information. The following presentation of line broadening mechanisms in Section 5.2.3 is taken from [23] adapted to include mechanisms relevant for spectroscopy in general:

The lineshape of the spectral profile is determined by several factors dictated by the spectroscopic properties of the species studied as well as the experimental conditions such as the pressure and temperature of the gas sample and the existence of electric and magnetic fields. These factors can be divided into homogenous and non-homogeneous categories, i.e., contributions to the lineshape that affect every element present in the sample in an identical way, or contributions which arise from an average effect that is not identical to every element, respectively. Here, the relevant mechanisms behind the lineshape of the spectral profile will be briefly discussed. They include natural broadening, collisional broadening (resonance and van der Waals broadening), Doppler broadening, Stark broadening, instrumental broadening, and saturation broadening. For a more detailed description and mathematical expressions, we refer the reader to [18,19,24–27].

Natural Line Broadening

One example of a homogenous contribution to the lineshape is natural broadening, which arises from the natural lifetime, τ, of the emitting state of the transition. From the Heisenberg uncertainty principle, $\tau \, \Delta E \geq \hbar$, it can be seen that the consequence of a finite lifetime of the upper state is an uncertainty in the corresponding energy and, therefore, the state is described by a range of energies, ΔE. The resulting natural broadening effect has a Lorentzian lineshape, however, this contribution to the width of the absorption lineshape is very small, being usually negligible in comparison with other broadening contributions.

Collisional Broadening

Another example of homogeneous broadening of the lineshape is collisional broadening. When collisions occur in a gaseous sample containing atoms and/or molecules, this effectively reduces the natural lifetime of the transition and thus increases the uncertainty in the energy of the transition, i.e., there is a broadening of the lineshape. The extent of the lineshape broadening depends on the nature of the energy transfer during a collision, and on the natural lifetime of the transition compared to the frequency of collisions. As the pressure increases and the interval between collisional events reduces, the extent of broadening of the lineshape increases. The collisional broadening has two components: pressure-induced broadening, also known as van der Waals broadening, and resonance broadening. The former is due to collisions of the absorbing particle with neutral perturbers that do not share a resonant transition with the radiating particle. The latter is due to collisions with perturbing particles with similar energy levels as the absorbing particle, which introduces the possibility of an energy exchange process (when both particles are from the same species, in literature this is often termed self-broadening). Collisional broadening leads to a Lorentzian lineshape, analytically described by

$$L(v) = \frac{2}{\pi} \frac{\Delta v_L}{\Delta v_L^2 + 4(v - v_0)^2},$$ (5.9)

where Δv_L is the Lorentzian linewidth given as a full width half maximum (FWHM) value and v_0 is the resonance frequency, i.e., the line center of the absorption profile. The Lorentzian linewidth due to collisional broadening is often conveniently expressed as

$$\Delta v_L = 2\gamma p,$$ (5.10)

with p being the operating buffer gas pressure [atm] and γ the pressure broadening coefficient dependent on the nature of the colliding species, generally given in half width half maximum (HWHM) [cm^{-1}/atm] at $T_{ref} = 296$ K and reference pressure $p_{ref} = 1$ atm. This parameter is transition dependent [28]. The broadening coefficient is often tabulated in literature in spectroscopic databases. If the broadening coefficients are not tabulated in the literature, which is often the case for atoms, then they have to be calculated using the analytical expressions of resonance and van der Waals broadening as one can find in [24]. Resonance broadening occurs if either the lower (l) or the upper (u) state of the radiative transition under consideration is the upper level of a resonance transition, i.e., if the level is connected to the ground state (g) by an allowed dipole transition [29]. When including all three perturbing transitions, the resonance broadening can be expressed as [18]

$$\Delta\lambda_{resonance} = \frac{3e^2}{16\pi^2\varepsilon_0 m_e c^2}\lambda_{ul}^2\left[\lambda_{lg}f_{gl}\sqrt{\frac{g_g}{g_l}}n_g + \lambda_{ug}f_{gu}\sqrt{\frac{g_g}{g_u}}n_g + \lambda_{ul}f_{lu}\sqrt{\frac{g_l}{g_u}}n_l\right].$$ (5.11)

Van der Waals broadening for radiator r colliding with perturber p is given by [18]

$$\Delta\lambda_{van\,der\,Waals} \approx \frac{\lambda_{ul}^2}{2c}\left(\frac{9\pi\hbar^5 R_\alpha^2}{16m_e^3 E_p^2}\right)^{\frac{2}{5}}v_{rp}^{3/5}N_p,$$ (5.12)

where E_p is the energy of the first excited state of the perturber connected with its ground state by an allowed transition, N_p the number density of the perturber, v_{rp} the relative speed of the radiating atom and the perturber, and the matrix element $\overline{R_\alpha^2}$ equals to

$$\overline{R_\alpha^2} \approx \frac{1}{2}\frac{E_H}{E_\infty - E_\alpha}\left[5\frac{z^2 E_H}{E_\infty - E_\alpha} + 1 - 3l_\alpha\left(l_\alpha + 1\right)\right]. \tag{5.13}$$

Here, E_H and E_∞ are the ionization energies of the hydrogen atom and of the radiating atom, respectively, E_α the term energy of the upper state of the transition, l_α its orbital quantum number, and z the number of effective charges ($z = 1$ for a neutral emitter, $z = 2$ for a singly ionized emitter, etc.) [18].

Collisional broadening of the lineshape tends to dominate at atmospheric pressure while Doppler broadening, an example of inhomogeneous broadening, is negligible especially for cold plasmas. At higher temperatures, however, contribution by Doppler broadening has to be taken into account, and from it the translational temperature of the species can be obtained.

Even if the profile is pure Lorentzian, due to the Doppler broadening being very small, in principle, a value for the temperature can still be obtained via the expression for the temperature and pressure correction of the pressure broadening coefficient. The pressure broadened coefficient $\gamma(p, T)$ for a gas at pressure p [atm], temperature T [K], and partial pressure p_s is given by the following expression [28]:

$$\gamma\left(p,T\right) = \left(\frac{T_{ref}}{T}\right)^n \left(\gamma_{air}\left(p_{ref}, T_{ref}\right)\left(p - p_s\right) + \gamma_{self}\left(p_{ref}, T_{ref}\right)p_s\right), \tag{5.14}$$

where n is the coefficient of temperature dependence of the pressure broadening coefficient. In the absence of other data, the coefficient of temperature dependence of the self-broadening coefficient is assumed to be equal to that of the pressure broadening coefficient. Alternatively, the classical value of 0.5 can be used by default.

Doppler Broadening

Doppler broadening results from the shift in the frequency of the absorbed radiation during a transition, and this shift is dependent on the relative velocity of the absorber with respect to the direction of propagation of the radiation. For an atom or molecule that absorbs at a resonant frequency of v_{res} when stationary, the actual frequency at which the transition is observed is shifted to v_a when the molecule travels with velocity v, away from the radiation source and is given by

$$v_a = \frac{v_{res}}{1 - v/c}. \tag{5.15}$$

As a result of thermal motion, the molecules within the sample exhibit a Maxwellian distribution of their velocities along the propagation direction of the radiation, leading to a distribution of absorbed frequencies according to the direction of the particle's motion relative to the source. This inhomogeneous mechanism gives rise to the following normalized Gaussian lineshape

$$G(v) = \frac{2}{\Delta v_{Dopp}} \sqrt{\frac{\ln 2}{\pi}} exp\left[-\frac{4\ln 2 (v - v_0)^2}{\Delta v_{Dopp}^2}\right], \qquad (5.16)$$

where Δv_{Dopp} is the FWHM Gaussian linewidth for Doppler broadening with its dependence on temperature T, given by

$$\Delta v_{Dopp} = 2v_0 \sqrt{\frac{2\ln 2 k_B T}{mc^2}} = 7.16 \times 10^{-7} v_0 \sqrt{\frac{T}{M}}, \qquad (5.17)$$

where m is the molecular mass and M the molecular weight. Additionally, as the equation above highlights, the FWHM Doppler contribution, Δv_{Dopp}, is an increasing function of the resonant frequency, v_0, and thus ranges from tens of MHz in the mid-infrared, up to a few GHz in the UV. Such dependence results in different instrumental resolutions required for conducting Doppler-resolved studies in different regions of the electromagnetic spectrum.

Stark Broadening

The interaction of a molecule with the electric field caused by the free electrons in the sample can be thought of as many weak collisions with the electrons. This results in a lifetime shortening (like in collisional broadening) of the molecular state under consideration. When the contribution of electrons is much higher than that of ions, the Stark broadening is characterized by a Lorentzian profile, whose FWHM is directly related to the electron density. OES measurements on the H_β-line are commonly used for the determination of electron densities ranging from 10^{14} to 10^{18} cm^{-3} [30]. However, absorption spectroscopy on this transition is not trivial. Alternatively, one could in principle use measurements on metastable argon or helium atoms. The Stark broadening, however, is generally much smaller than the Doppler and collisional broadening. Therefore, the contribution of Stark broadening to the line profile is hard to determine and is usually neglected [31]. Despite these problems, in [32], an estimation of the electron density was performed in an argon/oxygen micro structured electrode discharge up to a pressure of 400 mbar; although, this was not possible at higher pressures. Stark broadening has been used in several cases for an analysis of atmospheric pressure plasmas [27].

Resonance Broadening in Emission Spectra

Resonance broadening occurs, when an excited atom collides with a ground state atom of the same species and the excited state has a radiative transition

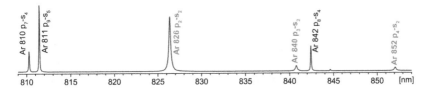

FIGURE 5.4 Broadening of resonant argon emission lines in an atmospheric pressure plasma jet observed with a medium resolution spectrograph. (From Pipa, A.V. et al., *Appl. Phys. Lett.*, 106, 244104, 2015. With the permission of AIP Publishing.)

to the ground state. The resulting resonance broadening of the spectral line (see **Figure 5.4**) can be expressed by [33]:

$$\Delta\lambda = 9.3492 \times 10^{-14} \sqrt{\frac{g_G}{g_R}} \lambda_R f_R N \tag{5.18}$$

where λ and λ_R are the wavelengths of the line under observation and the resonant line, respectively. The broadening described in Equation (5.18) allows to calculate the gas temperature through its dependence on the particle density. The advantage of such a temperature measurement is the noninvasiveness, the spatiotemporal resolution and the simplicity of the experiment setup.

In [33], the gas temperature from an atmospheric pressure plasma jet was determined with a moderate resolution 1200 g/mm grating 750 cm spectrometer with a 20 μm slit.

Instrumental and Saturation Broadening in Absorption Spectra

When deducing properties of the plasma from the lineshape of a spectral absorption profile, one has to take into account the influence of the spectrometer or the laser on the lineshapes, i.e., instrumental broadening. Depending on the components of the spectrometer (laser), this instrumental broadening can have a Gaussian or a Lorentzian lineshape with FWHM linewidth $\Delta\nu_{inst}$. Preferably, the instrumental linewidth of the spectrometer is much smaller than or similar to the spectral width of the investigated atomic or molecular transition. When the instrumental linewidth is comparable or larger than the width of the spectral linewidth, large errors can be made when deconvoluting the spectra profile to obtain information on parameters such as temperature or other broadening mechanisms. This is generally less of an issue for atmospheric pressure plasmas due to larger collisional broadening of the studied spectral profiles compared to instrumental broadening. Another broadening effect appears in case of power saturation. Due to the optical coupling of the lower and the upper states by the radiation field, the homogeneous line width of the transition is broadened and becomes $\Delta\nu_S = \Delta\nu_L \sqrt{1+S}$ [19], S being the saturation parameter. In this case, the absorption coefficient gets smaller in amplitude and larger in width.

5.3 Optical Emission Spectroscopy

Optical emission spectroscopy probes excited species in a plasma. As it is a passive technique, it is inherently a non-invasive measurement method. In low pressure or equilibrium plasmas, thermal equilibrium conditions are often present. In thermal equilibrium, and in local or partial local thermal equilibrium conditions, optical emission spectroscopy can probe the population density of the observed excited energy states and allows to deduce population of lower energy states via Saha's equation. Assuming Boltzmann distribution, also ground state concentrations can be determined. The strongly nonequilibrium character of atmospheric pressure plasma jets prohibits most approaches valid for plasmas in thermal equilibrium; however, with careful analysis of the plasma conditions, optical emission spectroscopy can yield a multitude of information. Techniques using optical emission spectroscopy are described in Chapters 6 and 7.

The line emission coefficient ε of a radiative electronic transition by spontaneous emission is given by:

$$\varepsilon_{line} = n_2 A_{21} \frac{h\nu}{4\pi} \tag{5.19}$$

ε_{line} is the integral of the spectral emission coefficient ε_ν over the spectral line.

$$\varepsilon_\nu = \varepsilon_{line} P_\nu \tag{5.20}$$

with the normalization $\int_{line} P_\nu d\nu = 1$ and P_ν the measured line profile.

Energy states can be populated by a variety of mechanisms. Most population processes in plasmas occur through electron excitation. The collision process of an atom or molecule with an electron is described by a cross section or rate coefficient. The cross section varies with the electrons' energy. With pressure, the number of collisions of atoms and molecules increases, so that at atmospheric pressure excitation through collisions with ions and excited species significantly contributes to population of excited energy states.

5.3.1 Molecular Spectroscopy

Molecular spectroscopy shows that a molecule offers more degrees of freedom than an atom. Specifically, energy states of vibrational and rotational energy can be accessed. **Figure 5.5** shows a sketch of the energy levels of a diatomic molecule and its first electronic state. Resolved are vibrational states and the rotational states for the first vibrational state of the electronically excited molecule.

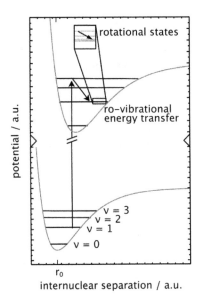

internuclear separation / a.u.

FIGURE 5.5 Molecular energy levels and energy transfer.

The potential curve for diatomic molecules follows the Morse potential that can be calculated by:

$$V_{Morse} = D_e e^{-2a(r-r_0)} - 2D_e e^{-a(r-r_0)}$$ (5.21)

D_e is the molecule dissociation energy, r_0 is the internuclear separation with the least potential energy and a describes the stiffness of the oscillator system represented by the two atoms in the molecule. The notation in Equation (5.21) is chosen so that the potential at $r = \infty$ is zero. The potential is therefore the negative of the dissociation energy at equilibrium distance r_0.

The vibrational energies in the Morse potential are approximated by a polynomial function:

$$E_{vib} = \omega_e \left(v + 1/2\right) - \omega_e x_e \left(v + 1/2\right)^2 - \omega_e y_e \left(v + 1/2\right)^3$$ (5.22)

The spectroscopic constants ω_e, $\omega_e x_e$, and $\omega_e y_e$ are tabulated or can be calculated from the Morse potentials' parameters.

Rotational energies can be approximated by

$$E_{rot} = hcB_0 \left(J(J+1)\right) + D_v \left(J(J+1)\right)^2$$ (5.23)

The spectroscopic constants B_0 and D_v can be found in spectral databases. The quantum numbers J follow the selection rules $\Delta J = 0, \pm 1$. The total energy of the energy state is calculated by $E = E_{electronic} + E_{vibrational} + E_{rotational}$.

A molecular band of a diatomic molecules is separated in Q-, R-, and P-branch which differ by the ΔJ-values at which the energy transition occurs.

For the Q-branch $J_{upper} = J_{lower}$, for the R-branch, $J_{upper} = J_{lower} + 1$ and for the P-branch, $J_{upper} = J_{lower} - 1$.

From the resulting molecular spectra, species can be identified, vibrational, rotational, and gas temperature can be determined, and species concentrations can be obtained.

5.4 Absorption Spectroscopy

The simplest and most direct way to determine absolute species densities is absorption spectroscopy. Absorption spectroscopy is a calibration free technique. Fundamentals of absorption spectroscopy are described in the following, which is an excerpt from [23].

5.4.1 Atomic and Molecular Absorption

Sensitive atomic and molecular absorption spectroscopy in atmospheric pressure plasmas commonly uses permitted dipole transitions (single-photon absorption) between eigenstates in the outermost shell. From quantum mechanics, eigenstate energies are obtained as solutions of the stationary Schrödinger equation and they are a unique set of energies for atoms and molecules. Transition dipole selection rules (e.g., parity change) limit the number of allowed transitions leading to specific spectral absorption patterns for each species. For electronic transitions, these patterns are typically observed in the VUV to near infrared (NIR) spectral region, and, for rotational-vibrational absorption transitions, in the mid infrared (MIR) spectral region.

Wavelength calibrated absorption spectra and thus absolute absorption line or absorption band positions are determined by various techniques: Usually in the VUV to NIR spectral domains, calibrated spectrometers or wavemeters are employed, while in the MIR, Fabry-Pérot etalons and reference gas cells are used.

The unique set of transition wavelengths for each species is generally available in spectroscopic databases. This allows the distinct identification of a particular species present in a plasma. Limited instrumental resolution and overlapping of multiple lines may hamper plasma species identification. This is relatively easily solved by the identification of several absorption lines of the same species. Note that a particular absorption transition in plasmas is observed only if there is a significant population density in the lower state. Consequently, absorption lines may be absent as a result of certain plasma conditions, e.g., the electron temperature. In such a case, a careful line identification procedure is required.

5.4.2 Beer–Lambert Law

Absorption spectroscopy relies on the Beer–Lambert law, which describes the absorption of light by a sample. The Beer–Lambert law links the intensity attenuation of optical radiation through a homogeneous sample to the density N of species present in it

$$I(\lambda) = I_0(\lambda)e^{-N\sigma(\lambda)L} = I_0(\lambda)e^{-k(\lambda)L}, \qquad (5.24)$$

where $I(\lambda)$ is the transmitted radiation, I_0 the incident radiation, $\sigma(\lambda)$ the wavelength dependent absorption cross-section, L the absorption path-length, and $k(\lambda)$ the wavelength dependent absorption coefficient. The degree to which light is absorbed is thus quantified, and, from the magnitude of the absorbance, the density of absorbing species in a sample can be determined

$$Absorbance(\lambda) = -\ln\left(\frac{I}{I_0}\right) = N\sigma(\lambda)L. \qquad (5.25)$$

Absorption features, however, are never strictly monochromatic and are generally spread over a range of wavelengths centered at λ_0, resulting in the line profile of the transition. Several processes contribute to the increase of the width of the line profile, including the lifetimes and thermal motion of the species, the collisions between species, and the influence of electric and magnetic fields on the species behaviour. Hence, the integrated absorption coefficient over an absorption line, k_λ, gives a more useful measure of the absorbance

$$\frac{1}{L}\int_{-\infty}^{\infty} -\ln\left(\frac{I}{I_0}\right) = \sigma_{int}N = k_\lambda = \int_{line} k(\lambda)d\lambda, \qquad (5.26)$$

where σ_{int} is the integrated cross-section and is defined as

$$\sigma_{int} = \int_{-\infty}^{\infty} \sigma(\lambda)d\lambda. \qquad (5.27)$$

As shown in the next section, by using the Beer–Lambert law, absolute densities of absorbing species in the plasma can be extracted directly from their spectral profiles. The spectral line positions usually provide species identification while line profiles are connected with the properties of the species in the plasma, for example their translational temperature.

5.4.3 Oscillator Strength: Atomic Density

The allowed dipole transitions from atomic ground states to excited states have typical energies in the VUV and UV range. When absorption spectroscopy is used to probe atomic densities in the electronic excited states (metastables, resonant levels, etc.), the corresponding transition wavelengths are usually in the visible (VIS) and NIR domain. In case of nonequilibrium atmospheric plasmas, excited electronic states of atoms are not in Boltzmann equilibrium with their ground states. Because the energy separation between electronic levels is large (eVs), the main population mechanisms are not the heavy particle collisions but electron kinetics, excitation transfer, and radiative processes. In this case, to relate excited state densities to electronic ground states, collisional-radiative models are required. However, the density of the lower state (which can be a ground state or an excited state) in the absorption process is easily determined by

$$\int_{line} k(\lambda)d\lambda = \frac{e^2\lambda_{ik}^2}{4\varepsilon_0 m_e c^2} f_{ik}\left(N_i - \frac{g_i}{g_k}N_k\right),\qquad(5.28)$$

where $k(\lambda)$ is the absorption coefficient integrated over the entire absorption line, λ_{ik} the resonance wavelength corresponding to the electronic transition $i{\rightarrow}k$, f_{ik} the oscillator strength (dimensionless), N_i and N_k the densities of the lower and upper states, ε_0 the vacuum permeability, e and m_e the electron charge and mass, respectively. If the two levels are in Boltzmann equilibrium, then the relation between the absorption coefficient and the lower level density is given by

$$\int_{line} k(\lambda)d\lambda = \frac{e^2\lambda_{ik}^2}{4\varepsilon_0 m_e c^2} f_{ik}N_i\left(1 - e^{-\frac{h\nu}{k_B T}}\right),\qquad(5.29)$$

where k_B is the Boltzmann constant, T the electron temperature [K], and $h\nu$ the energy of the absorbed photons [J] with $\nu = c/\lambda$. The bracket term in the right side of Equation (5.29) accounts for stimulated emission. In atmospheric pressure plasmas, the population of the upper state of most atoms is a very small fraction of the lower state, because of the large energy separation between levels. Therefore, the bracket term in Equation (5.29) normally equals to one. For simplicity, the above expressions are using atomic oscillator strengths. This spectroscopic property can be found for instance in the NIST Atomic Spectra Database [34]. We recommend this database, because it includes the statistical weights of the two levels, so it is easy to apply. Notice that in Equations (5.28) and (5.29) the absorption is described in wavelength space (as this is mainly being used for UV and VIS atomic transitions in literature). For practical purpose, Equation (5.29) can be written as

$$\int_{line} k(\lambda)d\lambda = 8.85\times10^{-13}\lambda_{ik}^2 f_{ik}N_i,\qquad(5.30)$$

where $k(\lambda)$ is in cm^{-1}, λ in cm, and N_i in cm^{-3}, and the constant in the right side of the equation is also in cm units.

Atomic lines in atmospheric plasmas often exhibit Lorentzian profiles due to collisional broadening dominating the broadening mechanisms of the line profile. A practical expression in that case is

$$N_i = \frac{k(\lambda_0)\Delta\lambda_L}{5.64\times10^{-13}\lambda_{ik}^2 f_{ik}}, \qquad (5.31)$$

where $k(\lambda_0)$ is the peak absorption coefficient ($\lambda_0 \equiv \lambda_{ik}$, the resonance wavelength), and $\Delta\lambda_L$ the FWHM value of the line profile [cm^{-1}]. Notice the expressions given above are for single line absorption; when line overlapping occurs, the contributions of different species (due to unresolved transitions of the same or of different atoms) need to be considered.

5.4.4 Line Strength: Molecular Density

Using MIR absorption spectroscopy, individual absorption lines of vibrational-rotational transitions in electronic ground state of molecules can be measured. Often used techniques are laser-based methods, e.g., tunable diode laser-absorption spectroscopy (TDLAS) and quantum cascade laser-absorption spectroscopy (QCLAS) or a broadband high-resolution apparatus, e.g., a high-resolution Fourier transform infrared (FTIR) spectrometer. The integrated absorption coefficient $k(v)$ in wavenumber space ($v = 1/\lambda$ in cm^{-1}) over an absorption line is related to the molecular species density by

$$\int_{line} k(v)\,dv = S(T)n, \qquad (5.32)$$

where $S(T)$ [cm^2 cm^{-1} molecule^{-1}] is the line strength (or the line intensity) of a specific transition at temperature T [K], and n [molecules cm^{-3}] is the total density of the molecular species in all internal states of the molecule. We choose to give here the integral of the absorption coefficient in wavenumber space because MIR spectroscopy is commonly presented in wavenumber units. It should be noted that the line strength defined in this way is temperature dependent via the Boltzmann relation between $N_{v,J}$ and n, the ro-vibronic level for which the absorption is measured. The information on the rotational and vibrational temperatures can be deduced from the relative densities in the rotational and vibrational levels. Assuming local thermodynamic equilibrium, the line strength of a particular rotational-vibrational transition is given by [28]

$$S_{v''J''}^{v'J'}(T) = \frac{8\pi^3}{3hc}v_{v''J''}^{v'J'} \frac{g_{v''J''}\,exp\!\left(-\dfrac{E_{v''J''}}{kT}\right)}{Q(T)} \Re_{v''J''}^{v'J'}\left[1-exp\!\left(-\frac{hcv_{v''J''}^{v'J'}}{kT}\right)\right], \qquad (5.33)$$

where v and J are the quantum numbers of the lower ($''$) and upper ($'$) ro-vibronic levels, $v_{v''J''}^{v'J'}$ [cm^{-1}] the spectral line transition wavenumber, $g_{v''J''}$ the statistical weight of the lower level, $E_{v''J''}$ the energy of the lower level, $Q(T)$ the total internal partition function, and $\mathfrak{R}_{v''J''}^{v'J'}$ the weighted transition-moment squared. The statistical weight and the total internal partition function include the electronic, vibration, rotation and nuclear spin terms (see for details [35,36]). The term in the squared brackets accounts for stimulated emission. While for electronic transitions this term can be neglected (large ΔE), in the case of rotational-vibrational transitions probed in the electronic ground state, an important population is present in the upper state. This occurs even in room temperature plasma environments. Therefore stimulated emission is often significant and needs to be considered. Most of the above molecular parameters can be found in the HITRAN database [37,38].

While the theoretical calculation of the line strength is complex, its measurement is easily obtained for stable molecules. Experimentally, a reference cell is filled with a molecular gas at ambient temperature and controlled pressure. The gas density is calculated based on the ideal gas law. Knowing the length of the cell, the integrated absorption coefficient is then measured and the $S(T)$ at room temperature is obtained using Equation (5.32). For radicals, however, this procedure becomes very challenging. Due to their short lifetimes, a complex apparatus is needed to generate and maintain high radical densities, and calibration methods are then used to measure absolute radical population (e.g., [36]).

5.4.5 Effective Absorption Cross-Section: Molecular Density

When using broadband absorption techniques (e.g., grey body lamps) or when unresolved molecular spectral features are measured by laser absorption (e.g., absorption of large molecules or absorption with pre-dissociation of the electronic excited state), effective absorption cross-sections are needed for species density measurements. In these cases, often the rotational, vibrational molecular spectral features are unresolved, resulting in a structureless absorption spectrum. Consequently, densities are obtained not from a single line absorption but from unresolved molecular bands, which are spectrally convolved with the instrumental functions. Absolute concentrations can then be obtained if not a distinct point in the spectrum is analysed, but a carefully defined spectral range Δv, which is in fact additional averaging. This gives the following relation between the molecular density and the effective absorption cross-section, $\sigma_{eff}(v)$ [39],

$$\sigma_{eff}\left(\langle v \rangle\right) = \frac{1}{nL}\ln\left(\frac{I_0(\langle v \rangle)}{I(\langle v \rangle)}\right) = \frac{1}{nL}\ln\left(\frac{\langle I_0(v) \rangle \cdot \Delta v}{\langle I(v) \rangle \cdot \Delta v}\right) = \frac{1}{nL}\ln\left(\frac{A_0}{A}\right). \quad (5.34)$$

In this way, the least error prone density determination is to not calculate the absorbance from the transmitted intensity I at v, but to use the area (or integral) $A \approx I\Delta v$. In the calculation of σ_{eff}, the contribution of Δv cancels out in Equation (5.34), i.e., $\sigma(\lambda)$ is in cm^2 units. The effective absorption cross-section can depend on temperature, pressure, and instrumental resolution.

When very high-resolution absorption devices are used, instrumentally independent cross-sections are needed. Otherwise, cross-section data need to be convolved with the instrumental function employed in the plasma diagnostic experiment. Underestimated densities are obtained if instrumental functions are disregarded. Absorption cross-sections from UV to MIR for molecular species can be found for example in the HITRAN database [38].

5.5 Scattering Techniques

Light scattering holds a plethora of information on the scattered medium. Diagnostic techniques that use scattering methods are accurate and yield the possibility to probe electron properties as well as gas properties such as density or temperature. Generally, elastic scattering and inelastic scattering are distinguished. Elastic scattering from heavy particles (Rayleigh scattering) is used in combustion or aerodynamics studies to probe shockwaves and temperatures [40]. Elastic scattering of light on electrons (Thomson scattering) is considered the gold standard of electron diagnostics in low temperature plasmas. Inelastic scattering of light on molecules (Raman scattering) yields information about temperature, species type, and sometimes structure of the scattering medium. If the scattering particles are larger than the wavelengths of light, elastic Mie scattering occurs, which has been used to study particle nucleation.

Scattering originates from inducing an oscillating dipole in the scattering medium [41]. An electromagnetic wave incident on an atom or molecule causes the particle's electrons to oscillate. The electron cloud is shifted with respect to the molecule and a dipole moment is induced. The resulting oscillating dipole is the source of an emitted electromagnetic wave. If the process is elastic, the emitted electromagnetic wave has the same frequency as the incident wave. The emitted wave is the scattered incident wave (see **Figure 5.6**).

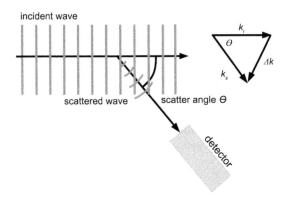

FIGURE 5.6 Scattering geometry.

5.5.1 Rayleigh Scattering

Rayleigh scattering is elastic scattering of electromagnetic radiation by atoms or molecules or other particles smaller than the wavelength of radiation. It can be used as simple diagnostic of temperature or density of gases [40]. Rayleigh scattering originates from the interaction of the oscillating electric field of a light wave and the localized charges of the probed particles. Due to the polarizability of the scattering particles, the charges move with the frequency of the incident light and result in dipole radiation emitted from the particle, which is observed as scattered light.

For the Rayleigh scattering cross sections, the following relations can be derived ([40] and references therein):

The Rayleigh scattering cross section for a spherical particle such as a noble gas atom can be gained from the classical expression for radiation from an infinitesimal small oscillating dipole. The Rayleigh scattering cross section for atomic gases with spherically symmetric scatterers is given by

$$\sigma_{sph} = \frac{8\pi^3\alpha^2}{3\varepsilon_0\lambda^4} = \frac{24\pi^3}{\lambda^4 N^2}\left(\frac{n^2-1}{n^2+2}\right)^2 \tag{5.35}$$

with the polarizability α written as a function of the refractive index n, N the scatterer density (particles per m^3). Equation (5.35) shows that Rayleigh scattering is inversely proportional to λ^4, which reveals that scattering of UV photons is far more efficient than that of IR photons.

For diatomic molecules such as nitrogen, orientation of the molecule toward the incident radiation needs to be considered by averaging over arbitrary molecule orientations. In contrast to spherically symmetric particles, the induced dipole moment must not be in the direction of the applied field. The Rayleigh scattering cross section can be expressed as a function of the ratio of horizontally-to-vertically polarized light ρ_0 propagating parallel to the horizontal plane [40]. ρ_0 is an expression depending on the mean polarizability α and the anisotropy γ of the molecule collective. The Rayleigh scattering cross section for diatomic molecules is equal to the scattering cross section for spherically symmetric particles with a correction factor called King correction factor:

$$\sigma = \sigma_{sph}\left(\frac{6+3\rho_0}{6-7\rho_0}\right) = \sigma_{sph}\left(1+\frac{10\gamma^2}{45a^2}\right) \tag{5.36}$$

This expression can be simplified for the condition typically used in laser induced fluorescence (LIF) setups: The incident laser radiation propagating along the x-axis, polarized along the z-axis. The light is only collected under a small solid angle and resultantly the total differential cross section is relevant for LIF measurements:

$$\frac{\partial \sigma_v}{\partial \Omega} = \frac{3\sigma}{8\pi} \frac{1}{2+\rho_0} \left(\rho_0 + \left(2 - 2\rho_0\right)\left(1 - \cos^2\theta_z\right) \right) \tag{5.37}$$

$$\frac{\partial \sigma_H}{\partial \Omega} = \frac{3\sigma}{8\pi} \frac{1}{2+\rho_0} \rho_0 \tag{5.38}$$

From the differential cross section, the scattering power which is collected over the solid angle observed by the detector $\Delta\Omega$ can be calculated. Further simplification of Equation (5.37) is possible, when the excitation is oriented at a 90° angle to the detection system along the y-axis and $\cos^2\theta_z$ becomes zero (see Equation 5.38).

The Rayleigh scattering signal is proportional to the observed scattering volume V_c, the differential cross section integrated over the observed solid angle, the number of scattering particles in the scattering volume and a constant ξ that contains the spectral and optical efficiencies of the complete detection optics, and the laser irradiance I_L:

$$S_R = \xi I_L n V_c \int \frac{\partial \sigma}{\partial \Omega} d\Omega \tag{5.39}$$

With the ideal gas law $n = P/k_B T_G$, the particle density in a gas or plasma volume can be determined at known gas temperature. k_B is the Boltzmann constant, T_G is the gas temperature at the pressure P.

5.5.2 Thomson and Raman Scattering

Thomson scattering describes the scattering of incident waves on electrons such as the free electrons in a plasma [41]. It has been successfully used on atmospheric pressure plasmas [42–44].

Thomson scattering is proportional to the density n of the scattering particle density. With the differential scattering cross section $d\sigma/d\Omega$, the measured power of scattered laser light can be calculated by:

$$P_\lambda = n \cdot \xi \cdot \Delta\Omega \cdot P_i \frac{d\sigma}{d\Omega} \cdot S_\lambda\left(\lambda\right) \tag{5.40}$$

P_λ is the measured power in wavelengths units, ξ combines all wavelengths dependent parameters of the detection setup as well as the length of the scattering volume, $\Delta\Omega$ is the solid angle of detection and S_λ includes the spectral distribution as a function of the wavelength.

From the Thomson scattered signal, electron parameters can be determined. The velocity of the electrons leads to a Doppler broadening of the Thomson signal. Thus, the electron temperature can be determined from the broadening according to:

$$T_e = \frac{m_e c^2}{4 k_B}\left(\frac{\Delta\lambda}{\lambda_l}\right)^2 \tag{5.41}$$

with $\Delta\lambda$ the FWHM of the Thomson scattering signal profile, λ_l the laser wavelength, m_e the electron mass, c the speed of light and k_B the Boltzmann constant. For non-Maxwellian electron energy distribution functions, the bulk electron temperature describing 95% of the electrons of medium electron temperatures are determined by Thomson scattering, whereas the high energy electron-tail temperatures are not determined.

The spectral distribution function of the scattered light directly reflects the one-dimensional velocity distribution function $F(v)$ of the electrons. It can be shown for $F_x(v_x)$ that

$$F_{\vec{v}} = -\frac{F_x'(v)}{2\pi v} \tag{5.42}$$

As the energy distribution function is linked to the velocity distribution function, it follows that the electron energy distribution function can be derived from the velocity distribution function and thus from the Thomson scattered signal according to:

$$f_E(E) = -\frac{2}{m_e} F_x'\left(\sqrt{2E/m_e}\right) \tag{5.43}$$

Cut off of the EDF at an energy E_0 results in an overall reduction of particles with a velocity lower than $v_0 = \sqrt{2E/m_e}$ and a complete depletion of particles with a velocity higher than v_0. This is due to the fact that electrons of one velocity v_0 contribute equally to all velocities lower than v_0 in one direction.

For a calibration of Thomson scattering signals, the setup dependent quantities given by ξ and $\Delta\Omega$ in Equation (5.40) need to be determined. Measurement of a different scattering signal such as, e.g., Raman scattering on a known density of Raman active medium can yield this calibration. With this, the concentrations of electrons from a comparison of the two scattering signals can be gained:

$$n_e = n_i \frac{P_\lambda^T \, d\sigma^R/d\Omega}{P^R \, d\sigma^T/d\Omega} \tag{5.44}$$

Raman scattering follows the same mechanism of light scattering on a dipole as Rayleigh and Thomson scattering. For Raman scattering, the electrons bound to a molecule are the origin of scattering. The direction of the dipole moment induced in a molecule depends on the orientation of the molecule towards the electric field. The molecular polarizability tensor describes this fact. When it is averaged over molecule orientation, it can be written in terms of its molecular polarizability α and its anisotropy γ. The rotational Raman scattering cross section in simple linear molecules such as O_2 or N_2, linear molecules with no electronic angular momentum coupled to the scattering, depends on γ only [45].

The Raman scattering cross section perpendicular to the incident beam for a transition from rotational quantum state J to J' is [45]:

$$\sigma_{J \to J'}^{\perp} = \frac{64\,\pi^2}{45} b_{J \to J'} \left(\omega_0 + \Delta\omega_{J \to J'} \right)^4 \gamma^2 \tag{5.45}$$

where ω_0 is the angular frequency of the incident light and $b_{J \to J'}$ are the Placzek-Teller coefficients. For simple linear molecules, b can be written as:

$$b_{J \to J+2} = \frac{3(J+1)(J+2)}{2(2J+1)(2J+3)} \tag{5.46}$$

$$b_{J \to J-2} = \frac{3J(J-1)}{2(2J+1)(2J-1)} \tag{5.47}$$

$$b_{J \to J} = \frac{J(J+1)}{(2J-1)(2J+3)} \tag{5.48}$$

Summing the polarization and space dependency of the scattering cross section over all polarization, and integrating over all directions, yields the relation for the total rotational Raman scattering cross section

$$\sigma_{tot} = \left(\frac{8\pi}{3} \right) \sigma^{\perp} (1 + 2p) \tag{5.49}$$

with p the depolarization.

5.6 Fluorescence Techniques

Properties of laser induced fluorescence (LIF) techniques make them ideally suited for diagnostics of atmospheric pressure plasmas. Especially the small dimensions and high spatial and temporal gradients of plasma jets used in medicine require diagnostic techniques that can study particle dynamics with high resolution both in space and time. Laser induced fluorescence, a measurement method widely used in combustion and flame diagnostics [46], provides the necessary resolution [11,47].

The general principle is that laser photons induce an electronic excitation of the probed atom or molecule, transferring the particle from a low-lying energy state <1> to a higher lying energy state <3>. The atom or molecule emits a fluorescence photon when it transfers from the higher lying energy state <3> to a lower lying energy state <2>. With calibrated photon flux and beam properties, population density of state <1> can be gained. Generally, a high photon flux is required for this technique. Therefore, as a light source, typically pulsed lasers are used. Most commonly laser induced fluorescence techniques are one-photon LIF (short LIF) and two-photon LIF (TALIF). Both techniques are described in the following section.

5.6.1 Single Photon Absorption Laser-Induced Fluorescence

LIF spectroscopy is the process in which absorption of one photon transfers an atom or molecule from a lower electronic state <1> (often the ground state) to a higher excited state <3>. After emission of a fluorescence photon, a lower state <2> is reached. De-excitation from state <3> back to state <1> is also possible for one-photon LIF. The use of one-photon excitation poses several experimental challenges that require consideration for an evaluation of the measurements. In optically thick media, the fluorescence photons can be reabsorbed, saturation effects can lead to a non-linear dependence of the fluorescence on the laser photon flux, and laser scattering obscures the fluorescence signal if absorbed and emitted photon are of the same frequency (de-excitation from state <3> back to state <1>). Especially, probing the concentration of ground state densities of light atoms such as atomic oxygen or hydrogen by LIF is difficult, since their electronic transition energies lie above 6.5 eV, which corresponds to laser wavelengths in the vacuum ultraviolet wavelength regime that require considerable experimental effort for generation and detection in atmospheric conditions. While these efforts are made in some cases [48], typically two UV photons are used to study light atoms using TALIF spectroscopy. LIF spectroscopy is typically used for detection of small molecules such as OH or NO. Indirect detection of species can be performed through an analysis of the fluorescence decay time. Collisional quenching of laser excited species reduces the effective fluorescent lifetime. With known quenching coefficients and quenching species, quencher-densities can be determined as is, e.g., the case for water concentration in plasma jets over a water surface [49].

LIF spectroscopy on molecular species, especially in cold atmospheric pressure plasmas, is complicated by energy dissipation processes. Rotational energy transfer and vibrational energy transfer dissipate the electronic energy over many ro-vibrational energy states, both in the excited and in the lower energy state (see **Figure 5.7**). When the lifetime of the electronic state is long enough, equilibrium with translational temperatures can be reached, and the energy level population distribution acquires Boltzmann characteristics.

In a simple two-level system without ro-vibrational energy transfer, where excitation from ground or low energy state <1> occurs to the excited energy state <2>, population depends on the photon absorption rate from state <1>

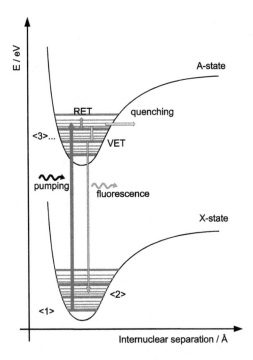

FIGURE 5.7 Energy scheme of LIF on molecules and energy dissipation processes involving rotational and vibrational energy transfer (RET & VET).

to state <2> given by the Einstein coefficient of absorption B_{12}. Depopulation of energy level <2> occurs through spontaneous emission, described by the Einstein coefficient A_2 and by quenching Q.

Generally, the population n_2 of level <2> follows the rate equation

$$\frac{dn_2(t)}{dt} = B_{12}I(t)f_B n_1(t) - \left[B_{21}I(t) - A_2 + Q\right]n_1(t) \qquad (5.50)$$

I is the time dependent laser intensity, B_{21} is the Einstein coefficient for stimulated emission. A_2 describes all processes of spontaneous emission from level <2>. The Boltzmann factor f_B describes the population ratio of two energy levels for a given temperature in case of levels in thermodynamic equilibrium. Collisional quenching is described by the quenching coefficient Q summed over all quenching species according to $Q = \Sigma_i\, n_i k_i$ with the density n and quench coefficient k of quenching species i. Quenching can simplify the rate equation if it includes all depopulating mechanisms including electronic and vibrational energy transfer (VET).

In certain cases, the following simplifications can be made so that quantitative results can be gained: For low laser intensities, the population of the lower energy state remains nearly constant and stimulated emission (B_{21}) can be neglected. Also, saturation of the LIF signal does not occur. Rotational energy transfer (RET) is very fast. If it is so fast that the lifetime of the excited energy state is dominated by the lifetime of the vibrational energy levels, LIF spectroscopy can be performed broadband to account for all rotational energy states. If vibrational energy transfer can be neglected and the ro-vibrational population of the ground state is in equilibrium with the gas temperature, the population of level <2> can be used to derive the ground state density by the Boltzmann equation.

Figure 5.8 shows a LIF setup with observation of the fluorescence emission at 90° angle. The LIF volume of observation is given by the intersection of the exciting laser beam and the optical path of the detection system. This yields high spatial resolution in the order of the Rayleigh range of the laser focus and the focusing optics. The LIF signal for a simplified two level energy system with above assumptions is [50] (cf. e.g., [51]):

$$S_{LIF} = \frac{\Omega}{4\pi} \int A_2 \xi n_2 \Gamma(\lambda) dV_c \qquad (5.51)$$

with Ω the solid angle of the detection optics, ξ the wavelength dependent efficiency of the detection system, Γ the line profile of the transition.

A dimensionless overlap fraction can be used to account for the spectrally broadened nature of both the laser radiation and the absorption transition [52]. The fraction represents the loss from optimum maximum interaction that is due to both detuning of the center frequencies and the

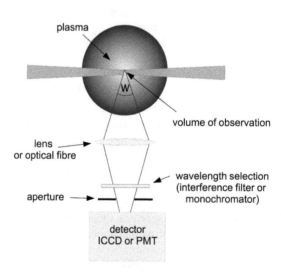

FIGURE 5.8 Typical LIF setup and intersecting volume element of excitation laser beam and optical path of detection.

different spectral distributions of the laser irradiance and the absorption transition [52].

For the simplified case, the solution to Equation (5.50) is:

$$n_2(t) = B_{12}n_1 \int_0^t I(t)e^{-(A_2+Q)(t-\tau)}d\tau \qquad (5.52)$$

Inserting Equation (5.52) in Equation (5.51) yields a fluorescence signal equation for which merely the quenching and optical parameters of the system have to be determined. These parameters need to be determined by calibration through Rayleigh scattering or absorption spectroscopy.

5.6.2 Two-Photon Absorption Laser-Induced Fluorescence

Using two photons for excitation of species has the advantage that UV photons can be used even for transitions of light atoms with excitation energy above 6.5 eV, which corresponds to single photon energies that reach VUV. VUV makes experimental approaches challenging due to more difficult generation and transport of light in the vacuum ultraviolet wavelength region. Two-photon absorption excites an atom from a lower state (e.g., ground state) <1> to an excited state <3> by simultaneous absorption of two photons via a virtual state <i> [53]. TALIF spectroscopy thus has a quadratic dependence on the laser energy and can thus not be calibrated with linear techniques such as Rayleigh scattering as is the case for one-photon LIF.

Figure 5.9 shows the population and depopulation mechanisms relevant for TALIF spectroscopy. From this, the rate equations can easily be derived, describing the processes of populating level <3> and depopulating level <1>.

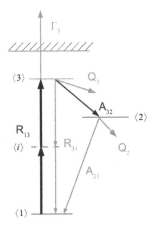

FIGURE 5.9 Population and depopulation of energy levels in TALIF spectroscopy on atomic species.

$$\frac{d}{dt}n_{\langle 1\rangle} = -R_{13}(t)n_{\langle 1\rangle}(t) \tag{5.53}$$

$$\frac{d}{dt}n_{\langle 3\rangle} = R_{13}(t)n_{\langle 1\rangle}(t) - n_{\langle 3\rangle}(t)\cdot\left(A_3 + Q_3 + \Gamma_3(t)\right) \tag{5.54}$$

Level <3> is depopulated by photoionization with the rate Γ, by quenching with the effective rate Q, and by spontaneous emission described by A.

The two-photon excitation rate R is given by the two-photon absorption cross section times the normalized line profile of the two-photon absorption and the squared laser intensity over hv.

The rate equations are valid under the assumption that the laser intensity is low enough to not significantly change the ground state density.

The dipole selection rules for two-photon transitions follow directly from the selection rules for a single photon transition, since the optical selection rules are a combination of two one-photon steps (see **Table 5.2**). The allowed changes for spin S, the orbital angular momentum L, and the total angular momentum J show that two-photon absorption can be used for transitions that are forbidden for single photon absorption. Not all transitions shown in **Table 5.2** are generally allowed. $\Delta J = 1$ is, for example, not allowed for photons of equal energy.

As described for absorption spectroscopy, TALIF spectroscopy also is affected by spectral broadening. For low pressure, Doppler shift and broadening can be used to determine temperatures and velocities of the probed species. Since the Doppler broadening results from velocity components with respect to the laser beam, using two photons in opposite directions allows to perform Doppler broadening free spectroscopy. For species at atmospheric pressure, collisional broadening will dominate the spectra.

The fluorescence photon number is given by

$$n_\phi(t) = A_{32}\int_0^\infty n_{\langle 3\rangle}(t)dt \tag{5.55}$$

With the population density $<3> = \int_0^t R(t')e^{-(A+Q)(t-t')}dt'$.

Table 5.2 Selection Rules for Allowed Changes for One and Two-Photon Transitions of an Atom		
Angular Momentum	One-Photon Selection Rule	Two-Photon Selection Rule
ΔS	0	0
ΔL	± 1	$0, \pm 2$
ΔJ	± 1	$0, \pm 1, \pm 2$

With the described dependence of R to the laser intensity, the number of fluorescence photons is given by:

$$n_\phi(t) = a_{32} \frac{\sigma^{(2)}}{(h\nu)^2} g(\nu) n_{\langle 1 \rangle} \int_0^\infty I_0^2 \, dt \qquad (5.56)$$

a_{32} is the reduced branching ratio $A_{32}/(A_3 + Q_3)$.

The fluorescence photon yield can be influenced by various effects: First, stimulated photoemission will occur at higher laser intensities and obscure the fluorescence signal. Second, amplified spontaneous emission (ASE) occurs mainly in direction of the laser beam reducing the detected photons. ASE results from a population inversion of states <3> and <2>, when depopulation by fluorescence or quenching of state <2> is slower than pumping of state <3>. Third, photodissociation can generate surplus atomic species that lead to an artificially higher species concentration measurement. All of these effects are a result from too high laser intensity. Therefore, LIF and TALIF spectroscopy require a careful evaluation of the applied laser intensity.

6

Diagnostics of the Core Plasma

THE CORE PLASMA IS characterized by electron dynamics and electron-ion and electron-atom interaction. Constantly a complex species generation and decomposition process takes place. Diagnostics need to be highly space and time resolved to yield information about neutral, charged, ground state, and excited species dynamics.

6.1 Electrical Diagnostics

Plasmas carry free electrons and are, therefore, conducting. This means that information about the plasma properties in the core plasma region can be gained from electrical measurements. Different techniques can be applied, depending on the type of excitation.

6.1.1 Measurement in kHz Discharges

In kHz plasma jets and in dielectric barrier discharges, voltage probe and current probe measurements yield reliable results. Still, separating the power dissipated in the plasma itself from, e.g., displacement current effects, is challenging and makes a thorough evaluation of measurements difficult. One approach is to use an equivalent circuit model [54,55]. For determination of power in dielectric barrier discharges, it has become the method of choice to measure the charge on the electrodes versus the applied voltage plot. First suggested by Manley [56], this method was soon performed on a variety of discharges [54,57–61].

Dielectric barrier discharges with a single dielectric on one electrode can be represented by a series of two capacitances. One is the capacitance of the

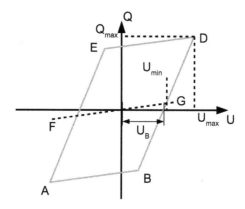

FIGURE 6.1 Charge voltage curve for sinusoidally excited dielectric barrier discharges.

gap C_G influenced by the dimensions and the gas type and the other is the capacitance of the dielectric C_D. The total capacitance can thus be calculated from the series of the two capacitances by

$$C_{tot} = \frac{C_G C_D}{C_G + C_D} \tag{6.1}$$

In a sinusoidally excited dielectric barrier discharge, the charge voltage diagram yields a parallelogram (see **Figure 6.1**) from which the dissipated electric energy can be derived. For model calculations based on equivalent circuit diagrams, the total capacitance C_{tot} of the electrode discharge assembly needs to be measured. This can be done either by measuring the current in time and determining the derivative of the time dependent voltage $U(t)$

$$i(t) = C_{tot} \frac{dU(t)}{dt} \tag{6.2}$$

or by measuring the accumulated charge, which is the integral of the time dependent current according to

$$\int_0^t i(t') dt' = Q(t) + Q(0) \tag{6.3}$$

Integrating both sides of **Equation (6.2)** yields

$$C_{tot} = \frac{Q(t) + Q(0)}{U(t) - U(0)} \tag{6.4}$$

This equals the slope of line $\overline{FG}$ in **Figure 6.1**. The accumulated charge in a dielectric barrier discharge can be measured by introducing a measuring capacitance in series with the electrodes and measuring the voltage drop across this capacitance [56]. The capacitance of the measuring capacitor C_M needs to be significantly larger than the capacitance of the dielectric and of the discharge gap. Measuring C_{tot} via **Equation (6.4)** has the advantage of being less sensitive to noise compared to taking the derivative of the measured voltage according to **Equation (6.2)**.

The enclosed area of the parallelogram measured in the charge-voltage diagram shown in **Figure 6.1** yields the power per excitation cycle deposited into the dielectric barrier discharge.

The discharge occurs at the paths $\overline{BD}$ and $\overline{EA}$. No discharge is present at the paths $\overline{DE}$ and $\overline{AB}$. The breakdown voltage U_B is the point where the Lissajous figure crosses the voltage axis. The area of the parallelogram (and thus the power per cycle) can be derived from

$$P_{cyc} = 4\, C_G U_0 \left(U_{max} - \frac{C_G}{C_{tot}} U_0 \right) \tag{6.5}$$

It should be noted that the minimal required voltage U_{min} for breakdown is slightly higher than the actual gap voltage at breakdown U_B and depends on the capacitance of the system. The proportionality factor between U_B and U_{min} is $\left(1 - C_{tot}/C_D\right)^{-1}$.

Determining the maximum voltage U_{max} and charge Q_{max} for different applied voltages will result in a straight line in the Q–V plot [61] of which the slope is C_D. This line can be used to determine C_D, even in short pulsed excited dielectric barrier discharges [61].

Figure 6.2 shows that this approach can be used to determine C_D also in more complex electrode geometries [62].

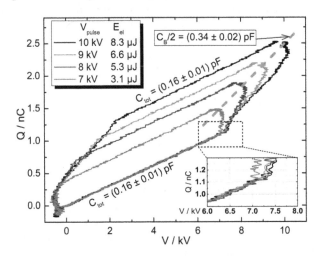

FIGURE 6.2 Q–V plot in a pin type dielectric barrier discharge to determine the electrical energy per pulse. (From Kettlitz, M. et al., *J. Phys. D Appl. Phys.* 45, 245201, 2012. © IOP Publishing. Reproduced with permission. All rights reserved.)

The approach to use Lissajous figure measurements for determination of power dissipation is most prominently used in dielectric barrier discharges, but has found its place in the electrical characterization of plasma jets (see e.g., [63]).

6.1.2 RF Discharges

Measuring dissipated power in radio frequency plasmas involves further challenges compared to power measurements at lower frequencies [64].

The average power in an electrical discharge is calculated by

$$P = f \int_0^{1/f} I(t) \cdot V(t) dt \tag{6.6}$$

$I(t)$ and $V(t)$ are the time dependent current and voltage signals. f is the frequency of the electric excitation. For a sinusoidal excitation, the power can simply be calculated from the mean values of the voltage ($V_{RMS} = V_0/\sqrt{2}$) and current ($I_{RMS} = I_0/\sqrt{2}$) signals as well as from their phase difference φ by

$$P = I_{RMS} \cdot V_{RMS} \cdot \cos(\varphi). \tag{6.7}$$

In radio frequency discharges, however, when using voltage probes for the voltage measurements, the probe's capacitance has a strong influence on the resonance matching circuit of the RF setup. This is due to the reactance as well as the resistance of the voltage probe typically being in the same order of magnitude as the ones for the matching unit and the plasma source [65]. Attaining reliable values for voltage and current in RF discharges is, therefore, difficult. For example, the capacitance of the μAPPJ studied in [65] is about 2 pF and the capacitance of the used voltage probe is 8 pF. Not necessarily the current and voltage amplitudes are the most problematic value, but the phase φ between voltage and amplitude results in large measurement errors, as an estimation by error propagation reveals. The error in the measured power from **Equation (6.7)** is calculated by

$$\frac{\Delta P^2}{P^2} = \left(\frac{\Delta V_{RMS}}{V_{RMS}}\right)^2 + \left(\frac{\Delta I_{RMS}}{I_{RMS}}\right)^2 + \left(-\tan(\varphi) \cdot (\cos(\varphi) - \cos(\varphi - \Delta\varphi))\right)^2 \tag{6.8}$$

An off-resonance phase-difference of 89° between current and voltage, and a measurement error of 1% in voltage, current, and phase, will result in an error in the power measurement of almost 90%.

One possible way to determine voltage and current in an RF discharge is to implement a full equivalent circuit model of the plasma source and the measuring probe. This approach leads to good results [66]. A simpler approach is to use a very basic matching unit consisting only of a coil. Measuring current and voltage between the power amplifier and the matching unit, instead of after the matching unit allows to operate the plasma at impedance matched conditions without the influence of the probes [64]. The power dissipated in the plasma can then be calculated from differences of the root mean square values of the power dissipated in the system with the plasma switched off and the dissipated power with the plasma switched on. This holds true, assuming that the power consumed within the matching coil remains constant, independent of the state of the plasma source. If simply a coil is used for matching, the power loss within the coil is from the resistive features of the coil leading to Joule heating. A temperature monitoring reveals any changes in power loss within the coil.

While the latter approach yields reliable results for the dissipated power, it has to be taken into account that in RF discharges, especially at frequencies above 13.56 MHz, the RF wavelengths lie in the meter range within the dimensions of the used cables. Thus, the positioning of the voltage and current probes has an influence on the measurement due to local minima or maxima of the wave. Furthermore, a distinction between forward propagating and backward propagating signal needs to be made.

Instead of voltage and current probes, directional couplers can be employed [65]. A calibration will yield the coupling factors for the forward directed signal (k_F) and for the backward directed signal (k_B). The dissipated power (with plasma switched on and plasma switched off, respectively) can then be calculated from the forward voltage V_F and the backward voltage V_B according to

$$P_{off/on} = \frac{V_{F,\ off/on}^2 \cdot k_F^2 - V_{R,\ off/on}^2 \cdot k_R^2}{50\,\Omega} \tag{6.9}$$

Figure 6.3 shows an exemplary power measurement performed as described. The power of the RF discharge with the plasma switched off is proportional to I^2 with the matching coil's resistance as proportionality factor. The difference

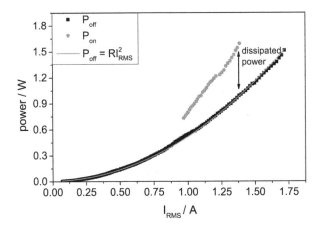

FIGURE 6.3 Power measurements in a capacitively coupled RF-argon discharge. (From Dünnbier, M. et al., *Plasma Sour. Sci. Technol.*, 24, 065018, 2015; Attribution 3.0 Unported (Creative Commons Attribution 3.0 licence in Ref. [67]), http://creativecommons.org/licenses/by/3.0/, 2017.)

between the power with the plasma off (black squares and solid line fit) and the power with the plasma switched on (grey circles) is the power dissipated within the plasma.

Error propagation performed on **Equation (6.9)** with the same error estimation of 1% used for **Equation (6.8)** yields a maximum error of 10%.

6.1.3 Short Pulsed Discharges

Short pulsed discharges make use of nanosecond excitation to avoid arcing and heating processes that lead to plasma instabilities. In nanosecond excitation, accurate measurements of the pulse amplitude are intricate. One technique yielding reliable current measurements is the use of a current shunt [68,69]. Current shunts are introduced into the power line. They have a low resistance and allow measuring the current through measuring the small voltage drop without strong influence on the circuit. In a forward current shunt, the resistors are inserted in the high-power line. A problem of the forward current shunt is that the high voltage lies also at the oscilloscope and in the measuring equipment. To avoid possible destruction of measurement equipment, in pulsed excitation, the reflected pulse can be used and the shunt can be placed in the grounded shielding of the coaxial cable used to connect the pulser to the plasma source. **Figure 6.4** shows a drawing of a respective arrangement. In [68], 10 low inductance 3 Ω resistors are soldered into the shielding of the coaxial cable (see **Figure 6.4**). The total shunt impedance Z_{shunt} amounts to 0.3 Ω.

The line current $I_{line}(z,t)$ can be calculated by [68]:

$$I_{line}(z,t)=\gamma \cdot \frac{Z_{out}+Z_{scope}}{Z_{scope}} \cdot \frac{V_{scope}}{Z_{shunt}} \tag{6.10}$$

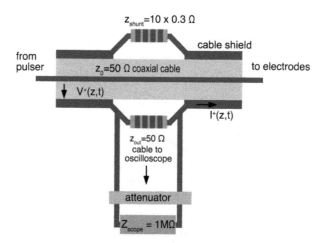

FIGURE 6.4 Schematics of the back current shunt. A comparison of forward and reflected pulse allows to determine the dissipated electric energy.

Z_{out} is the output impedance of 50 Ω of the shunt connector, Z_{scope} the impedance of the oscilloscope. The voltage across the shunt measured at the oscilloscope is V_{scope}. The voltage is attenuated by three 10 DB attenuators, which in
the equation is reflected by the factor γ of $10^{2/3}$.

6.2 Plasma Dynamics—Fast Imaging

Excited species in a plasma can be monitored by fast imaging. From fast imaging results, excitation processes in electrical discharges can be analyzed. For
a quantitative evaluation of the measurement results, model calculations are
usually required. When a wavelength selective element is included into the
setup, spectroscopic information can be gained. In repetitive discharges,
time resolution can be gained from imaging at specific times of the excitation period: Phase-resolved imaging or spectroscopic techniques can capture
dynamic processes with picosecond time resolution at MHz acquisition.

Figure 6.5a shows a typical setup for phase resolved optical emission spectroscopy (PROES). Essential for the measurement is a gateable and triggered
intensified charged couple device (iCCD) camera. Typically, an image intensi-
fier consisting of a photoelectric surface, a multichannel plate, and a fluorescent screen, is observed with a fast CCD camera. The photoelectric surface
generates electrons from the incident photons that are amplified pixel by pixel
through the multichannel plate. On the fluorescent screen, the amplified electrons result in a signal that can be recorded with the CCD camera. The multichannel plates' amplification is regulated by a high voltage across the plate.
The amplification is proportional to V^2. This high voltage can be switched on
and off on a picosecond time scale. By gating the amplifier, only photons from
a certain time frame generate a signal on the fluorescent screen. The camera

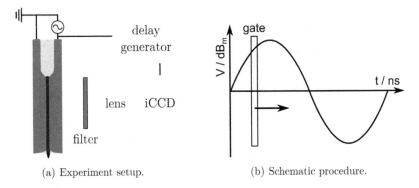

(a) Experiment setup.   (b) Schematic procedure.

FIGURE 6.5 Typical setup of filtered iCCD camera phase resolved optical
emission spectroscopy. (a) Experiment setup and (b) Schematic procedure.
(From Dünnbier, M. et al., *Plasma Sour. Sci. Technol.*, 24, 065018, 2015;
Attribution 3.0 Unported (Creative Commons Attribution 3.0 licence in Ref. [67]),
http://creativecommons.org/licenses/by/3.0/, 2017.)

averages multiple gate cycles in one image frame. If the gate is synchronized with the excitation cycle (see **Figure 6.5b**), an average image of the discharge optical emission at a certain position in the excitation phase can be acquired. Recording several images at increasing phase positions within the discharge excitation cycle produces an image sequence of the excitation dynamics of the plasma. With this technique, excitation dynamics were recorded, that coined the term plasma bullets [70]. The technique can be applied, when optical access allows the use of a camera. From the recorded images, a phase plot can be extracted, showing excitation dynamic on the y-axis and time on the x-axis (**Figure 6.6a** and **b**). Each data column is either a slice of an image at a certain point in time, or a binned pixel column in case of uniform plasmas such as the capacitively coupled μAPPJ shown in **Figure 6.5**. Binning reduces the noise of the images.

A wavelength selective element such as a filter allows observing only the emission from selected energy transitions (**Figure 6.7**). In [65], a filter at around 750.5 nm for the Ar($2p_1$) energy state and another filter at around 811 nm for Ar($2p_7$) and Ar($2p_9$) are used (see **Figure 6.6a** and **b**). These excited argon states are dominantly populated by electron impact excitation. For the studied discharge, the emission intensity of the Ar($2p_7$) at 810.37 nm is about 20% of the emission intensity of the Ar($2p_9$) line at 811.53 nm.

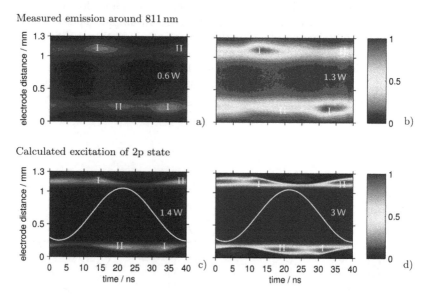

FIGURE 6.6 Spatio-temporal variation of the measured argon emission around 811 nm (a), (b) and calculated excitation rate of the 2p state (c), (d) for two different power settings at 27 MHz. The measured emission and calculated excitation rates are normalized to their respective maximum values at higher power. The white line marks the voltage course for better readability. (From Dünnbier, M. et al., *Plasma Sour. Sci. Technol.*, 24, 065018, 2015; Attribution 3.0 Unported (Creative Commons Attribution 3.0 licence in Ref. [67]), http://creativecommons.org/licenses/by/3.0/, 2017.)

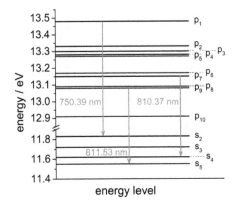

energy level

FIGURE 6.7 Three selected line transitions used for PROES measurements. (From Dünnbier, M. et al., *Plasma Sour. Sci. Technol.*, 24, 065018, 2015; Attribution 3.0 Unported (Creative Commons Attribution 3.0 licence in Ref. [67]), http://creativecommons.org/licenses/by/3.0/, 2017.)

A clear advantage of PROES is that it contains a multitude of time and space information as well as energy level information about the discharge. It can pose as validation and input for, e.g., fluid modeling of plasma processes. **Figure 6.6c** and **d** show the result from a fluid modelling of the capacitively coupled μAPPJ operated in argon. The excitation rate of the 2p state is shown in time. It can be seen that the excitation features match the measured emission profile well. Two power settings are studied.

Stochastic processes in plasmas, however, cannot be studied by this technique, as the image averaging process requires a trigger point that correlates to a specific emission feature in the discharge. In semi stochastic processes, e.g., in streamer type plasma jets where the plasma filaments propagate in open surroundings and the filament position is determined by the turbulent flow field, information about time development can be gained. Information about the special expansion of the filament is obscured by the averaging process [71], though. This loss in information depth makes the development of further single shot techniques necessary.

Is the plasma breakdown time point stochastic in time, measurement techniques other than phase resolved emission need to be applied. One such technique is cross correlation spectroscopy, where an optical event is correlated with an electric event, in order to record stochastic breakdown processes.

Streak cameras represent one of the earliest methods to record fast emission processes: Solar corona images were recorded using a slit moving quickly in front of a photo plate. Thus, depending on the velocity of the slit, time dynamics of fast processes could be recorded along the moving direction of the slit.

Modern streak cameras use a CCD camera for light detection. In the same way as with the multichannel plate, incident photons are transformed to electrons. The electrons' direction is deflected along the x-axis with time as in a fluorescent oscilloscope. The recorded image shows space resolution along one

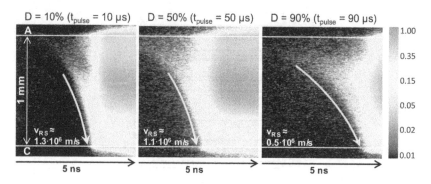

FIGURE 6.8 Breakdown process in a microdischarge as a function of duty cycle. (From Kettlitz, M. et al., *J. Phys. D Appl. Phys.*, 45, 245201, 2012. © IOP Publishing. Reproduced with permission. All rights reserved.)

axis and time resolution along the other axis. With this setup, fast processes of, e.g., streamer breakdown can be recorded [62] (see **Figure 6.8**). If an imaging spectrometer is mounted in front of the streak camera, spectral development in time can be recorded.

6.3 Electron Properties

Electrons are the key species of plasma and therefore determination of electron properties is fundamental to understanding plasma processes. At atmospheric pressure, however, electron properties of a plasma are most intricate to diagnose. From low-pressure plasma research, a number of electron property diagnostic methods are known, the most well-known of which is the use of a Langmuir probe. This technique uses an electric probe to which an electric potential is applied. From the second derivative of the *I–V* characteristics, the electron energy distribution of a plasma with isotropic electron velocity distribution can be directly measured. Although very sensitive to noise due to the derivation process, the simplicity of the approach and the direct access to the electron energy has made Langmuir probe measurements an integral technique for plasma diagnostics (see e.g., [72]). Nevertheless: At high pressure in regimes that are collision dominated, and in nonequilibrium plasma situations with plasma sheath regions in the range of μm, as well as in atmospheric pressure plasmas of sub millimeter dimensions, many of the assumptions fundamental to Langmuir probe data analysis do not apply. Reliable probe measurements of nonequilibrium plasmas at atmospheric pressure have yet to be made.

Most electron property measurement techniques that have been successfully applied to atmospheric pressure plasma jets can be applied in the core plasma region as well as in the plasma effluent (for diagnostics specific to the effluent only, see Chapter 7).

The main approaches to electron diagnostics are based either on scattering of light at the electrons, on measurement of the change in refractive index by

Table 6.1   Comparison of Electron Property Measurement Techniques

	OES (+ Model)	OAS (+ Model)	Langmuir Probe	Thomson Scattering
Space resolution	− −	−	+	+ +
Non-invasive	+ +	+	− −	+
Self sufficient	− −	− −	+ +	+ +
Experimental Simplicity	+ +	+	+ +	− −
Evaluation simplicity	− −	− −	+	+
Collisional plasmas	o	o	− −	+ +
Range of n_e	o	o	+ +	−
n_e and T_e	+	+	+ +	+ +

++ ≙ excellently suited, o ≙ neutral, − − ≙ not suited.

different electron densities, or on change in emission or absorption properties such as line broadening due to light electron interaction. These diagnostic techniques vary considerably in their ease of use and applicability. **Table 6.1** shows four measurement approaches in comparison. Optical techniques are mostly non-invasive. Methods based on optical emission or absorption such as line-broadening techniques are characterized by a relatively simple measurement setup, while requiring a rather complex evaluation procedure: Most of these techniques require atomic or collisional data as input for the evaluation. For example, the entirety of all broadening mechanisms needs to be accounted for in an evaluation of the small line broadening due to the electrons. Furthermore, optical emission spectroscopic techniques based on atomic transition lines in a collisional plasma require the use of a collisional radiative model that takes the relevant plasma processes into account. Laser based electron diagnostic methods, on the other hand, especially Thomson scattering, require a thorough and sometimes intricate experiment setup. The diagnostic results, however, are worth the experimental effort. Thomson scattering experiments have proven to be most reliable to date to determine electron energy distribution and temperature of atmospheric pressure plasma jets.

6.3.1 Scattering Techniques

Thomson Scattering

Thomson scattering is the elastic scattering of light on a free charged particle such as an electron. From the Thomson scattering signal, electron properties can be gained. **Equation (5.41)** shows that the electron temperature can be

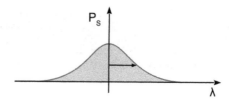

FIGURE 6.9 Broadened line profile of Thomson scattered radiation. The FWHM is proportional to the electron temperature.

determined from the width of the profile of the scattered light (see **Figure 6.9**). The broadening results from electron temperature dependent Doppler broadening.

Electron density can be determined from the scattered power of the Thomson scatter signal. After calibration, e.g., with a Raman scattered signal on a known gas concentration, the electron density is calculated from the ratio of the scattered signal powers and from the scattering cross sections multiplied by the Raman scatter species densities according to **Equation (5.44)**. Although a simple and universal technique, the following aspect of Thomson scattering complicates measurements significantly. The Thomson scattering signal is extremely weak. An estimation of the expected Thomson signal reveals the sensitivity required for a measurement setup [43]: Of a 20 W laser with a photon flux of several 10^{19} photons/s, only several 10^5 photons/s are Thomson scattered (at an electron density of 10^{18} cm^{-3}). From these photons, only 10^3 photons/s are collected by an optics with 10 cm diameter and a focal length of 60 cm. Taking the transmission probability of a detection system with lenses, spectrograph, and detector, into account: of the initial 10^{19} photons/s, only some 10 photons/s are detected. The Thomson scattered signal is 19 orders of magnitude lower than the incident laser beam. Any other scattering of the laser radiation, mainly Rayleigh scattering, will overcast the Thomson signal. This has important implications for the diagnostic setup. The very low intensity of the scattered signal means that highly sensitive detectors are required. The high noise from the laser radiation requires the use of filters with extremely high aspect ratios. Steep filter edges and strong optical density in the filter range can block out most of the direct unscattered laser light. Typically, triple grating spectrographs are used, in which the laser light is filtered by inverse apertures after the first grating. Additionally, scattered light is blocked by beam shaping apertures and walls inside the spectrographic system (see **Figure 6.10**). Further approaches make use of atomic filters that consist of an optical cell containing a gas that absorbs the central wavelength of the laser radiation.

Provided that attention is paid to the following aspects, Thomson scattering can be a non-invasive technique. The laser power must be high enough to yield a large enough scattering signal. At the same time, the laser power must be low enough to not disturb the plasma itself. Disturbances can, for example, occur through photoionization, multi-photon excitation,

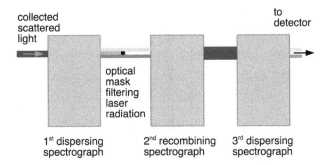

FIGURE 6.10 Triple grating used for filtering noise from scattered light at the laser wavelength. 2nd and 3rd spectrograph are for additional stray light reduction.

dissociation or ionization, and through laser heating of electrons [42,44]. To observe the correct laser energy regime, the linear dependence of the Thomson signal to the laser power as well as a constant Thomson line width must be ensured [42].

The scattered laser light's low power makes further reduction of noise by averaging measurements necessary. Thomson scattering, therefore, was so far only employed on repetitive plasmas, or plasmas with a high electron density [42]. Imaging of the scattered laser light along a laser line allows space resolved Thomson measurements. In plasma jets operated in air, this yields a mixed signal of Raman and Thomson scattering. Noble feed gas of the plasma jet will deplete the ambient species from the plasma region. Thus, in the centre, no Raman scattering on air species occurs. Only Thomson scattering on the electrons can be detected. Outside the plasma region, no electrons are present, and only Raman scattered light can be detected. In a single measurement, air species or temperature can be determined from the Raman scattered signal and electron properties can be deduced from the Thomson scattered signal (see **Figure 6.11**).

Scattering of Microwave Radiation

Electron densities can also be determined by scattering of microwave radiation on the oscillating plasma dipole. Typically, microwave scattering in weakly ionized plasmas is used in conditions where the plasma dimensions are considerably larger than the microwave wavelength. In [73] it could, however, be shown that when the length scale of the plasma is in fact much smaller than the wavelength of the incident microwave radiation, the plasma can be regarded as a point source for induced dipole radiation. Is the free electron distribution smaller than the wavelength of the microwave radiation, the wave is Rayleigh scattered. Time resolved plasma parameters such as time dependence of electron density and information about loss rates can be gained. The single shot measurement yields a highly time resolved signal which has, however, low space resolution, since it is acquired throughout the entire observed scattering volume.

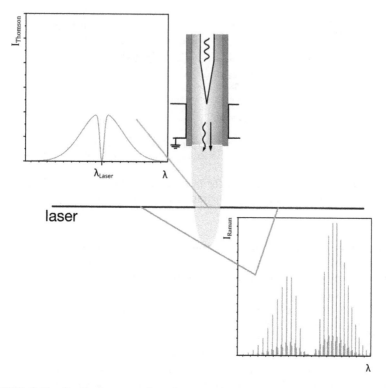

FIGURE 6.11 Scattering experiment on a plasma jet. Imaging spectroscopy allows 1D spectral imaging discerning Thomson scattering in the plasma jet with noble feed gas (no Raman scattering) and Raman scattering signal from ambient air for calibration and temperature measurements. From both spectra, the Rayleigh and Mie scattered signal at the laser wavelength is filtered.

With an incident microwave, ideally of linear polarization in the plane of the plasma column, the current is distributed throughout the plasma volume, when the skin thickness of the microwave-plasma interaction is $\delta = 2/\sqrt{2\mu_0\sigma\omega} > r$. r is the radius of the plasma column, ω is the angular frequency of the microwave, and σ is the plasma conductivity. The microwave electric field polarizes the plasma column and induces a dipole moment [73] leading to an oscillation of the electrons. The scattered electric field amplitude is proportional to the scattering volume (of the plasma column) and the plasma conductivity.

With $\sigma = e^2 n / m v_m$, a function of the electron charge e, mass m, and electron neutral transport collision frequency v_m, the electron density n can be derived from the calibrated scattering signal.

This technique has been applied to filamentary plasmas in various diagnostic configurations [74,75]. An exemplary setup is shown in **Figure 6.12**.

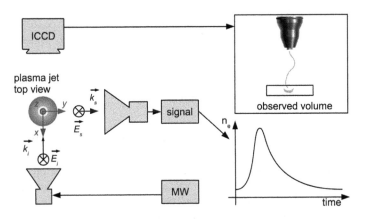

FIGURE 6.12 Setup of microwave Rayleigh scattering.

6.3.2 Electron Properties through Optical Emission Spectroscopy

Optical emission spectroscopy is used for determining electron densities in low pressure plasmas and plasmas in thermal equilibrium. In atmospheric pressure plasmas, where the electron energy distribution deviates from a Maxwellian distribution, and where assumptions made from local thermal equilibrium considerations like the Saha-Boltzmann relation require careful evaluation of their applicability, additional steps need to be taken to ensure reliable results.

In the following, the three most commonly used optical emission methods are presented and examples of their application for atmospheric pressure plasmas will be given.

Stark Broadening

As discussed in Section 5.2.3, the Stark broadening of an absorption or emission line is caused by interaction of the local electric field with the electromagnetic wave. Are electrons dominantly responsible for changes in the local electric field, the Stark broadened line profile is Lorentzian. For an evaluation of Stark broadening, all other broadening effects of the spectral line need to be accounted for. With sufficient spectral resolution of the detection system and high enough accuracy of the broadening parameter, Stark broadening of optical emission lines can be used to determine electron densities in nonequilibrium atmospheric pressure plasmas [10,11,18,20,23,27,42,76–78]. Typically, a calculation of the electron density is complex. Different approaches are possible [24]. It is advisable to use lines of those atoms that are already well studied. Commonly, Stark broadening of hydrogen H_α and H_β lines is investigated. In atmospheric pressure plasma jets, water impurities present in the feed gas can often be sufficient for hydrogen line emission. Otherwise, small admixtures of hydrogen or water [79] need to be added, with the drawback that either plasma properties or line broadening mechanisms are influenced.

Several approximations for the broadening of H_α and H_β (see (6.11)) have been published (see e.g., [80]) and have been applied for an evaluation of atmospheric pressure plasmas [27].

$$\Delta\lambda_{FWHM} = 4.800 \,\text{nm} \times \left(\frac{n_e}{10^{23}\,m^{-3}}\right)^{0.68116} \tag{6.11}$$

Stark broadening has prominently been used for dense plasmas with electron densities of greater than 10^{16} cm^{-3}. For these plasma conditions, theoretical calculations yield tabulated data that can be used for an evaluation. Electron densities below 10^{14} cm^{-3} as they are typical for atmospheric pressure plasmas make it necessary to consider fine structure levels of the Balmer line and each level has to be considered with separate Stark broadening [81].

The above description only takes into account Stark broadening assuming electron distribution homogeneous in space. Atmospheric pressure plasmas are characterized by high gradients and small dimensions. Thus, evaluation of the Stark broadening profile from line of sight measurements will only be an averaging over electron density values from different regions of the discharge. This gives only an averaging over different densities. A superposition of two electron concentrations together with a superposition of the resulting profiles will result in a better fit of the observed line profile [82] (see **Figure 6.13**).

While predominantly hydrogen lines are used for Stark broadening investigations, also argon emission lines as well as helium emission lines have been studied for an evaluation of electron properties in atmospheric pressure plasmas [82–84]. Since Stark broadening is observed in optical emission spectra, this technique can be applied in a great variety of discharges, including small scale discharges, e.g., filamentary DBDs [85], as well as micro gap discharges of 200 µm size [86]. In addition to atmospheric pressure plasmas, also plasmas in liquid have been studied with Stark broadening [76].

Bremsstrahlung Continuum

A technique that has been readily employed in fully ionized plasmas is the use of continuum radiation for electron diagnostics [27,87,88]. The Bremsstrahlung continuum originates from deceleration of electrons at ions and atoms. Three processes contribute to the emission continuum observed in plasmas: electron-atom free-free Bremsstrahlung, electron-ion free-free Bremsstrahlung, and electron-ion free-bound recombination radiation. The respective emissivities (ε_{ea}^{ff}, ε_{ei}^{ff}, ε_{ei}^{fb}) with ff denoting free-free and fb denoting free-bound are given by [89]:

$$\varepsilon_{ea}^{ff} = \sqrt{\frac{2}{m_e}}\,\frac{n_e n_a}{\lambda^2}\,\frac{hc}{4\pi}\int_{hv}^{\infty} Q_{ea}(\lambda,E)\sqrt{E}f(E)dE \tag{6.12}$$

$$\varepsilon_{ei}^{ff} = C_{ei}\xi_{ei}^{ff}\,\frac{n_e n_i}{\lambda^2}\,\frac{c}{\sqrt{kT_e}}\,e^{-hv/kT_e} \tag{6.13}$$

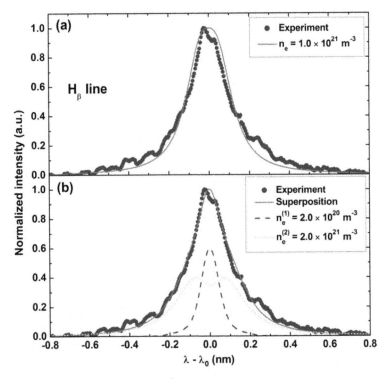

FIGURE 6.13 Single temperature fit (a) and superposition of two temperature fits (b) of the H_β line for a helium dielectric barrier jet. (From Xiong, Q. et al., *Plasma Sour. Sci. Technol.*, 22, 015011, 2013. © IOP Publishing. Reproduced with permission. All rights reserved.)

$$\varepsilon_{ei}^{fb} = C_{ei} \xi_{ei}^{fb} \frac{n_e n_i}{\lambda^2} \frac{c}{\sqrt{kT_e}} \left[1 - e^{-h\nu/kT_e} \right] \tag{6.14}$$

with $C_{ei} = \dfrac{1}{(4\pi\varepsilon_0)^3} \dfrac{32\pi^2 e^6}{3\sqrt{3}c^3 (2\pi m_e)^{3/2}} = 2.023 \times 10^{-63} \left[J^{3/2} m^3 sr^{-1} \right]$

ξ_{ei} are the Biberman factors and a function of wavelength, temperature, and ionization degree. For wavelengths <800 nm, they greatly deviate from 1 (see [89] and references therein).

In [87], continuum radiation was used to study electron density in a sub-atmospheric pressure arc plasma: At low pressure conditions and electron densities of nearly 10^{20} m^{-3}, electron-ion free-bound recombination radiation dominated the continuum.

In [90], the Bremsstrahlung continuum was investigated in a 13.56 MHz capacitively coupled argon RF discharge at atmospheric pressure. At 1.5 eV electron temperature with ratio of electron density and atom density smaller

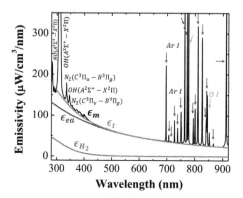

FIGURE 6.14 Fitted emissivity of the Bremsstrahlung continuum with correction for water dissociative radiation continuum in an argon atmospheric pressure capacitively coupled plasma. (From Park, S. et al., *Appl. Phys. Lett.*, 104, 084103, 2014, with the permission of AIP Publishing.)

than 10^{-3}, meaning an electron density of less than 10^{16}cm^{-3}, it was found that $\varepsilon_{ea}^{ff} \gg \varepsilon_{ei}^{ff} + \varepsilon_{ei}^{fb}$. Including dissociative continuum radiation from hydrogen impurities, the theoretical continuum radiation fits the measured spectrum well (see **Figure 6.14**).

It has to be noted that in [90] a Maxwellian electron energy distribution is assumed. In regions where no line emission from the plasma is present, ratios from two different wavelengths are used to take 2D images of the electron density distribution. With a comparative evaluation of the validity regime, this method may prove valuable for a simple determination of the electron density in atmospheric pressure plasmas.

Teaming Optical Emission Spectroscopy with Modelling

Electron properties can be determined using a combination of optical emission spectroscopy and collisional radiative models (CRM). From optical emission spectroscopy, line ratios of select emission lines are used [91]. This method is non-invasive and allows space resolved line of sight measurements. Frequently applied in plasmas where local thermal equilibrium can be assumed [92], a number of CRM approaches in connection with optical emission measurements have been applied for different plasmas. The common procedure is to establish a model of those atom (and molecule) energy states, which are identified to be relevant for the studied plasma parameters. For each energy level, a balance equation involving all population and depopulation mechanisms is implemented and the resulting rate equation system is solved.

At atmospheric pressure, energy states are typically not in Boltzmann equilibrium. Instead of pooling several energy states into an effective state, it is more realistic to resolve those energy states that can be expected to be not in equilibrium. This approach was used in [93], where the collisional radiative model includes kinetic processes that determine population densities of the first four ($1s_{5-2}$—Paschen notation) and next ten ($2p_{10-1}$—Paschen notation) excited levels, belonging to the $3p^54s$ and $3p^54p$ Ar level configurations, respectively. Emission lines from two different energy levels are compared. The ratio of the population of the selected 2p levels is strongly dependent on the electron density. The emission is proportional to the level population. By taking the ratio of two transitions, other constants are eliminated. The authors have used the ratio of $R_{13}(n_{2p1}/n_{2p3})$ for the electron density regime $n_e < 10^{13}$ cm^{-3} and the ratio of $R_{36}(n_{2p3}/n_{2p6})$ for the electron density regime $n_e > 10^{13}$ cm^{-3} (**Figure 6.15**).

[93] compared three different methods for measurement of electron density in three different types of plasma at atmospheric pressure. Stark broadening was compared to the line ratio measurement method and to an electrical model. All electron density results were in excellent agreement with each other (**Figure 6.16**).

It is important to validate model results with measurements to gain reliable electron property information. Phase and space resolved optical emission spectroscopy, for example, yields plenty data points for a validation of a model. Teaming PROES measurements with 1D, or 2D fluid simulations makes it possible to determine time and space resolved electron energy and density values. An example is shown in **Figure 6.17**. In [65], a fluid-Poisson model is employed in combination with PROES measurements. For the capacitively coupled argon plasma, seven excited atomic argon species are included. The argon energy levels used are shown in **Figure 6.7**. Energy levels $1s_5 - 1s_2$ are resolved. 2p levels are not resolved for each level, as in the previously described approach, but are separated into three groups, namely $2p = 2p_{10}$ to $2p_5$, $2p' = 2p_4$ to $2p_1$, and hl = all higher lying states. With an improved drift-diffusion approximation for the electron particle and energy fluxes [94],

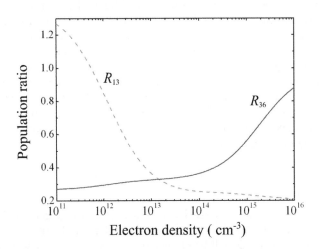

FIGURE 6.15 Population ratio of $R_{13}(n_{2p1}/n_{2p3})$ and $R_{36}(n_{2p3}/n_{2p6})$ as a function of electron density. (From Zhu, X.M. et al., *J. Phys. D Appl. Phys.*, 42, 142003, 2009. © IOP Publishing. Reproduced with permission. All rights reserved.)

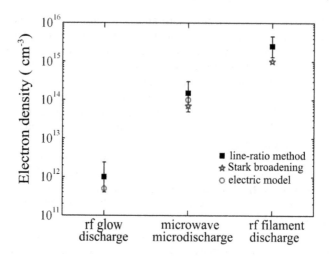

FIGURE 6.16 Comparison of electron energy measurements on three different plasma types with three different measurement techniques. (From Zhu, X.M. et al., *J. Phys. D Appl. Phys.*, 42, 142003, 2009. © IOP Publishing. Reproduced with permission. All rights reserved.)

the transport of the electrons in time and space can be accurately described. Input for the model is the power dissipated in the plasma that was measured according to Section 6.1.2. Further input is the gas temperature measured by resonant line broadening of an argon emission line.

Figure 6.17 shows that the spatio-temporal features of the argon 2p1 emission follow the ground state excitation of the 2p′ modelled energy state.

The teamed modelling and OES approach yields detailed information about electron properties and plasma mode. The model accurately describes the transition from α- to γ-mode in the atmospheric pressure discharge. The α-mode is dominated by volume processes where bulk electrons sustain the plasma. In the γ-mode discharge secondary electrons contribute to ionization and the discharge becomes hotter with a higher electron density. It is found that the critical voltage for the α- to γ-mode transition follows the excitation frequency with a $1/f$ law. Also the $f^{3/2}$ scaling law of the α- to γ-mode transition found for low pressure RF capacitively coupled glow discharges described by Raizer et al. [95] is confirmed for atmospheric pressure capacitively coupled RF discharges (see **Figure 6.18b**).

6.3.3 Electron Density Measurement Techniques Based on the Refractive Index of Plasma

Several diagnostic techniques based on measuring the change in refractive index by plasma have been readily employed to study electron density. A few of the most intriguing methods will be presented, the reason for their selection here being: (a) they have yet not been extended to atmospheric pressure plasmas, but show potential to be in future, (b) they have (tentatively) been used for atmospheric pressure plasmas and need to be studied further for their validity regime,

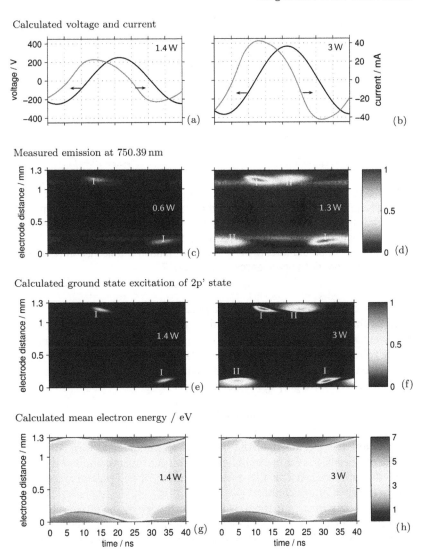

FIGURE 6.17 Temporal variation of calculated discharge voltage and current (a), (b), measured argon emission at 750.39 nm (c), (d), calculated ground state excitation rate of 2p′ state (e), (f) and calculated mean electron energy (g), (h) for two different power settings at 27 MHz. The measured emission and calculated excitation rates are normalized to their respective maximum values at higher power. (From Dünnbier, M. et al., *Plasma Sour. Sci. Technol.*, 24, 065018, 2015; Attribution 3.0 Unported (Creative Commons Attribution 3.0 licence in Ref. [67]), http://creativecommons.org/licenses/by/3.0/, 2017.)

Nonequilibrium Atmospheric Pressure Plasma Jets (N-APPJs)

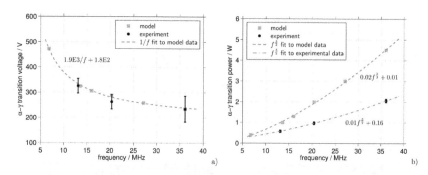

FIGURE 6.18 α- to γ-mode transition voltage with $1/f$ dependence on the frequency (a) and $f^{3/2}$ scaling law for the transition of α- to γ-mode with the frequency in a capacitively coupled argon atmospheric pressure plasma jet (b). (From Dünnbier, M. et al., *Plasma Sour. Sci. Technol.*, 24, 065018, 2015; Attribution 3.0 Unported (Creative Commons Attribution 3.0 licence in Ref. [67]), http://creativecommons.org/licenses/by/3.0/, 2017.)

or (c) they can be used for atmospheric pressure plasmas that are either larger or hotter than plasma jets typically used for plasma medicine and need further improvement to meet the requirements of small scale, high gradient plasmas.

The refractive index of a collisionless plasma can be expressed by the cut-off electron density and the plasma frequency. The dispersion relation of an electromagnetic wave propagating in a plasma is

$$\omega^2 = \omega_p^2 + \left(c \cdot k\right)^2 \tag{6.15}$$

ω is the angular frequency of the incident electromagnetic wave, ω_p is the plasma frequency, c is the speed of light and k is the wave vector. For diagnostic methods making use of this relation at atmospheric pressure, collisions have to be taken into account.

The refractive index n is the speed of light divided by the phase velocity of an electromagnetic wave. The phase velocity is the angular frequency divided by the wave vector. Thus, with $n^2 = \left(c \cdot k\right)^2/\omega^2$ results from **Equation (6.15)**

$$n^2 = 1 - \frac{\omega_p^2}{\omega^2} \tag{6.16}$$

Figure 6.19 shows the square of refractive index n as a function of angular frequency of the electromagnetic wave according to **Equation (6.16)**.

It is evident that the refractive index becomes zero, when the angular frequency of the incident electromagnetic wave is equal to the plasma frequency. The plasma frequency is related to the electron density n_e according to

$$\omega_p^2 = \frac{n_e e^2}{\varepsilon_0 m} \tag{6.17}$$

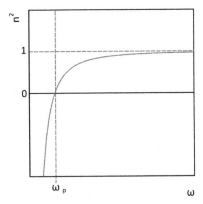

FIGURE 6.19 Square of the refractive index as a function of frequency.

with e the electron charge, m the electron mass, and ε_0 the permittivity of free space. When the refractive index becomes zero, an electromagnetic wave of the plasma frequency is reflected on the plasma.

The refractive index (**Equation 6.16**) can be rewritten with the cut-off electron density $^c n_e$ as a function of electron density to

$$n = \sqrt{1 - n_e \Big/_{^c n_e}} \qquad (6.18)$$

In a collisional plasma, the real part of the refractive index can be derived from Appleton's equation [96] for the unmagnetized case with a magnetic field $B = 0$ to

$$n' = \sqrt{\frac{\Xi}{2} + \frac{1}{2}\sqrt{\Xi^2 + \frac{v_c^2}{\omega^2}\left(\frac{\omega_p^2}{\omega^2 + v_c^2}\right)^2}} \qquad (6.19)$$

with v_c^2 the electron neutral collision frequency and $\Xi = 1 - \dfrac{\omega_p^2}{\omega^2 + v_c^2}$

In a collisional plasma, microwave radiation is attenuated with the attenuation index

$$\chi = \sqrt{-\frac{\Xi}{2} + \frac{1}{2}\sqrt{\Xi^2 + \frac{v_c^2}{\omega^2}\left(\frac{\omega_p^2}{\omega^2 + v_c^2}\right)^2}} \qquad (6.20)$$

The collision frequency is related to the electron temperature T_e via

$$v = n_0 \sqrt{2kT_e \Big/_{m_e}} \, \sigma_0 \qquad (6.21)$$

n_0 is the neutral particle density, m_e is the electron mass, k is the Boltzmann constant.

Equations (6.20) and **(6.21)** can be simplified, if $v \gg \omega_p$ [97,98].

Changes in refractive index can be studied using interferometry. Microwave interferometry has been used in low-pressure plasmas (see e.g., [99,100]). The above relations can be utilized to measure electron densities as well as electron temperatures in collisional plasmas, i.e., plasmas at atmospheric pressure. In a helium dielectric barrier discharge at atmospheric pressure, microwave interferometry has been used to determine electron density and electron temperature as a function of time [101,102].

As alternative to microwave interferometry, THz time domain spectroscopy was used to determine electron densities [103]. The high absorption of radiation in the THz regime and the fact that THz frequency is close to typical plasma frequencies allows measuring low plasma densities from 10^{11} to 10^{13} cm^{-3}. In [104], THz spectroscopy was compared to microwave interferometry and excellent agreement was found for a variety of discharge gases and parameters with pressures up to 20 Pa. THz spectroscopy is less sensitive to vibrations. Furthermore, THz time domain spectroscopy measures the phase shift of an electromagnetic pulse and is not limited by the plasma frequency.

Microwave transmission measurements (without a second interferometry arm) were performed on a dielectric barrier discharge array in nitrogen [105]. The results were compared to an interferometry method based on CO$_2$-laser heterodyne interferometry that has special resolution in the µm regime [106]. Heterodyne infrared interferometry uses a reference laser beam that is shifted in the order of MHz with respect to the probing beam. This allows performing interferometry with the beat frequency of the two laser beams.

It needs to be mentioned that in plasmas of low ionization degree at atmospheric pressure that are optically thin, the refractive index also has contributions of heavy particles. The electron density can be gained from an equation similar to Equation (6.16) with heavy particles taken into account. [107] also used heterodyne interferometry in the infrared to study electron densities, separating the contribution of the heavy particles to the refractive index from the contribution of the electrons by pulsing the discharge. Fast changes in refractive index are attributed to electron contribution and slow changes to heavy particles. The refractive index depends on the electron density according to [107]

$$n-1 = -\frac{e^2}{2\left(c^2 m_e \varepsilon_0 4\pi^2\right)}\lambda^2 n_e + \left(A + \frac{B}{\lambda^2}\right)\frac{n_h}{n_h^o} \qquad (6.22)$$

λ is the wavelength of the probing radiation, n_h is the heavy particle density at the given conditions, and n_h^o at normal conditions of $T = 273$ K and 1 bar pressure.

For air at 1 bar and 273,15 K (as studied in [107]), A is 2.871×10^{-4} and B is 1.63×10^{-18} m^2. A similar approach of phase-modulated dispersion interferometry in [108] uses second harmonic generation on a CO_2 laser for electron density determination.

A non-interferometry method for measuring electron density in cold atmospheric pressure plasmas makes use of beam deflection by changes in the refractive index similar to gas temperature measurements in laser Schlieren deflectometry [109], described in Section 7.3.2. Employing a Shack-Hartmann sensor allows to image the beam deflection of an expanded laser beam in a 2D-map [110]. Shack-Hartmann sensors measure the wave front of an incident light-wave with a lenslet array in front of a CCD chip. Fundamental for this measurement technique is the assumption that, while the refractive index depends on the densities of electrons, neutrals, and ions, only the contribution of the electrons exhibits a wavelength dependency [111]. Using different wavelengths for the measurement thus allows determining electron densities. The technique was employed to measure 2D electron density distribution in a positive pulsed streamer in air [112]. The sensitivity for the electron density of a laser produced plasma has been determined to be 5×10^{17} cm^{-3} [113].

6.4 Excited and Ground State Atomic, Molecular, and Ion Species

6.4.1 Absorption

Gaining absolute densities of ground state and excited state species in the core plasma region of plasma jets is difficult. A calibration free method is absorption spectroscopy. Several studies investigate large area plane parallel electrode assemblies in DBD or jet geometry with kHz or RF-excitation. Typical plasma jets are, however, very small, so that applying absorption spectroscopy to the core plasma region is in most cases impossible due to the lack of optical access or absorption length. A small number of jet geometries have been developed that allow a line of sight optical access through the discharge region to perform absorption spectroscopy.

There are only few plasma geometries that allow optical access from two sides in order to perform core plasma absorption spectroscopy.

The μAPPJ (**Figure 6.20c**) is a plasma jet specifically designed to allow access to the discharge region. This capacitively coupled plasma jet has a 1×1 mm cross-section and a 20–30 mm long plasma channel between two plane parallel electrodes. Its sides are capped with glass to allow optical access from both sides and from an angle. Absorption spectroscopy inside the core plasma on a broader electrode dimension was performed for ozone [123] by UV LED absorption spectroscopy. The setup is shown in **Figure 6.21**. Furthermore, VUV absorption spectroscopy on O(^{1}D) was performed by

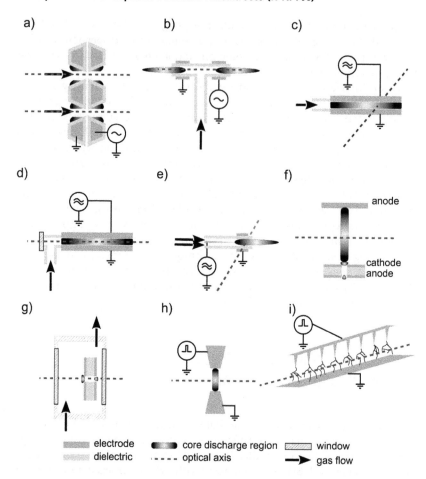

| electrode | core discharge region | window |
| dielectric | optical axis | gas flow |

FIGURE 6.20 Nonthermal atmospheric pressure plasma setups allowing absorption spectroscopy inside the core discharge region. Possible geometries are (a) grid-based dielectric barrier discharges [105] (b) T-shaped tubular jets [114] (c) and (d) parallel plate capacitive jets [115,116] (e) tubular jets with pin centered electrode and open ring outer electrode [117] (f) microhollow cathode with extended plasma region [118] (g) microhollow cathode with optical path through the hollow cathode [119] (h) pin to pin geometry without dielectric [120] or with dielectric [62] (i) long absorption lengths using several pin electrodes in a row [121] or by using a knife edge electrode in a DBD type discharge [122].

synchrotron radiation [124]. Highly spectrally resolved measurements yielded results on the atomic oxygen concentration inside the discharge. Additionally, OH UV-absorption spectroscopy with a laser-stabilized arc lamp [125] was performed [126].

Access to the plasma discharge region can also be gained in a T-shaped jet [114]. Typical kHz streamer type jets with double ring electrodes around dielectric tubing usually do not allow two-sided optical access because the gas

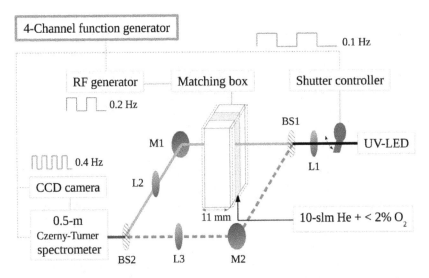

FIGURE 6.21 Schematic of the two-beam UV-LED absorption setup. The *probe beam* and *reference beam* are represented by the grey-solid line and the dashed line, respectively. (From Wijaikhum, A. et al., *Plasma Sour. Sci. Technol.*, 26, 115004, 2017; Attribution 3.0 Unported (Creative Commons Attribution 3.0 licence in Ref. [67]), http://creativecommons.org/licenses/by/3.0/, 2017.)

line blocks one side. In a T-shaped jet, two effluents emerge on either side of the T-exits (**Figure 6.20b**). Using an L-shaped plasma jet geometry (shown in **Figure 6.20d**) and capping one side with a window represents a further solution to avoid having the feed gas line block the view. These geometries make absorption spectroscopy possible within the discharge region albeit with the drawback that all plasma regions are probed along the line of sight.

Having to probe core plasma and effluent of a plasma jet simultaneously by line of sight absorption along the jet axis can be avoided when setting the optical axis perpendicular to the plasma jet. In jet effluents, this approach has been used to study species concentrations in numerous investigations as discussed in Sections 7.2.1 and 7.2.2. A predominant reason why absorption spectroscopy can usually not be performed within the core plasma is that the electrodes block the view. By using mesh electrodes, or—in a pin type plasma jet with outer, grounded electrode—splitting the grounded electrode as shown in **Figure 6.20e**, even the onset of the streamer formation within the dielectric tubing can be observed [117].

Non-jet-like nonthermal plasmas often allow for a better optical access to the discharge region. Examples that have relevance to the kinetics of plasma jets used in plasma medicine are microhollow cathode discharges that can be probed either directly through the hollow cathode region (**Figure 6.20g**) [119], or through an extended plasma via a third electrode (**Figure 6.20f**). Grid electrode (**Figure 6.20a**) configurations (in [105] also with a micro sized gap between the electrodes covered with a dielectric) allow absorption spectroscopy through the discharge region in the grid cavities. A gas flow through the grid generates a blown-out dielectric barrier discharge jet.

Most accessible for an analysis of the discharge core region is a pin to pin electrode configuration—with metal electrodes [120] (see **Figure 6.20h**) or dielectric covered electrodes [62]. In [21], cavity ring-down spectroscopy was used for space resolved absorption spectroscopy on nitrogen in a pin to pin electrode discharge. Pin to pin electrode configurations are rarely used for plasma jets. They are nevertheless worthwhile to mention, since breakdown mechanisms and reaction kinetics are very similar to jet-like streamer discharges. The same argument can be made for pin to plane geometries. These point-like sources of plasma generation all exhibit the drawback that small-scale geometries result in small absorption lengths. When long averaging or integration times are not possible, long absorption path lengths are required to achieve low signal to noise ratio. With pin to pin or pin to plane geometries, this can be achieved by using several pin electrodes in a row [121] or by using a knife edge electrode in a DBD type discharge [122] (see **Figure 6.20i**).

In plasma jets providing optical access to the core plasma region, tunable laser diode absorption spectroscopy can be successfully used to determine metastable species [127,128]. The absorption cross section is large enough and the stability of diode lasers allows a high signal to noise ratio. The space resolution allows to measure spatial profiles of the metastable density and to compare these profiles to plasma models [129]. Should the absorption length be not sufficient, cavity enhanced methods such as cavity ring down absorption spectroscopy can enhance the sensitivity of the measurement system [120].

6.4.2 Laser Fluorescence and Scattering Methods

Laser scattering or laser-induced fluorescence spectroscopy methods allow high spatial resolution down to the beam focus diameter. Regarding the experiment setup, scattering, and fluorescence methods are more complex than absorption techniques. Scattering and fluorescence setups require access (entrance and exit path) for the laser beam and a third optical access for observation of scattered light or fluorescent radiation. Is the respective access given, scattering and fluorescent methods yield highly space and time resolved measurement results. With sufficient calibration, laser-induced fluorescent techniques have long been used to study atmospheric pressure plasmas [47,130]. Pin to pin geometry offers a broad diagnostic access possibility as shown by Rayleigh, LIF and TALIF measurements at lower pressure determining H, and OH concentrations along with gas temperature [131]. Fluorescence measurements are not limited to ground state species but allow the study of metastable species as demonstrated in a pin to plate geometry at atmospheric pressure [132]. Vital to the reaction chemistry of atmospheric pressure discharges is the generation of atoms. Atoms of light elements have their electronic transition from the ground state at energies that require vacuum ultraviolet photons for single photon excitation. Two-photon excitation allows using ultraviolet radiation for detection of hydrogen [133], nitrogen [134], and oxygen atoms [115] in the discharge region of cold

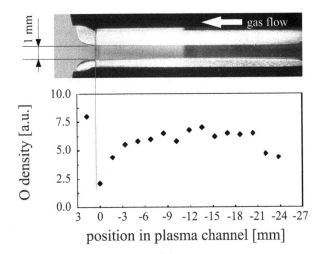

FIGURE 6.22 Atomic oxygen concentration along the discharge of a capacitive coupled atmospheric pressure plasma jet in helium measured with TALIF spectroscopy. The concentrations of atomic oxygen reached 3×10^{16} cm^{-3} with a linear dependence on input power. (From Knake, N. et al., *Appl. Phys. Lett.*, 93, 131503, 2008, with the permission of AIP Publishing.)

atmospheric pressure plasmas. TALIF spectroscopy was used for atomic oxygen detection in the discharge region of a capacitive coupled μAPPJ operated in helium. **Figure 6.22** shows the parallel plate electrode geometry of the plasma jet and the discharge channel in which the TALIF measurements were performed [115]. From the concentration development along the jet axis, production rates can be deduced.

TALIF spectroscopy requires observation preferably at a 90° angle to the exciting laser beam. In [135], nitrogen atom measurements were performed in a DBD in air with a glass dielectric plate on top of a pin array. Observation of the fluorescence radiation in the streamer located on the center pin of the array was performed through a high-voltage water electrode on top of the glass dielectric.

Molecular concentrations in ground state and in excited state can be determined by one-photon LIF spectroscopy. $N_2(A^3\Sigma_u^+)$ concentration, for example, have been determined by LIF spectroscopy in a pin to plate geometry [132] as well as in a DBD arrangement [136]. [136] applied so-called optical-optical double resonance (OODR) LIF developed in [137]. OODR LIF measurements of $N_2(A^3\Sigma_u^+)$ makes use of the fact that at high pressure, the transition probabilities of the second positive system of nitrogen are about 100 times larger than those of the first positive system [138]. Using two separate lasers thus probes

$$N_2\left(A^3\Sigma_u^+, v''\right) + h\nu_{L_1} \to N_2\left(B^3\Pi_g, v'\right) + h\nu_{L_2} \to N_2\left(C^3\Pi_g, v\right) + h\nu_E \quad \textbf{(6.23)}$$

While LIF and TALIF spectroscopy yield atomic and molecular densities, the most reliable method to get information about the electron properties in an atmospheric pressure nonequilibrium discharge is Thomson scattering. Few reports of measurements inside the discharge region of plasma jets have been published. In the core plasma of a capacitive coupled helium discharge, Thomson scattering has been performed in [128] where, additionally, Thomson scattering proved to be sensitive for highly excited electronic states in Rydberg molecules.

6.4.3 Atomic and Molecular Densities from Optical Emission Spectroscopy

Optical emission spectroscopy remains the least complex experimental technique, the data evaluation set aside. At atmospheric pressure, excitation of atoms and molecules to radiative states does not occur purely through electron impact excitation. Pooling reactions and heavy particle collisions that lead to energy transfer populate radiative states in a non-negligible manner. An analysis of optical emission spectra has to take this into account. With the correct approaches, atomic densities can be determined inside the discharge region of plasma jets by optical emission spectroscopy.

One approach is to observe the emission in the time afterglow of the plasma, where electron impact excitation is negligible. For nitrogen containing plasmas, the time development of the emission intensity signal of $N_2(B, v)$ in the plasma afterglow [139] can be used to determine the nitrogen atom density in the plasma through extrapolation.

Nitrogen emission originates dominantly from the recombination of nitrogen atoms according to

$$N + N + M \xrightarrow{k_1} N_2\left(B\,{}^3\Pi_g, v'\right) + M \tag{6.24}$$

where v' denotes the vibrational level. The $N_2\left(B\,{}^3\Pi_g, v'\right)$ state is depleted by relaxation into the A-state through spontaneous emission

$$N_2\left(B\,{}^3\Pi_g, v'\right) \xrightarrow{k_2(v')} N_2\left(A\,{}^3\Sigma_u, v''\right) + h\nu \tag{6.25}$$

and by quenching through the main operating gas (He in [141])

$$N_2\left(B\,{}^3\Pi_g, v'\right) + M \xrightarrow{k_3(v')} N_2\left(\neq B\right) + M \tag{6.26}$$

From the resulting rate equations, with known $k_i(v')$, the following relation of the N-atom density to the $N_2\left(B\,{}^3\Pi_g, v'\right)$ emission intensity can be derived [139]

$$n_{N(t_0)} = \frac{-d\ln\left(I_{N_2\left(B\,{}^3\Pi_g, v'\right)}\right)/dt}{4k_1(v')n_M} \tag{6.27}$$

Diagnostics of the Core Plasma

This method is valid, if only **Reactions (6.24)** through **(6.26)** are contributing to the nitrogen atom kinetics, in a regime, where $1/\left(d\ln\left(I_{N_2(B^3\Pi_g, v')}\right)\right)/dt$ is linear with t. Extrapolating this function to $t = 0$ yields the nitrogen atom concentration in the discharge. In a capacitive atmospheric pressure plasma jet, this method can be applied in α- as well as in γ-mode [140].

Metastable molecular nitrogen densities can be obtained in atmospheric pressure nitrogen discharges by optical emission spectroscopy, when small admixtures of NO are added to nitrogen. In these mixtures at atmospheric pressure, ground state NO is dominantly excited by metastable nitrogen $N_2(A^3\Sigma_u^+)$ molecules [141].

$$N_2\left(A^3\Sigma_u^+\right) + NO\left(X^2\Pi\right) \rightarrow NO\left(A^2\Sigma^+\right) + N_2\left(X^1\Sigma_g\right) \qquad (6.28)$$

The emission of NO-γ $\left[X^2\Pi \rightarrow A^2\Sigma^+\right]$ can thus be taken as a measure for $N_2(A^3\Sigma_u^+)$ metastable concentrations as shown e.g., in pulsed corona discharges [142] and in DBD discharges [136].

Teaming optical emission spectroscopy with modelling yields absolute atomic oxygen densities [143]. Investigation of a helium RF atmospheric pressure plasma jet was performed by so-called diagnostic based modelling (DBM). While TALIF spectroscopy also yields reliable density results of atomic oxygen (see above), it requires optical access for laser beam as well as for observation optics. DBM requires only one optical access for spectroscopic observation and can, therefore, be used in plasmas where TALIF measurements are impossible. For DBM, a model based on one-dimensional numerical fluid simulation combined with kinetic treatment of the electrons [144–145] is used. The DBM method is a modified actinometry approach. In actinometry, two optical emission lines are compared in their intensities: one originating from the investigated species, the other from an actinometer species of known density. It is assumed that (1) excitation of the two emission lines arises from electron impact excitation from the ground state and (2) primary decay results from photon emission. Given that (3) both transitions' electron impact excitation cross sections match in threshold and shape as a function of electron energy, the two species' line intensity ratio will be proportional to their density ratio [146]. These assumptions, however, cannot be made for plasmas at atmospheric pressure. Since at atmospheric pressure collisional de-excitation can outbalance the optical transition rate, condition (2) will not hold true. Neither is condition (3) given, since the electron impact excitation thresholds show a significant difference of 2.5 eV for O(3p ^{3}P) and Ar(2p$_1$), the states observed in [143]. By using spatio-temporal-characteristic results from modelling (see Section 6.3.2) time and space resolved electron dynamics can be gained. From this, effective excitation rate coefficients can be determined for the observed transitions and thus the atomic oxygen density can be derived, taking relevant processes including dissociative attachment into account. **Figure 6.23** shows the atomic oxygen concentration measured inside the discharge region of a capacitive coupled atmospheric pressure plasma jet operated in helium with small argon admixtures as actinometer gas and

301

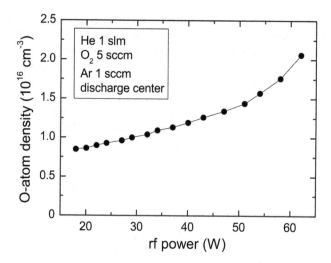

FIGURE 6.23 Atomic oxygen density measured in the μAPPJ as a function of input power. (From Niemi, K. et al., *Appl. Phys. Lett.,* 95, 151504, 2009, with the permission of AIP Publishing.)

oxygen as precursor gas. The overlinear increase in oxygen concentration can be attributed to the onset of γ-mode discharge with a higher electron density.

6.5 Energy Dissipation in the Core Plasma Region

Energy dissipation in the core plasma region of atmospheric pressure plasma jets strongly depends on the type of gas and admixture used, on the modality of the electric energy supply, or the electrode configuration. Furthermore, it is important to know time scales and species abundances in order to assess possible energy dissipation mechanisms (**Figure 6.24**).

Inside the discharge region, electric energy is dissipated through the electric field. Characteristic time for interaction is given by the polarizability of

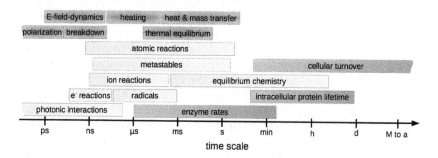

FIGURE 6.24 Timescales in atmospheric pressure plasmas. (From Reuter, S. et al., *J. Phys. D Appl. Phys.,* 51, 2018; Attribution 3.0 Unported (Creative Commons Attribution 3.0 licence in Ref. [67]), http://creativecommons.org/licenses/by/3.0/, 2017.)

the medium which is calculated by $\tau = \frac{\varepsilon_0}{\sigma}$, ε_0 is the vacuum permittivity and σ is the plasma conductivity. τ is also called Maxwell time. The plasma conductivity is given by $\sigma = \frac{e^2 n_e}{m_e \nu_e}$. Here, $\frac{e}{m_e \nu_e}$ is the electron mobility, which is in the order of 10^{-12} s^{-1} at atmospheric pressure and room temperature. Local E-field dynamics are influenced by the space charges that depend on the electron collision frequency (see Equation 6.21).

At atmospheric pressure, collisions are dominant and heavy particle collisions contribute significantly to a dissipation of the electric energy. Collisions are ruled by energy and charge conservation. This leads to energy and charge exchange processes. Atom collision processes relevant for plasmas are, apart from elastic scattering with no exchange, ionization, associative ionization, charge transfer, attachment, and chemical reaction.

The type of electric excitation determines chemical pathways in the plasma, since it strongly influences the electron energy distribution function. At the same mean power input, nanosecond excitation will generate a much higher peak electron-energy than DC-power input. With nanosecond excitation, simulations for the kINPen showed a peak power density of 4500 W cm^{-3} [147], compared to typical values around 200 W cm^{-3}. This results in about 20% higher electron temperature.

Energy transfer reactions in atmospheric pressure plasmas are strongly influenced by metastable species (see **Table 6.2**). In helium plasmas, metastables are especially relevant due to their high internal energy of 20.6 and 19.82 eV. But also in argon plasmas, metastables with an energy of 11.72 and 11.55 eV play a vital role. Electron impact generation of metastables as well as collisions of electrons with metastable species, feeding energy back to the electrons, influences the electron energy distribution function.

Additionally, formation of argon excimer species is relevant for energy transfer to the atoms and molecules in an argon plasma jet. Model calculations show that argon excimers are the most abundant excited argon species in a pure argon discharge with nanosecond excitation [147] (see **Figure 6.25**). Ar$_2^+$ are the dominant ions with concentrations matching those of the electrons.

Argon excimer excited energy states lie in the energy regime of electronic excitation of small atoms (see **Figure 6.26**). Argon excimer has a resonant energy transfer to atomic oxygen populating the 3s^{3}S^0 state of atomic oxygen at 9.54 eV that decays to 2p^4 ^{3}P energy state emitting the line triplet at 130.2,

Table 6.2 Metastable Species Energy		
Species	Energy State	Energy
He	^{1}S	20.6 eV
He	^{3}S	19.82 eV
Ar	^{3}P$_2$	11.72 eV
Ar	^{3}P$_0$	11.55 eV
N$_2$	A$^3\Sigma_u^+$	6.17 eV
O$_2$	a$^1\Delta_g$	0.98 eV

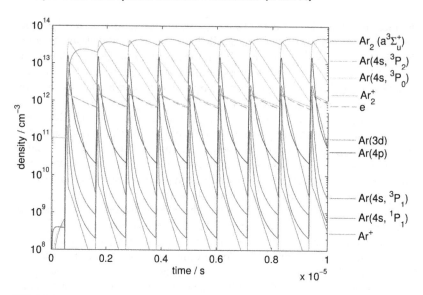

FIGURE 6.25 Species concentration from modelling results of a 1 MHz argon plasma jet. (From Schmidt-Bleker, A. et al., *Plasma Sour. Sci. Technol.*, 25, 015005, 2016; Attribution 3.0 Unported (Creative Commons Attribution 3.0 licence in Ref. [67]), http://creativecommons.org/licenses/by/3.0/, 2017.)

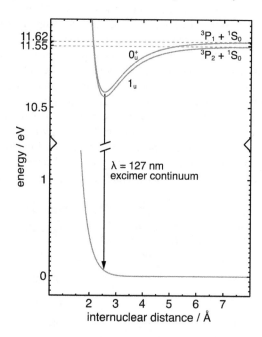

FIGURE 6.26 Excimer continuum potential curve for argon. (Data taken from Bretagne, J. et al., *Beiträge aus der Plasmaphysik*, 23, 295–312, 1983 and Michaelson, R.C. and Smith, A.L., *J. Chem. Phys.*, 61, 2566–2574, 1974.)

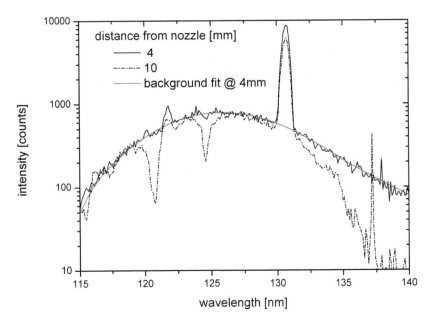

FIGURE 6.27 Argon excimer emission. (From Reuter, S. et al., *Plasma Sour. Sci. Technol.*, 21, 024005, 2012. © IOP Publishing. Reproduced with permission. All rights reserved.)

130.5, and 130.6 nm [148]. This resonant energy transfer can result in a high population of the excited atomic oxygen, even at low oxygen concentrations [54]. **Figure 6.27** shows the argon excimer continuum emitted from a 1 MHz atmospheric pressure plasma jet (kINPen) in ambient air. The atomic oxygen emission can be seen at 130 nm. With greater distance to the nozzle, parts of the argon excimer continuum are absorbed by air and plasma generated species.

With molecular admixture, the plasma composition changes and initial chemical species are produced. Metastable species are relevant for dissociation and excitation of many species. Oxygen atoms and metastable oxygen atoms, nitrogen atoms, and hydrogen or OH are dominantly generated by excited argon species. Metastable molecular nitrogen is for higher air concentrations generated through electron impact and for lower air concentrations through excited argon species. Singlet oxygen metastable molecules are dominantly generated through electron impact.

Water admixture leads to increasing generation of OH [149]. In helium discharges with water admixtures, OH can be generated by metastable species. Namely $O(^1D)$ and helium metastables will lead to generation of OH by **Reactions (6.29)** and **(6.30)**, respectively [124–126]

$$O\left(^1D\right)+H_2O \rightarrow 2\,OH \tag{6.29}$$

$$He_m + 6H_2O \rightarrow HH_{11}O_5^+ + OH + He + e^- \tag{6.30}$$

Nonequilibrium Atmospheric Pressure Plasma Jets (N-APPJs)

Especially for plasma jets used in plasma medicine, metastable excitation processes of humid air species become relevant even inside the plasma region: diffusion along the discharge walls upstream the gas flow will introduce air species into the discharge region. This diffusion process will influence the generation efficiency of reactive species such as atomic oxygen [115]. Admixture of O_2 to a nitrogen containing argon plasma will result in a reduction of atomic nitrogen due to formation of NO [134]. In atmospheric pressure plasmas this is a more dominant reaction than charge transfer of, e.g., NO_2^+ and N_2O^+ leading to excited NO species [150], since cations are neutralized by electrons faster than charge transfer reactions can occur.

In a nanosecond pulsed helium discharge, Thomson scattering allowed to detect Rydberg molecules of helium that significantly contribute to the energy dissipation in elevated pressure helium discharges and are an intermediate to helium metastable species [128].

7

Diagnostics of the Plasma Effluent (in Jets)—From Near Field to Far Field

THE EFFLUENT ZONE IS much more accessible to optical and other diagnostic methods compared to the core plasma region. This is one reason for the multitude of studies regarding the plasma effluent zone (see [10] and references therein). It presents an intricate region of transition from plasma chemistry to equilibrium reaction chemistry. This zone is characterized by flow interaction with the ambient. In plasma jets, flow processes can be very influential on the resulting reaction products, however, also in DBD type discharges, through ion wind, flow affects reactive species generation mechanisms.

Chapter 7 describes flow measurement and effects in cold atmospheric pressure plasma jets followed by a section on measurements of short living atoms and molecules. Gas temperature has a strong effect on reaction products. Respective measurement techniques will be described. Charged species and electric field measurements will be discussed. A description of energy dissipation processes and transport and reaction chemical processes reaching into the far field gas phase region without presence of plasma species concludes the chapter.

7.1 Flow Field Diagnostics

While in low pressure plasmas gas flow effects do not dominate processes, this is drastically different at atmospheric pressure. In plasma jets operated in ambient air, air species quickly diffuse into the effluent. In [151] an analytical solution for the diffusion equation has been developed, assuming that

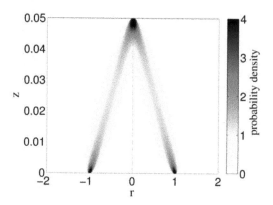

FIGURE 7.1 Probability density of particles from the nozzle edge of a rectangular gas flow. Particles travel predominantly on straight lines to a given point in the near field of the nozzle. At $z = 0$, particle paths start at a normalized radial distance of 1, or −1 respectively. The center is $r = 0$. (From Schmidt-Bleker, A. et al., *Phys. Fluids*, 26, 083603, 2014, with the permission of AIP Publishing.)

particles diffusing in from the side of the jet nozzle dominantly reach a given point in a straight line (see **Figure 7.1**). By mapping these species' pathways to a given flow field, diffusion patterns for arbitrary flow fields can be easily calculated. This path mapping approximation can be applied, for example, to analyze absorption measurements [9] or to evaluate heat transport [152] or diffusion based quenching of LIF signals [153]. However, as the situation usually is more complex involving turbulences and interaction of plasma and flow, measurement of the flow situation is preferred.

The simplest approach to determine the flow pattern is by Schlieren diagnostics. Schlieren uses the change in a light pathway caused by a change in refractive index. The latter can be induced either by change in particle composition or by density changes e.g., through temperature changes. As described in [15], a Schlieren setup usually consists of a point light source collimated to a beam of parallel light which is imaged onto a 2D detector; sometimes a laser is used as light source. A knife-edge blocks the light that passes through the focal point (see **Figure 7.2**). Is the light deflected by changes in the index of refraction as, for example, in a plasma jet effluent, the deflected light will appear on the detector. A background-subtracted image will show the change in optical density and can be used to determine particle density or gas temperature. For plasma Schlieren measurements, green light is recommended since cold atmospheric pressure plasmas for most cases emit little light in this wavelength.

Results of Schlieren measurements contributed to the validation of computational reaction kinetics models and to an understanding of the diffusion of ambient species into plasma jet effluents.

Although Schlieren images are only line of sight images—unless tomographic techniques are employed—they nevertheless reveal important information (**Figure 7.3**).

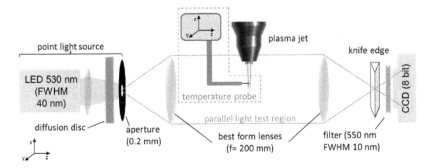

FIGURE 7.2 Typical Schlieren setup as used for diagnostics on an atmospheric pressure plasma jet. (From Schmidt-Bleker, A. et al., *J. Phys. D Appl. Phys.*, 48, 175202, 2015. © IOP Publishing. Reproduced with permission. All rights reserved.)

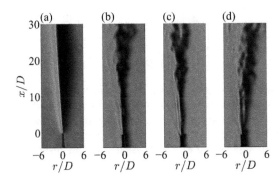

FIGURE 7.3 Schlieren photography of (a) helium jet flow at $Re = 135$ and (b–d) helium jet flow at $Re = 135$ with plasma ignited at (b) 10 kV, (c) 13 kV, and (d) 16 kV. (From Whalley, R.D., and Walsh, J.L., *Sci. Rep.*, 6, 31756, 2016; Attribution 4.0 International (Creative Commons Attribution 4.0 licence in Ref. [155]), http://creativecommons.org/licenses/by/4.0/, 2017.)

In [156] it was found by Schlieren imaging and iCCD imaging that the gas flow in a helium atmospheric pressure plasma jet follows the discharge. These fast Schlieren measurements can observe fast changes in the flow pattern. Similar results have been found in [157], where it was shown that different voltage pulses generate local changes in the flow pattern.

In [158] shadowgraphy and iCCD imaging are also used to determine the energy dissipation processes in a plasma jet's effluent. Here, thermal effects are deemed to be responsible for varying diffusion of air species affecting the effluent shape.

The intrinsic drawback of the above-described techniques is that they are line of sight methods and thus in most cases give a qualitative impression only.

A new diagnostic approach for nonthermal plasma jets is planar LIF combined with emission imaging. In [153], this approach is used to determine the

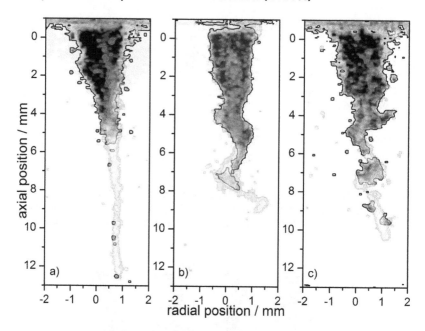

FIGURE 7.4 Flow pattern using OH as tracer molecule indicating the Ar air boundary visualized by quenching of OH fluorescence (black and white) for Air densities >0.1% to 1% overlaid with Ar emission (light grey – red and yellow in original) stereoscopically recorded. (From Iseni, S. et al., *J. Phys D Appl. Phys.*, 47, 152001, 2014. © IOP Publishing. Reproduced with permission. All rights reserved.)

flow structure without plasma by using a fluorescent tracer and the flow pattern when the plasma is switched on by using plasma generated OH as tracer molecule (**Figure 7.4**).

The onset of turbulence is characterized by Kelvin-Helmholtz instabilities. These are attributed to very localized heating at the powered electrode that produces an acoustic wave. This wave disturbs the shear layer between the feed gas and the ambient air. The disturbance propagates with local gas speed [159].

7.1.1 Particle Imaging Velocimetry

While the above methods can give an impression of the flow pattern, they do not determine the flow field with its velocity vectors. For this, particle-imaging velocimetry (PIV) needs to be employed. This technique is known from aerospace engineering, where particles in a flow situation are illuminated by a laser sheath and tracked with a camera. Image analysis over time yields the velocity fields of the studied systems. With the rise of plasma actuators in aerospace engineering applications, PIV is increasingly used in plasma research (e.g., [160]) (**Figure 7.5**).

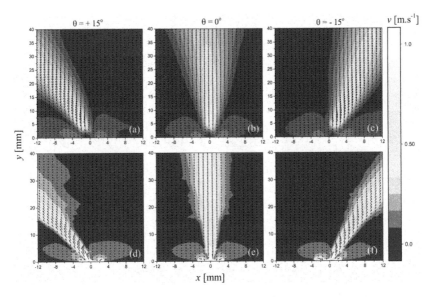

FIGURE 7.5 PIV on an induced jet flow induced by a surface barrier discharge (a–c) Measured velocity vector maps showing impact of a +15°, 0°, and −15° phase difference between applied voltages, respectively. (d–f) Calculated velocity vector maps showing impact of +15°, 0°, and −15° phase difference between applied voltages, respectively. All cases correspond to an electrode separation distance of 5 mm. (From Dickenson, A. et al., *Sci. Rep.*, 7, 14003, 2017; Attribution 4.0 International (Creative Commons Attribution 4.0 licence in Ref. [155]), http://creativecommons.org/licenses/by/4.0/, 2017.)

7.2 Short-Living Atoms and Molecules

Short-living atoms and molecules are key species in plasma reactivity and set cold plasmas apart from equilibrium chemistry. Electronic energy is converted into chemically highly active atoms or molecules. Their characterization and detection is vital to gain an understanding of effects of plasmas on biological systems.

7.2.1 Absorption Spectroscopy

Light atoms absorb from the ground state to the first electronically excited state in the vacuum ultraviolet spectral region. Vacuum ultraviolet radiation is absorbed by air species. This makes detection setups more complex than UV- or visible light absorption spectroscopy. Radiation source and detection system need to be operated in an air free environment. A setup described to study plasma jets uses two window extensions with a small gap in between them, where the plasma jet is placed (see **Figure 7.6**). The arms are evacuated

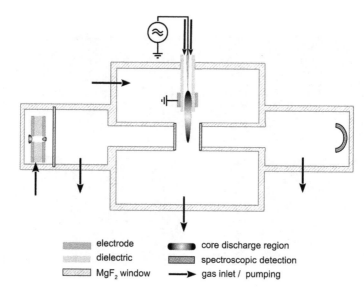

electrode
dielectric
MgF$_2$ window
core discharge region
spectroscopic detection
gas inlet / pumping

FIGURE 7.6 VUV absorption spectroscopic setup as described in [162]. The VUV-radiation carrying arms are evacuated. A microhollow cathode lamp generates resonant radiation when fed with the studied species in a helium buffer. The plasma jet is in a controlled environment with nitrogen buffer gas.

and connected to a light source and to a spectrometer [162]. As light source, a micro hollow cathode discharge (MHCD) lamp is used and the operating gas was varied to contain the probed species in a helium buffer gas. The MHCD generates the desired radiation at low pressure yielding a small widths emission line of the probed species that can be used to measure the pressure broadened absorption profile in the plasma jet.

Since plasma jets emit VUV radiation, self-absorption can be used as means to detect species in the plasma jet effluent. For example, from the argon excimer continuum around 125 nm, absorption by molecular oxygen and by singlet oxygen can be observed. A spectrum can be seen in **Figure 6.27**. From absorption measurements at greater distances, oxygen molecule concentration can be determined and ambient species diffusion can be measured.

Absorption spectroscopy in the UV-spectral range is useful for detection of light molecules. Typically, ozone concentration is determined by absorption at 254 nm [163]. Absorption spectroscopy in the UV spectral range has the advantage that cross sections are typically very high. A challenge of multicomponent plasma effluents is that absorption cross sections of relevant molecules in the UV spectral region are overlapping and it needs to be ensured that the correct molecular absorption band is observed (see **Figure 7.7**). For species of high-enough concentration or with sufficient absorption length, absorption spectroscopy in the infrared, where molecular absorption spectra are separated, can be used. For ozone, both, Fourier transform infrared absorption spectroscopy [164] and infrared laser-absorption spectroscopy are used [163].

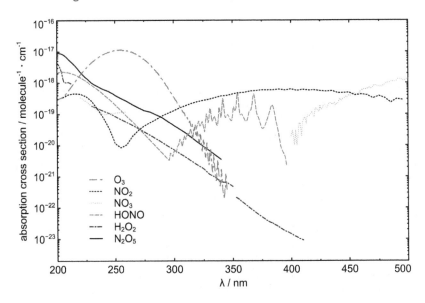

FIGURE 7.7 Absorption cross-sections of dominant species in humid air plasmas. (Data taken from [165]; Modified from Reuter, S. et al., *Plasma Sour. Sci. Technol.*, 24, 054001, 2015; Attribution 3.0 Unported (Creative Commons Attribution 3.0 licence in Ref. [67]), http://creativecommons.org/licenses/by/3.0/, 2017.)

Due to the hydroxyl radical's relevance in plasma medicine, many studies put special focus on its detection. OH is used as a tracer molecule for flow pattern recognition (see Section 7.1) by laser induced fluorescence spectroscopy. Quantitative detection is, nevertheless, often performed by absorption spectroscopy in the ultraviolet spectral range. OH originates from water dissociation in plasma and leads to generation of hydrogen peroxide via OH + OH -> H_2O_2. OH is known to have several biological effects [166]. The small absorption length of plasma jet effluents as well as the comparatively small absorption cross section of OH in the UV spectral range [167] requires that experiments need to be very precise or other measures have to be taken to ensure a high signal to noise ratio as for example by using cavity ring down setups [168]. Otherwise, the plasma dimension has to be enlarged to ensure sufficient absorption lengths for determining OH [169]. Performing sensitive absorption spectroscopy requires stable light sources. This is a property of light emitting diodes that are now available also in the UV-spectral range. With a spectrally resolving detection system, highly sensitive absorption measurements can be performed [169]. Absorption measurements can be used to calibrate LIF measurements [170,171].

For sensitive detection, laser-absorption spectroscopy is often used. Lasers, especially tuneable diode lasers (TDL), are an excellent light source due to their stability, their coherence, and their small divergence. TDL absorption spectroscopy in the near infrared is frequently used to study metastable species such as argon or helium metastables [117,127,129,172].

7.2.2 Multi-Pass Absorption and Cavity Enhanced Techniques

The problem of insufficient detection sensitivity in absorption spectroscopy measurements can be solved by the use of longer absorption lengths. Since plasma jets are typically small and species distribution in plasma jet efflu-ents can be as small as several tens of micrometers depending on the species lifetime, measurement techniques need to be adapted to provide sufficient absorption lengths. Multi-pass arrangements and cavity enhanced absorp-tion spectroscopy techniques provide absorption lengths that can be orders of magnitude higher than that of single pass techniques. Conventional multi-pass arrangements provide several ten to hundred passes, but typically do not allow spatial resolution [173,174]. A novel focal multi-pass cell which does pro-vide spatial resolution uses retroreflectors and two parabolic mirrors to focus the beam pathways to one point where a plasma jet can be positioned [175]. As with regular absorption spectroscopy, cavity, enhanced absorption methods rely on the Beer–Lambert Law of absorption, which connects the incident light intensity of a probing lamp with the transmitted light through the absorption coefficient of the absorbing medium, namely the plasma. For cavity enhanced methods, the absorbing medium is placed inside the cavity. The probing light is reflected back and forth inside the cavity, passing the absorbing medium multiple times. Cavity mirrors have a high reflectivity of greater than 0.999 so that up to tens of thousands passes can be achieved. A small portion of the light exits the cavity each pass through the exit mirror which has a slightly lower reflectivity. There are a large number of techniques that make use of cav-ity enhanced absorption spectroscopy [176]. One of the most commonly used techniques is cavity ring down spectroscopy (CEAS) [177].

Until the development of pulsed cavity ring down spectroscopy, absorp-tion spectroscopy used continuum light sources, e.g., by measuring the phase difference between a modulated cavity beam and a reference beam [178]. In cavity ring down spectroscopy, a pulsed laser is used and the decay time of the pulse is measured [179]. The decay time τ_0 of the light intensity at the cavity exit depends on the exit mirror reflectivity and the absorption coefficient of the medium inside the cavity. The pulse will be reflected mul-tiple times inside the cavity. Each pass, photons are absorbed by the medium inside the cavity and every other pass, a part of the photons exit the cavity through the exit mirror. The development of the intensity as a function of time follows:

$$I(t) = I_0 \cdot e^{-\frac{c}{L}|\ln(R)|t} = I_0 e^{-\frac{t}{\tau_0}} \tag{7.1}$$

with L the cavity length and R the mirror reflectivity. In case the pulse length is smaller than twice the cavity length, or the pulse number is small, the inten-sity at the exit of the cavity is a decaying chain of pulses with the pulse peak intensity following **Equation (7.1)**. With greater pulse length or higher pulse number entering the cavity, the exit signal follows an exponential function as shown in the cavity ring down setup sketch in **Figure 7.8**.

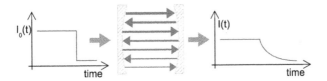

FIGURE 7.8 In cavity enhanced diagnostic techniques, the absorption length is drastically increased by multiple transitions of the light through the absorbing medium inside the optical cavity formed by two highly reflective mirrors. (From Lu, X. et al., *Phys. Rep.*, 630, 1–84, 2016, Copyright 2016, with permission from Elsevier.)

Is an absorbing medium of length l introduced into the cavity, the effective pulse lifetime is reduced and the absorber density n can be calculated by [177]:

$$n = 16\pi^2 \frac{\tau_{21}}{\lambda_{12}^2} \frac{w_1}{w_2} \frac{L}{l} \int \frac{1}{\lambda^2} \left(\frac{1}{\tau(\lambda)} - \frac{1}{\tau_0} \right) d\lambda \qquad (7.2)$$

With τ_{21} as lifetime of the excited state, λ_{12} as wavelength of the resonance transition, w_i the statistical weight of state i. The absorption line is then probed by measuring the wavelength dependent lifetime $\tau(\lambda)$.

Typically, cavity ring down spectroscopy is performed with monochromatic laser light, but also broadband white light cavity ring down spectroscopy that can probe multiple species at the same time has been reported [180]. Cavity enhanced methods, especially cavity ring down spectroscopy, are used to study low-pressure reactive plasmas [181–183]. In recent years, cavity ring down spectroscopy and other cavity enhanced detection techniques have been applied to atmospheric pressure plasmas [120,177]: The combination of high sensitivity with spatial resolution and high time resolution are ideal to study the high gradient small-scale atmospheric pressure plasma jets [18]. Like with other absorption spectroscopic techniques, not only ground state species can be detected, but also metastable species can be studied in a time-resolved manner [120].

A further development of cavity enhanced absorption spectroscopy uses an optical feedback (OF) dye laser in combination with a V-shaped cavity to study HO_2 generation in an atmospheric pressure plasma jet. The setup is shown in **Figure 7.9**. The V-shaped cavity allows for some light that enters the cavity to exit at the regular exit coupling mirror as well as at a third mirror, the cavity folding mirror (FM). This light is fed back to the laser diode. This leads to a strong reduction of the linewidth of the laser from MHz down to kHz. Additionally, the laser is mode locked to the respective cavity mode.

Figure 7.10 shows a comparison of a mode locked and a free running laser line profile. When the laser is frequency tuned, the laser frequency locks to subsequent laser modes remaining in each mode for as long as mode

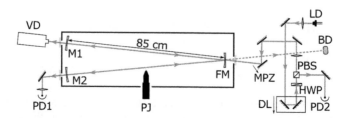

FIGURE 7.9 Schematic of the OF-CEAS experiment setup with the atmospheric pressure plasma jet kINPen effluent placed in one of the arms of the cavity. LD laser diode; DL delay line; HWP half-wave plate; PBS polarizing beam splitter; MPZ mirror on piezo-electric transducer; FM folding cavity mirror; M1, M2 cavity mirrors; VD Vidicon camera; PD1, PD2 photodiodes; BD beam dump; PJ kINPen plasma jet. (From Gianella, M. et al., *New J. Phys.*, 18, 113027, 2016; Attribution 3.0 Unported (Creative Commons Attribution 3.0 licence in Ref. [67]), http://creativecommons.org/licenses/by/3.0/, 2017.)

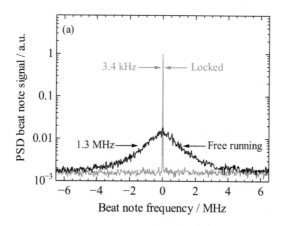

FIGURE 7.10 Comparison of a laser profile locked to a cavity mode (grey) to a free running laser profile (black). Intensity is orders of magnitudes higher and linewidth is reduced by three orders of magnitude. (From Gianella, M. et al., *New J. Phys.*, 18, 113027, 2016; Attribution 3.0 Unported (Creative Commons Attribution 3.0 licence in Ref. [67]), http://creativecommons.org/licenses/by/3.0/, 2017.)

matching conditions are met. For mode matching, the distance d between the laser diode facet and the entrance mirror to the cavity must meet the requirement $d = l_{FM-M2} + q(l_{FM-M2} + l_{FM-M1})$ with an integer $q \geq 0$ and l_{i-j} the distance between mirrors i and j [185].

The absorption coefficient of the kth mode can be derived from $\alpha_k = T/(L\sqrt{A_k}) + 2\ln(R)/L$ with A_k the amplitude of the kth mode, R and T the reflectivity and transmittivity of the mirrors, and L the cavity length. The factor T/L can be derived from the decay time of the cavity.

The species density n of a species can be derived from

$$n = \frac{\int \alpha(v)\, dv}{\int \sigma(v)\, dv} \tag{7.3}$$

Cavity ring down and cavity enhanced absorption spectroscopy are sensitive enough to detect minority species such as the above-mentioned HO_2. The detection sensitivity of the described V-shaped cavity enhanced detection system is in the order of $10^{12}\,cm^{-3}$ in a 4 mm wide plasma effluent.

7.2.3 Molecular Beam Mass Spectrometry

In low-pressure reactive plasmas, mass spectrometry are commonly used to study the plasma generated molecular and atomic charged and neutral reactive species. Advantages of mass spectrometry are the wide range of detectable species, space and—with some effort—also time resolved measurements of reactive species with a high sensitivity. Mass spectrometry detects particles separating them with respect to their mass over charge ratio. Several filter techniques can provide further selectivity such as energy selectivity. To select molecules and atoms, magnetic and electric fields are employed to alter the trajectory of the particle to either reach the detector or be filtered from the detection pathway. Fundamental for mass spectrometry is, therefore, a low pressure that inhibits particle collisions. A particle entering a mass spectrometer will typically pass four major components: An ion source will ionize neutral particles to prepare them for detection. Ion optics will form a beam of the ions to an energy analyzer that can select charged species with respect to their energy. The mass over charge analyzer subsequently selects ions of a specific m/z ratio to guide them to the ion detector. Energy filtering can be done, e.g., by a sector field energy filter or by a Bessel box [186].

For the mean free path of the particles to be long enough, the pressure has to be in the order of 10^{-2} to $10^{-4}\,Pa$. This is a challenge for atmospheric pressure plasma jet diagnostics. However, with differential pumping using a small orifice for the first pressure chamber, the pressure can be reduced stepwise to reach the required pressure inside the mass spectrometer. The aim is to generate a molecular beam that inhibits collisions of atoms (or molecules) with each other or with walls so that the reactive species composition remains frozen at the conditions present in the plasma. This technique has been used for several decades, e.g., to study flame chemistry [187].

For a sampling of plasma-generated species two distinct sampling schemes are used, their main difference being the pressure gradient between initial pressure and pressure inside the first sampling chamber. One scheme uses a rotating skimmer disk that opens the sampling orifice only briefly and thus allows to achieve a low background pressure, the other one uses strong pumping units to build up high pressure gradients. Depending on the sampling orifice diameter and pressure conditions, different beam forming mechanisms occur. The sampling orifice diameter typically has to be orders of magnitude higher than the mean free path at atmospheric pressure, which is in the order of 100 nm. Sampling orifices can be 10 to several 100 µm.

For diagnostics of atmospheric pressure plasmas, the plasma sheath thickness has to be considered. It lies in the order of 100 µm. Is the orifice in this order of magnitude or larger, the discharge can enter the sampling orifice and change the species composition and plasma-generated ions can be accelerated. In general, the sampling orifice allows for a pressure gradient to build up. A high-enough pressure gradient from atmospheric to first chamber pressure will accelerate the sampled particles to the speed of sound. After the sampling orifice, expansion will lead to a further acceleration of the particles and a supersonic shock wave will develop. To gain a molecular beam from the expansion zone, a skimmer needs to sample the beams in the region bounded by the frontal Mach disk and the side-wall barrel shock-walls. This so-called zone of silence is a region completely shielded from the outside conditions. Undisturbed sampling inside this region (in the order of a few mm distance from the orifice exit) will result in the formation of a molecular beam in the second sampling chamber.

Is the pressure in the first chamber too low for the development of a supersonic shock wave and the mean free path in the first chamber is longer than the characteristic lengths of the system, the continuum flow region near the orifice exit will gradually develop to a free flow region due to expansion. A molecular beam can then be sampled in the free flow region. To achieve such high-pressure gradients, either strong pumping with low orifice diameter or background pumping with a shuttered sampling orifice can be performed [188,189].

In general, molecular beam mass spectroscopic measurement needs to guarantee a low signal to noise ratio, a low distortion of the beam in the low-pressure regime inside the mass spectrometer, and an ionization of the neutral species that allows species detection but does not generate surplus species by dissociation.

Molecular beam mass spectrometry for atmospheric pressure plasmas jets has been used by several groups (see e.g., [188]). To measure neutral species with molecular beam mass spectrometry, the ionization stage of the spectrometer has to be used. Typically, electron impact ionization is used, since it is least disturbing on the species composition. Nevertheless, the ionization process can result in formation of secondary species through dissociative ionization. To distinguish plasma-generated species from species generated through electron impact dissociation inside the mass spectrometer, an electron energy dependent signal of the mass to charge ratio of the studied species needs to be performed [190]. **Figure 7.11** shows a so-called threshold ionization mass spectroscopic measurement of atomic nitrogen.

Varying the ionization stage's electron energy will yield a different signal from plasma generated nitrogen atoms compared to nitrogen atoms generated through dissociative ionization of molecular nitrogen in the mass spectrometer [192].

To gain absolute density values from molecular beam mass spectroscopic data, background subtraction and calibrating measurements have to be performed on a known reference species concentration. Species composition distortion has to

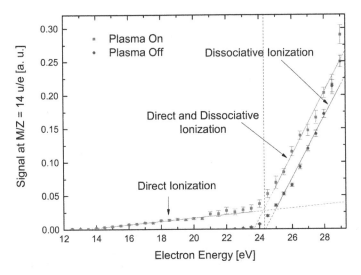

FIGURE 7.11 Signal of mass $M/Z = 14$ u/e for plasma on and plasma off as function of the electron energy in the ionizer. The helium flow was 1.4 slm and 0.25% N_2 was admixed at a distance of 4 mm between jet nozzle and sampling orifice. (From Schneider, S. et al., *J. Phys. D Appl. Phys.*, 47, 505203, 2014. Attribution 3.0 Unported (Creative Commons Attribution 3.0 licence in Ref. [67]), http://creativecommons.org/licenses/by/3.0/, 2017.)

be taken into account. The reference species, therefore, needs to be comparable in mass to the species that is being investigated.

When all conditions are met, molecular beam mass spectrometry indeed represents a versatile tool for the study of atmospheric pressure plasma generated neutral reactive species [189,193–199].

7.2.4 Laser Induced Fluorescence Spectroscopy

As described for diagnostics of the core plasma, laser induced fluorescence spectroscopy is one of the most sensitive diagnostic techniques to study atoms and small molecules also in the effluent. In the plasma jet effluent, accessibility is excellent and fluorescence methods can be readily applied. The geometry is simple: A laser beam is focused into the effluent region and the fluorescent radiation is observed typically at a 90° angle (see **Figure 7.12**).

Using a laser sheath and an intensified CCD camera, PLIF- or LIF-imaging yields two-dimensional fluorescence maps (see **Figure 7.4**) from which—given sufficient calibration—can be derived a concentration distribution of fluorescent species and quenchers [115].

Laser induced fluorescence spectroscopy can yield spatial and time resolved measurements as well as absolutely calibrated atom and molecule density measurements. For plasma jets, several approaches have been used and are described in the following.

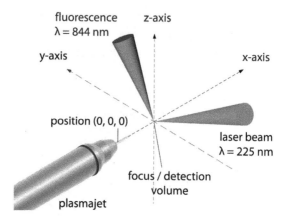

FIGURE 7.12 Laser and observing beam geometry in laser induced fluorescence spectroscopy on plasma jet effluents. (From Reuter, S. et al., *Plasma Sour. Sci. Technol.*, 21, 024005, 2012. © IOP Publishing. Reproduced with permission. All rights reserved.)

A LIF signal is given by Equation (5.51) in Section 5.6.1. For calibration, the spectral response function of the excitation and detection system ξ needs to be known. Furthermore, the solid angle $\Omega/4\pi$ of detection and the excitation volume influence the fluorescent signal strength. These factors can be determined by a calibration measurement that generates a photon response from a known species concentration. For single photon laser induced fluorescence, a process that is linear with laser intensity can be utilized. Typically, Rayleigh scattering is used in the same experiment setup. This method was first established by Salmon and Laurendeau [200]. The Rayleigh scattering signal is performed on a known scattering particle concentration. The Rayleigh signal—according to Equation (5.39)—contains all spectral and spatial response functions of the system. By measuring scattering as a function of the laser intensity the calibration of the measurement system can be performed.

The Rayleigh scattering signal is proportional to the scattering volume V, the differential cross section integrated over the observed solid angle, the number of scattering particles in the scattering volume, and a constant ξ that contains the spectral and optical efficiencies of the complete detection optics, and the laser irradiance I_L, as given by Equation (5.39). ξ can be determined from the slope of the Rayleigh signal as a function of laser energy per pulse.

When Rayleigh and LIF signal are measured in the same setup with the same collection volume V_c, the species concentration probed by LIF can be derived from Equation (5.51).

An alternative calibration approach is to generate or introduce a known concentration of fluorescing species. In this case, the LIF signal itself can be used to calibrate the measurement setup. In nitric oxide LIF measurements on atmospheric pressure plasma jets, this can, e.g., be done by adding a known amount of NO to the feed gas. In a region where ambient diffusion can be neglected, the LIF signal on the NO reference is then used to calibrate the measurement setup [201].

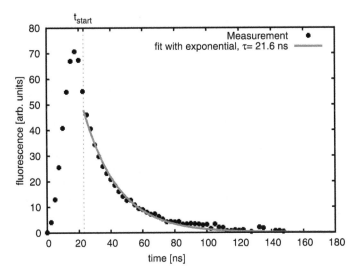

FIGURE 7.13 Lifetime measurement of OH-LIF signal on an argon atmospheric pressure plasma jet. (From Vorac, J. et al., *Plasma Sour. Sci. Technol.*, 22, 025016, 2013. © IOP Publishing. Reproduced with permission. All rights reserved.)

Depopulation of the excited states by collisional quenching reduces the fluorescent signal. At atmospheric pressure, quenching becomes relevant and needs to be accounted for. A straightforward approach is to measure the effective lifetime of the fluorescence. In measurements where the laser system has a pulse length of several nanoseconds, only transitions with long lifetimes can be evaluated (see **Figure 7.13**). When, for example, laser excitation is tuned to populate the vibrational ground state of the electronic A-state of OH, vibrational energy transfer is avoided, since excitation to higher vibrational levels is unlikely. The fluorescence to the electronic ground state is observed. The fluorescent radiation is, therefore, of the same wavelength as the laser light and cannot be filtered out. Scattering of laser light thus contributes significantly to the noise level. Evaluating after laser excitation only will still yield a good LIF measurement.

In cases where the short fluorescence lifetime inhibits a lifetime measurement, the quencher concentration and the respective quenching rates need to be known. When quenching cannot be measured, quencher concentration can be estimated by fluid modelling [203].

For example, for OH LIF measurements in post discharge time, a simple reaction model would involve only the dominant destruction pathways as generation occurs through electron molecule (ion and neutral) interaction [171].

OH is dominantly depleted through reaction with itself

$$OH + OH \rightarrow H_2O + O \qquad (7.4)$$

$$OH + OH + M \rightarrow H_2O_2 + M \qquad (7.5)$$

$$OH + H + M \rightarrow H_2O + M \qquad (7.6)$$

Under these simple assumptions, LIF signal decay can be fitted with a rate equation including the reaction rate coefficients of **Reaction (7.4)** through **(7.6)** as well as a lumped quenching coefficient. The result is an estimation of the OH concentration that proved accurate within a factor of 5 for various situations [169,171,204,205].

A full dynamic analysis of the LIF signal is much more complex. A six-level energy state model can reflect the situation precisely enough and has been applied in several studies for OH LIF measurement analysis (see e.g., [207]). **Figure 7.14** shows a schematic within the potential energy diagram for OH(X $^2\Pi$) and OH(A $^2\Sigma^+$). The analysis takes into account time and space variant excitation by the 2D laser sheet (for planar LIF) and stimulated emission characterized by the respective Einstein coefficients (B_{25} and B_{16}, respectively),

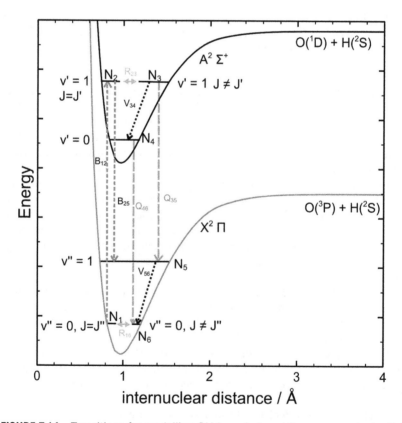

FIGURE 7.14 Transitions for modelling OH laser induced fluorescence (potential curve data taken from [206]—energy level data not to scale) as described e.g., in [207].

rotational energy transfer from the probed rotational level J'' of ground state vibrational level $v'' = 0$ of OH(X $^2\Pi$), and the excited rotational level J' of vibrational level $v' = 1$ from OH(A $^2\Sigma^+$) each to a respective lumped rotational level $J_{final} \neq J_{initial}$ within the respective vibrational level. **Figure 7.14** shows these rotational energy transfers with R_{ij}. Vibrational energy transfer occurs within the electronic energy state (marked in dotted by V_{ij}). Collisional quenching depopulates both vibrational energy states of OH(A $^2\Sigma^+$) to ground state vibrational states $v = 1$ and $v = 0$.

While this six-level model is giving accurate enough results, its numerical analysis is computational intensive and an analysis of time and space resolved measurements is not always feasible. Therefore, ways to reduce analysis efforts are required and found in various approaches to reduce the number of levels and energy transfer processes taken into account for the LIF dynamic analysis. Simplifications can be made when the laser energy is low and the LIF signal is linearly dependent on the laser energy. The depletion of the ground state and repopulation by stimulated emission can be neglected, which also means that the rotational energy distribution is thermalized and rotational energy transfer can be neglected. Additionally, the laser profile can be approximated to be homogeneous. The strongest simplification assumes that OH(X $^2\Pi$) is constant and only $v = 0$ and $v = 1$ of the electronic A-state are populated. With a laser pulse approximated by a rectangular profile, an analytical solution for the rate equations can be found [203,208]. Solving the rate equations numerically allows to add a rate equation for the ground state level and an experimentally gained laser pulse [209].

The six-level model can be reduced to a four-level model under the assumption that rotational energy transfer occurs on a time scale that is orders of magnitude smaller than absorption and vibrational energy transfer. Levels N_2 and N_3 as well as N_1 and N_6 are then coupled through thermalization with the gas temperature by the Boltzmann distribution [207,210]. When $v = 3$ of OH(A $^2\Sigma^+$) is populated by laser excitation, quenching can be neglected, as for OH(A $^2\Sigma^+$, $v = 3$) predissociation dominates all other depopulation mechanisms [211] circumventing the necessity to determine the influence of quenching.

For atmospheric pressure discharges, a linear regime LIF measurement can result in a very low signal. Increasing the signal by increasing the laser energy will result in partially saturated LIF measurements. Resultantly, stimulated emission, rotational energy transfer, as well as laser inhomogeneity need to be taken into account resulting in a five-level model [47]. LIF has been used in various studies on atmospheric pressure plasmas for several small molecules, e.g., on NO [30,201,212], N_2(A$^2\Sigma_u^+$) [136,213], and extensively—due to its relevance for plasma medicine—OH molecules [49,169,171,202,203,205,208,211,214–216]. LIF can also be used to study charged species such as N_2^+ ions [217]. In non-plasma environments, LIF has been used to study, e.g., NH(a$^1\Delta$) [48] (photolysis), or CH (in flames) [218].

Extension of laser induced fluorescence spectroscopy to atomic species without using vacuum ultraviolet photons leads to two-photon absorption laser induced fluorescence (TALIF) spectroscopy (see Section 5.6.2). Using two photons to excite the large energy gap of light atoms from ground to first electronic excited state means that calibration by one-photon techniques is not applicable. Therefore, Rayleigh scattering or linear LIF cannot be used to calibrate the wavelength and spatial properties of the TALIF setup. As with LIF spectroscopy, several approaches for a calibration scheme can be employed. The simplest approach is to calibrate TALIF signals on the studied plasma source by a TALIF signal on a known species concentration of a reference species. Only noble gas atoms are stable and can serve as reference species without further experimental efforts. Calibration using noble gas reference species is described below.

Taking the species that is investigated also as calibration species, requires a calibration species generation through a known process such as photodissociation, or through a so-called flow tube reactor. Here, a microwave plasma generates the desired species (e.g., atomic oxygen), which is then guided in an inert tube to the detection volume. The species concentration can be calibrated by titration [219]. For atomic oxygen a well-known titration scheme uses NO_2:

$$O + NO_2 \rightarrow NO + O_2 \qquad (7.7)$$

The flow tube reactor is constructed so that at the detection volume all NO_2 has reacted with the atomic oxygen. The TALIF signal of atomic oxygen is measured as a function of NO_2 admixture. When the TALIF signal disappears, the atomic oxygen concentration stochiometrically equals the NO_2 admixture.

In a pure oxygen plasma, collisional quenching of the atomic oxygen ($3p^3P_2$) state can be used as calibration measurement [220].

Most commonly, noble gas calibration schemes are used [221]. The concentration of the noble gas species is known from mass flow and from ideal gas law. The measured TALIF signal is thus attributed to a known species concentration and is proportional to the optical and geometric properties of the experiment setup. The density of the plasma generated species x can thus be derived from:

$$n_x = \gamma \frac{a_{21}(r) \, \hat{\sigma}^{(2)}(r) \, I(r)}{a_{21}(x) \, \hat{\sigma}^{(2)}(x) \, I(x)} n_r. \qquad (7.8)$$

$\hat{\sigma}^{(2)}$ are the two-photon absorption cross sections of the studied species x and the reference (noble gas) species r with known density n_r. I is the normalized fluorescence intensity. γ is the ratio of all wavelength dependent system properties of the probed and the reference species.

Calibration schemes aim to have excitation and fluorescence wavelength of the probed and reference species as close as possible in order to have as little influence of the wavelength dependent properties on both fluorescent signals. If the excitation and fluorescent wavelength of probed and reference species are close enough, γ is close to unity. The branching ratio a_{ik} of the calibration gas and the studied species x is the ratio of the Einstein coefficients for

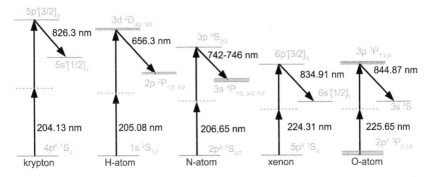

FIGURE 7.15 TALIF calibration schemes for hydrogen, nitrogen, and oxygen.

spontaneous emission A_{ik} of the observed transition and the sum of all other processes depleting the energy level, from which the fluorescence is observed. This includes spontaneous emission to other lower energy levels as well as quench coefficients k of species q at density n_q. These quench coefficients k are temperature dependent.

$$a_{ik} = \frac{A_{ik}}{\sum_{k<i} A_{ik} + \sum_{q} k_q^i n_q} \qquad (7.9)$$

To analyze quenching, quencher concentrations need to be determined [9]. Several TALIF energy level schemes have been applied in the past. **Figure 7.15** shows a selection of noble gas TALIF schemes and the respective non-noble gas species that are calibrated by the noble gases [222]. To minimize the effect of wavelengths dependent properties of the experiment setup, excitation and fluorescence wavelength of the reference noble gas species have to be as close as possible to the ones of the probed species.

TALIF spectroscopy has been applied to study atomic oxygen in atmospheric pressure plasmas [6,16,78,115, 221–227] also time resolved [224], atomic nitrogen [228,229], as well as atomic hydrogen [230]. TALIF measurements have been compared to electrochemical probes for atomic oxygen and nitrogen [231].

7.2.5 Calibrated Optical Emission Spectroscopy

Optical emission spectroscopy yields valuable insight into species present in a plasma. A challenge lies in gaining absolute concentration values from optical emission spectroscopy.

The spectrally integrated emission coefficient ε_{ik} of a spontaneous emission from state $<k>$ to state $<i>$ is connected to the Einstein coefficient for spontaneous emission A_{ki} and the population density n_k via

$$\varepsilon_{ik} = \frac{h\nu}{4\pi} n_k A_{ki} \qquad (7.10)$$

325

It denotes the emitted power per time of a radiative transition. With an absolutely calibrated detection system the number of photons collected per solid angle can be transferred to absolute concentration values. It is thus possible to determine excited species concentrations such as singlet oxygen [232].

7.3 Gas Temperature

Knowledge of the gas temperature is fundamental for an analysis of the reaction chemistry. In nonequilibrium plasmas, time scales of collisional energy transfer determine whether thermal equilibrium between energy states can be reached. Some degrees of freedom can be in thermal equilibrium, while others are not. Gas temperature, the translational temperature of a gas, can be measured by techniques that directly access the translational motion of particles (one of these is, e.g., Doppler broadening spectroscopy see Section 5.2.3) or by other techniques accessing related quantities including rotational and vibrational temperature gained from molecular spectroscopy, provided that these are in equilibrium with the gas temperature.

7.3.1 Schlieren

Since the refractive index of a gas is related to the gas density, refractive index measurements yield information about the gas temperature. Thus, Schlieren measurements can be used for an accurate temperature determination. Combined with computational fluid dynamics (CFD) simulations, the method not only yields the temporally averaged ambient-air density and temperature in the effluent of the fully turbulent jet (see **Figure 7.16**), but also allows for an estimation of the calorimetric power deposited by the plasma [154].

The following excerpt from [15] describes how Schlieren measurements can be used to determine the gas temperature. A typical setup for Schlieren measurements is shown in **Figure 7.2** in Section 7.1. The studied region is illuminated with parallel light. The light intensity is measured with a detector after focusing on a razor blade. From a measurement of the intensity with (I) and without Schlieren (I_K), the contrast c can be determined from **Equation (7.11)** [152]:

$$c = \frac{I - I_K}{I_K} \tag{7.11}$$

It was shown that by measuring the contrast $\hat{c}_{fl}$ of the gas flow with the plasma switched off and $\hat{c}_{pl}$, the contrast with gas flow and plasma on, the temperature can be determined via:

$$T = T_0 \frac{n_0 + \hat{c}_{fl}/S - 1}{n_0 + \hat{c}_{pl}/S - 1} F \tag{7.12}$$

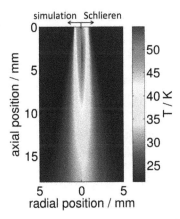

FIGURE 7.16 Comparison of temperature distribution from CFD simulations (left) and Schlieren measurements (right) in a 1 MHz argon atmospheric pressure plasma jet effluent. (From Schmidt-Bleker, A. et al., *J. Phys. D Appl. Phys.*, 48, 175202, 2015. © IOP Publishing. Reproduced with permission. All rights reserved.)

with

$$F = \frac{\epsilon_{r,air} + x_{Ar,pl}\left(n_{Ar} - n_{air}\right)}{\epsilon_{r,air} + x_{Ar,fl}\left(n_{Ar} - n_{air}\right)} \tag{7.13}$$

T_0 is the ambient gas temperature, S the Schlieren setup sensitivity, n denotes the refractive indices and x_{Ar} is the molar fraction of argon, while $1 + \epsilon_{r,i}$ denotes the reference index of refraction for species i. From the Schlieren measurements, the mole fraction of argon can be derived as well as the gas temperature (**Figure 7.16**).

7.3.2 Laser Schlieren Deflectometry

Fast temperature processes can be measured by time resolved measurements of the deflection of a laser beam through a plasma filament. This point-wise technique is called laser Schlieren deflectometry [109]. A laser beam is deflected in a radially symmetric optical milieu with a radially declining refractive index based on Fermat's principle of light paths that denotes that a beam of light travels that path between two points which allows it to travel the quickest. One such optical milieu is represented by an axially symmetric plasma filament, that locally heats gas, changing the refractive index. Proportional to the temperature gradient is the refractive index changing from center to outer regions. The deflection pathway of a probing laser beam is shown in **Figure 7.17**.

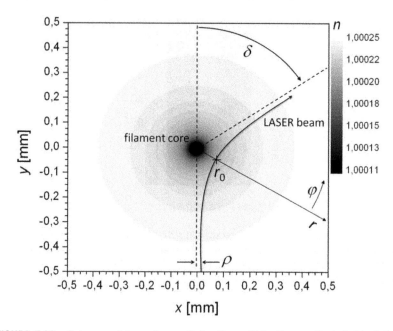

FIGURE 7.17 Scheme of laser beam deflection within the gradient field of the refractive index *n* caused by an axially symmetric plasma filament. The notation of trajectory parameters used in the equations is shown in the diagram. The trajectory is plotted schematically. (Reprinted from Schäfer, J. et al., *Rev. Sci. Instrum.* 83, 103506 (2012), with the permission of AIP Publishing.)

From Fermat's principle, the trajectory of the laser beam can be written in polar coordinates as [109]:

$$\varphi = \int \left(\frac{dr}{r\sqrt{\left(\frac{r}{\rho}f\right)^2 - 1}} \right) \tag{7.14}$$

with $f = f(r)$ as profile function connecting the refractive index $n = n_0 f$ to the gradient function with n_0 the refractive index of the surrounding. ρ is the lateral beam position. The angular deflection δ results from $\pi - 2 \cdot \varphi$:

$$\delta = \pi - 2\int_{r_0}^{\infty} \left(\frac{dr}{r\sqrt{\left(\frac{r}{\rho}f\right)^2 - 1}} \right) \tag{7.15}$$

with the integration boundaries going from r_0 to ∞. r_0 is the minimal distance of the laser beam from the central point (see **Figure 7.17**).

Measurement of the deflection δ is thus a measure for the local temperature profile in time. With an assumption of a hyperbolic refractive index profile outside the discharge filament and a constant profile inside the filament, an analytical solution for the maximum temperature profile as a function of deflection amplitude is:

$$T_1 = T_0 \left(1 - \frac{1}{2(n_0 - 1)} \delta_1 \right)^{-1} \qquad (7.16)$$

with δ_1 as maximum deflection angle, T_0 ambient temperature, and n_0 the refractive index inside the filament. The method is valid for regimes, where the refractive index is not dominated by electrons at electron concentrations of 10^{14} cm^{-3} as in [109].

7.3.3 Molecule Spectroscopy for Temperature Determination

Gaining temperature values from molecular spectra can be achieved by two methods. The first method applies a so-called Boltzmann plot. For the second method, a measured spectrum is fitted with a theoretical spectrum.

Both methods assume that the derived rotational temperature is in equilibrium with the translational temperature. This assumption needs careful evaluation for each respective situation. While the simplest way to measure molecular spectra is by optical emission spectroscopy, this technique probes only electronically excited levels. Reaching an equilibrium in these, in most cases, short lived energy states is much less likely than for electronic ground state levels.

For the determination of the rotational temperature from a Boltzmann plot, relative line intensities are plotted according to the following relation derived from the Boltzmann distribution (Equation 5.4):

$$Y = -\frac{\alpha}{k_B T} E_J \qquad (7.17)$$

where

$$Y = \ln\left(\frac{I\lambda_{JJ'}}{A_{JJ'}(2J+1)} \right) \qquad (7.18)$$

and α a proportionality constant and $A_{JJ'}$ are the Einstein coefficients. I are relative line intensities and E_J is the energy of the upper level. Plotting Y as a function of E_J yields a linear Boltzmann plot, where the slope yields the rotational temperature. Rotational temperature typically reaches thermal equilibrium quickly. In some cases, where population of rotational states through, e.g., metastables occurs, higher rotational energy states can be at nonequilibrium, while low lying states are already at equilibrium. In this case, the gas temperature can be derived from a two-temperature fit for high and low energy states, respectively.

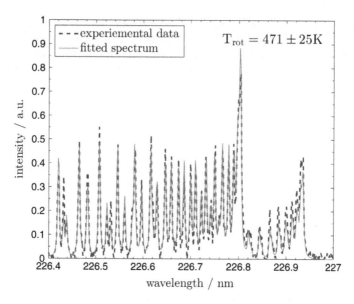

FIGURE 7.18 Rotational temperature determined by LIF spectroscopy on NO. (From Iseni, S. et al., *New J. Phys.*, 16, 123011, 2014; Attribution 3.0 Unported (Creative Commons Attribution 3.0 licence in Ref. [67]), http://creative commons.org/licenses/by/3.0/, 2017.)

The second method is to determine rotational temperature by fitting an emission spectrum to a theoretical spectrum gained from the molecular constants. Emission spectroscopy, however, only probes excited states. The most reliable option to determine gas temperature from rotational spectra is to use spectroscopy that probes the ground state, which typically is in thermal equilibrium with the gas temperature. Techniques are Raman spectroscopy, or LIF spectroscopy. With these techniques, the rotational temperature of the ground state of a molecule can be determined. **Figure 7.18** shows a rotational spectrum of NO determined by LIF spectroscopy with a respective theoretical fit.

Rotational spectroscopy to determine temperature in nonthermal plasmas has been applied for different plasmas (see e.g., [77,233,234]).

7.3.4 Rayleigh Scattering

To receive the gas temperature from Rayleigh scattering measurements, a reference measurement at a known gas temperature needs to be taken. From the relation $T_{gas} I_{plas} = T_{ref} I_{ref}$, the gas temperature can easily be determined. If a stable species such as for example helium is probed, I_{plas} is the intensity of the Rayleigh scattering on the species with the plasma switched on and I_{ref} is the intensity of the scattering on the species with the plasma switched off. Assuming room temperature in the latter case, the gas temperature in the plasma can be determined. Different gases have a different differential

Rayleigh cross-section [235]. Ideally, therefore, a reference spectrum should be performed in the same gas composition.

The neutral gas density is gained from the Rayleigh signal according to:

$$n_g = \frac{S_{Pl+St} - S_{St}}{S_{ref+St} - S_{St}} n_{ref} \qquad (7.19)$$

where n_{ref} is a known gas density injected into the plasma chamber determined by a fixed pressure and (room) temperature. S_{Pl} is the plasma Rayleigh signal. S_{St} is stray light determined with a reference measurement at low pressure. S_{ref} is the signal of a reference gas at a given temperature and pressure. For plasma jets operated in ambient air, the reference signal is that of air at room temperature.

7.4 Charged Species

While in low-pressure plasmas negative ions can, e.g., be detected by photo detachment measurements [236,237], at atmospheric pressure collisionality makes detection intricate. Electron detachment spectroscopy on a dielectric barrier discharge has been performed in [238]. More common methods to detect charged species in plasma jets are molecular beam mass spectroscopy and indirect measurements through techniques determining surface charges as described in the following.

7.4.1 Molecular Beam Mass Spectrometry

A well-suited technique for determination of charged species is molecular beam mass spectrometry, described in Section 7.2.3. Compared to measuring neutral species by molecular beam mass spectrometry, measuring ions is simpler, since there is no need for ionization of the probed species. This approach has the benefit of avoiding unintended ionization and potential dissociation of molecules within the mass spectrometer. To collect ions, the ion optics of the mass spectrometer has to be in line of sight with the sampling orifice. A variety of plasma jets have been studied by molecular beam mass spectrometry.

Both positive and negative ions can be sampled by setting up polarity of the mass spectrometers' electrodes in reverse.

Ionization occurs through different processes listed in **Table 7.1**. Most negative ions are generated through electron attachment either of a molecule (**Reaction 7.20**) or an atom (**Reaction 7.22**) as well as by charge-exchange of an ion with a neutral (**Reaction 7.21**). Ionization of a molecule by electron capture in vibrational ground state can lead to molecular ion formation (**Reaction 7.23**) which is unstable and after autodetachment forms vibrationally (and also rotationally) excited molecules. **Process (7.23)** is especially relevant for electron energies $\leq 2eV$ typical for the low temperature plasmas in plasma medicine.

Table 7.1   Dominant Ionization Processes

Reaction		Process
$AB + e^- \rightarrow A^- + B$	(7.20)	Dissociative electron attachment
$A^- + B \rightarrow A + B^-$	(7.21)	Charge exchange
$A + e^- + M \rightarrow A^- + M$	(7.22)	Electron attachment
$AB(v = 0) + e^- \rightarrow AB^- \rightarrow AB(v > 0) + e^-$	(7.23)	Electron capture leading to ro-vib excitation
$AB + e^- \rightarrow AB^+ + 2e^-$	(7.24)	Electron impact ionization
$A + B^* \rightarrow A^+ + e^- + B$	(7.25)	Penning ionization
$AB + e^- \rightarrow A + B^+ + 2e^-$	(7.26)	Dissociative ionization

Positive ions are mostly formed through electron impact ionization (**Reaction 7.24**) with electron energies above the ionization thresholds shown in **Figure 7.19**. From molecules, dissociative ionization leads to formation of positive ions (**Reaction 7.26**). In atmospheric pressure plasma jets with high collision rates, especially when operated with noble gases, Penning ionization, i.e., ionization by collision with electronically excited species, is a significant process (**Reaction 7.25**).

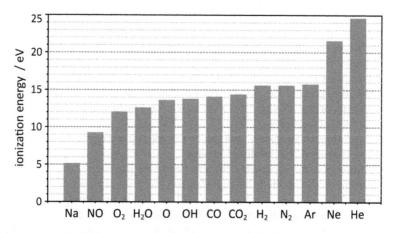

FIGURE 7.19   Ionization threshold potentials of species relevant for plasma medicine. (Data taken from [239].)

Especially processes **(7.28)** and **(7.30)** are relevant for plasma jets operated in air. Electron attachment leads to negative ion formation in electronegative gases with a large electron affinity, where the potential minimum lies below the ground state of the neutral molecule. This is the case for oxygen. The electronegativity of oxygen leads to the following effect in plasma jets: electrons generated by the propagating streamer front generate negative oxygen ions from atomic and molecular oxygen in its wake within the diffusion zone between ambient air and noble feed gas. The subsequent positive plasma pulse allows the ionization front to propagate within this channel of negative ions. Electron detachment by photons or metastables leads to a confined and far propagating plasma plume. When this electronegative channel is missing (such as in pure nitrogen surrounding), the propagation is not pronounced and as a result, the plasma plume is confined to the near nozzle region [240] (**Figure 7.20**).

Concentrations of negative ions can be estimated from the quasi neutrality of the plasma and from a measurement of the ion decay from the visible plasma tip [189].

A relevant process in atmospheric pressure plasmas operated in a humid environment is the formation of water clusters. Several processes lead to water cluster formation. Typically, water clusters are generated by hydration of parent ions in a step-by-step hydration forming larger and larger water clusters [241].

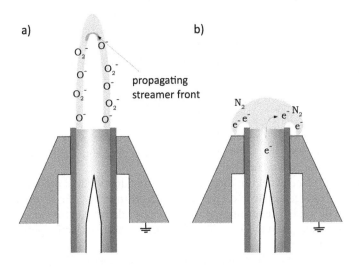

FIGURE 7.20 Effect of electronegative gas surrounding the plasma effluent resulting in a negative channel (a) and radially directed loss of electrons in case of missing electronegative gas (b). (From Reuter, S. et al., *J. Phys. D Appl. Phys.*, 51, 2018; Attribution 3.0 Unported (Creative Commons Attribution 3.0 licence in Ref. [67]), http://creativecommons.org/licenses/by/3.0/, 2017; according to the study published in [240].)

Initial ions for positive cluster formation are O_2^+, H_3O^+, and H^+, where O_2^+ leads to formation of $H_3O^+H_2O$, yielding $H_3O^+(H_2O)_n$ water clusters [242]. Development of water clusters of size n depends on discharge power and humidity admixture in helium discharges [243] and in argon discharges [244].

Depending on different downstream addition of different gases, ion composition in plasma effluents is changed [196]. Ions are a minority species in plasma effluents. Nevertheless, their respective role in medicine is relevant due to their high reactivity in biological systems.

7.4.2 Surface Charges

A different approach to measure charges in atmospheric pressure plasmas is by determining plasma deposition of surface charges on dielectric surfaces. This can be measured via the Pockels effect, which changes the polarization of light passing through a birefringent crystal (Pockels crystal) depending on the electric potential over the crystal. To measure light passing through the probing crystal, either a transparent electrode, such as an Indium Tin Oxide (ITO)-coating [245], or a grounded metal mirror, measuring the light passing the crystal both in transmission and reflection [246] have been used (see **Figure 7.21**). Different crystal materials can be applied. Typically, Bismuth Silicon Oxide (BSO) crystals are used. With an applied electric field, the BSO crystal effectively behaves like a wave-plate. Illumination and observation of the BSO crystal are parted by a polarizer analyser combination with a wave-plate in between. The Pockels effect states that the electric field affects the retardation in the Pockels crystal linearly. The retardation is the phase difference $\Delta\phi(E)$ between the two perpendicular electric field components of the incident light after passing the crystal. It is given by:

$$\Delta\phi(E) = \frac{2\pi}{\lambda} n_0^3 p_{BSO} dE \qquad (7.27)$$

λ is the incident lights wavelength, n_0 the refraction index of the crystal, p_{BSO} the Pockels coefficient, and d the thickness of the crystal. The Pockels coefficient depends on the crystal material and orientation.

Since the electric field in direction of the incident light is proportional to the surface charge density σ on the crystal according to $\sigma = E\varepsilon_0\varepsilon_r$, the Pockels effect can be used to determine the surface charge density on the surface of the Pockels crystal. ε_0 and ε_r are the vacuum and the relative electric permittivity of the crystal, respectively.

The intensity of the light detected by the camera follows [248]:

$$I = I_0 \left(\frac{1}{2} \left(e^{i\left(\frac{\pi}{2} + 2\Delta\phi(E)\right)} - 1 \right) \right)^2 \qquad (7.28)$$

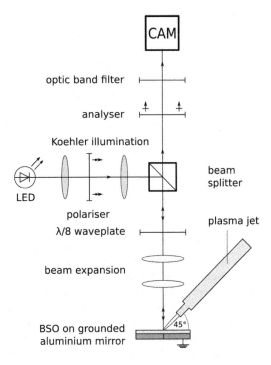

FIGURE 7.21 Surface charge diagnostics via the electro-optic Pockels effect. (From Wild, R. et al., *J. Phys. D Appl. Phys.*, 47, 042001, 2014; Attribution 3.0 Unported (Creative Commons Attribution 3.0 licence in Ref. [67]), http://creative commons.org/licenses/by/3.0/, 2017.)

First measurements of surface charges of a plasma via the electro-optic Pockels effect were made in a dielectric barrier discharge. The same principle can be used for measurements of charge deposition by plasma jets. Either observed in transmission through the crystal [249], or in reflection [247] as shown in the setup of **Figure 7.21**.

In both cases, the plasma jet is at an angle of 45° to the crystal surface to not obstruct the optical path, resulting in an asymmetric observed field profile. The measurements show that the charge deposition on the surface result in a backwards directed excitation [249] and that the negative charges are spread over a larger surface than the positive charges, which agrees with the electric field focusing mechanism described in [240]. At less than 0.4 W dissipated plasma power, the electric field is in the order of $10^5\,\mathrm{Vm}^{-1}$ [249]. When plasma jets are studied, a uniform field distribution cannot necessarily be assumed so that for an evaluation of the electric field from the Pockels effect a look up table is useful [250]. A charge deposit of 200 pC is required for generation of an extended discharge [250]. Depending on the electric field on the surface and gas flow properties, surface streamers develop on the crystal surface with a higher velocity [251].

The Pockels effect can also be used to measure the electric field (as superimposition of surface charges and electric field) on a small crystal attached to a fibre optics in an electric field probe system [252]. In plasma diagnostics, surface charge accumulations have to be taken into account. The Pockels probe allows to measure electric field changes spatially resolved within the probe diameter of one millimeter and time resolved up in the MHz regime [253].

7.5 Electric Field Measurements

Apart from indirect electric field measurements through the Pockels effect described in the previous section, or by a combination of optical emission spectroscopy with discharge modelling—e.g., on nitrogen second positive emission [254]—direct electric field measurements can be performed through optical spectroscopy. Several spectroscopic techniques can be used to determine electric field values. Among them are laser collision induced fluorescence (LCIF) [255] as well as laser induced fluorescence dip spectroscopy [256]. For atmospheric pressure plasmas, three methods have been applied:

Four-wave mixing with a picosecond laser on a hydrogen surface wave discharge has demonstrated a time and space resolved electric field measurement technique [257] on a high pressure discharge. The technique is based on a modified coherent anti-Stokes Raman scattering (CARS) scheme. In CARS, four wavelengths are mixed according to the energy scheme shown in **Figure 7.22a**. Three waves with the frequency ω_i generate the resulting anti-Stokes ω_a photons through a wave mixing process. The wavelength of $\omega_a = \omega_{Pump} + \omega_{Probe} - \omega_{Stokes}$. An electric field can replace one of the wavelengths resembling an electromagnetic wave of frequency 0 [258]. The resulting energy level scheme is shown in **Figure 7.22b**. Basically, the applied field induces a dipole moment in the normally non-polar molecules, proportional to the field, which allows the molecules to produce coherent emission at their Raman frequency [257].

The resulting emitted light intensity of the anti-Stokes beam I_A is proportional to the square of the electric field E and the intensity of the pump beam I_{Pump} and the probe beam I_{Probe} [259]:

$$I(\omega_a) \propto \left|\chi_{IR}^{(3)}\right|^2 E^2 I_{Pump} I_{Stokes} \tag{7.29}$$

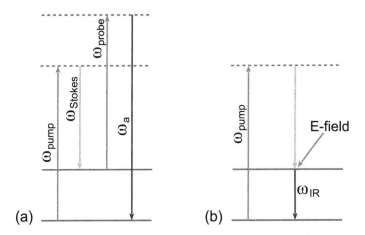

FIGURE 7.22 (a) CARS scheme and (b) electric field vector measurements scheme.

$\chi^{(3)}$ is the third order susceptibility for the four-wave mixing process in the medium.

Calibration is performed by a measurement in a known static electric field.

The direction of the electric field can be derived from measurements with parallel or perpendicularly oriented polarization of the pump and Stokes beam according to [257]:

$$I(\omega_a) \propto \left| \chi_{\parallel}^{(3)} \right|^2 E_{\parallel}^2 I_{Pump} I_{Stokes} \tag{7.30}$$

$$I(\omega_a) \propto \left| \chi_{\perp}^{(3)} \right|^2 E_{\perp}^2 I_{Pump} I_{Stokes} \tag{7.31}$$

The space and time resolution makes this measurement technique highly suitable for the electric field determination in atmospheric pressure plasma jets [260]. Although an excellent diagnostic technique, the method comes with two drawbacks: First, CARS-like spectroscopy for determination of the electric field requires a complex experiment setup with two different laser wavelengths. Second, the measurement technique is restricted to environments which allow Raman scattering.

A similar technique [261], also relying on wave mixing, requires a much simpler experimental effort and can be used independent of the medium the electric field is measured in: Laser induced second harmonic generation by electric fields has been known for a while (see e.g., [262]) but has only recently been used for plasma diagnostics [263]. The underlying principle is similar to E-field measurements by the CARS-like process, described above. As described in [261], fundamentally, three-wave nonlinear mixing processes such as second-harmonic generation are forbidden in symmetric and homogeneous media. The electric field acts as fourth wave—with zero wavelength—and breaks the symmetry. This makes generation of second harmonic light possible. The second harmonic light intensity follows an E^2 dependency. Advantages of this technique are that it is a non-resonant detection method requiring only a single input wavelength and detection of only one single output wavelength. With a model electric field from two rod electrodes, the second harmonic signal is quadratic to the electric field [264].

The polarization P at the second harmonic (2ω) is described by a third order non-linear process [261]:

$$P_i^{(2\omega)} = \frac{3}{2} N \chi_{i,j,k,l}^{(3)} \left(-2\omega,0,\omega,\omega\right) E_j^{(F)} E_k^{(\omega)} E_l^{(\omega)} \qquad (7.32)$$

$E^{(F)}$ is the electric field, $E_{k,l}$ are the electric fields of the incident laser radiation, which are of identical wavelength. N is the number density of the medium and χ the third order susceptibility. χ depends on the dipole moments of the medium and the orientations of the electric fields. The dependence of two χ components on the orientation of the polarization of the laser ((a) parallel to the electric field and (b) perpendicular to the electric field) makes it possible to measure the orientation of the electric field. Case (a) and (b) differ by a factor of 9. The second harmonic signal is proportional to the square of the electric field and the square of the laser intensity. It is additionally proportional to the interaction length and a phase-matching factor.

A different method for electric field measurements in plasmas utilizes emission spectroscopy. Stark polarization spectroscopy makes use of the Stark effect influencing the emission spectra depending on the local electric field. Electric fields induce a splitting of energy levels and a shift of these levels depending on the electric field [265]. Depending on the polarization of the observation, emission lines can be selected. For helium, various transitions have been studied. Empirical dependencies of the wavelength separation of the upper (allowed) and lower (forbidden) energy level for electric fields from 0 to 20 kV are given in **Table 7.2** [266].

This method has been applied for helium glow discharges [266], in dielectric barrier discharges [267,268], and in atmospheric pressure plasma jets [269,270].

λcentre / nm	Allowed Transition	Forbidden Transition	Polynomial for Separation λAllowed·λForbidden
402.6	2p ^{3}P^0–5d ^{3}D	2p ^{3}P^0–5f ^{3}F	$1.9 \cdot 10^{-6} \cdot E^4 - 9.8 \cdot 10^{-5} E^3 + 1.86 \cdot 10^{-3} E^2 - 2.6 \cdot 10^{-3} E + 0.0638$
447.1	2p ^{3}P^0–4d ^{3}D	2p ^{3}P^0–4f ^{3}F	$-1.06 \cdot 10^5 E^3 + 5.95 \cdot 10^{-4} E^2 + 2.5 \cdot 10^{-4} E + 0.1479$
492.1	2p ^{1}P^0–4d ^{1}D	2p ^{1}P^0–4f ^{1}F	$-1.87 \cdot 10^5 E^3 + 8.8 \cdot 10^{-4} E^2 + 1.4 \cdot 10^{-3} E + 0.1316$

Table 7.2   Helium Transitions for Stark Polarization Spectroscopy

Source: [266].

7.6 Energy Dissipation in the Plasma Jet Effluent and Energy Transport to the Far Field Gas Phase

The respective plasma jet type determines the energy dissipation mechanisms from jet effluent to the far field gas phase. While in plasma jets with an electric field perpendicular to the gas flow [116], reactive species and photons dominate the energy transfer in the jet effluent, in jets with ionization waves, primary plasma species are also significantly produced in the effluent region of the jet. Emission features of plasma jets can easily be measured, and allow conclusions about dominant energy transfer mechanisms.

Ionization fronts travel at velocities of some tens of thousand m/s in the effluent (see **Figure 7.23**). In air, the ionization waves are guided by residual electrons attached to electronegative species. These "stored" charges also act as energy reservoir allowing an easier ionization wave propagation with every successive plasma pulse similar to dielectric barrier surface wave discharges.

Figure 7.23 also shows a backwards directed feature (at a relative time point of 0.6 µs). This feature is also observed in a helium discharge [240]. The mechanism proposed for the backwards directed excitation feature is similar to the ones observed from dielectric surfaces [271]. Deposited charges pose as nucleation for a backwards directed ionization wave at high-enough electric

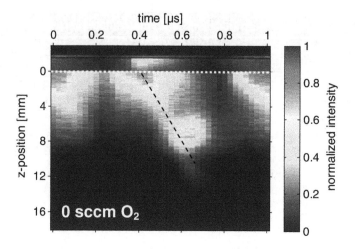

FIGURE 7.23 Phase plot of a phase resolved optical emission spectrum [163] of the effluent of a ~1 MHz argon plasma jet z-position of O shows the nozzle exit. (From Reuter, S. et al., *IEEE T. Plasma Sci.*, 43, 3185–3192, 2015. © IOP Publishing. Reproduced with permission. All rights reserved.)

fields. In open air, these charges are deposited on the electronegative oxygen. Potential release mechanisms are for example reactions of metastables with ions according to **Reaction (7.33)** [240]:

$$O^- + He_m; He^* \rightarrow e^- + products, k = 3 \cdot 10^{-10} cm^{-3}s^{-1} \qquad (7.33)$$

At room temperature, this reaction is probable enough to be a potential source for electrons leading to backwards directed excitation, other than Penning ionization, or direct ionization through ground state noble gas atom collisions. In helium, sufficient metastable species have been detected to possibly explain the observed feature [172].

Backwards-directed excitation dynamics is consequently only observed with oxygen in the surrounding atmosphere. Interestingly, helium metastable concentration drops below detection limit when the ambient atmosphere composition is changed from oxygen containing gas to pure nitrogen or argon [172] (see **Figure 7.24**).

Generation of helium metastables in oxygen containing surrounding is related to the presence of oxygen in the effluent surrounding. It was found by simulation and ICCD images that oxygen focuses the electric field lines at a distance to the nozzle (see **Figure 7.25**).

In noble gas ionization wave plasma jets, metastables pose an important pathway to dissipate electric energy to the molecules of ambient air. As described in Section 6.5, electron energy is transferred to noble gas species forming argon excimers and metastable species. Excitation occurs rapidly

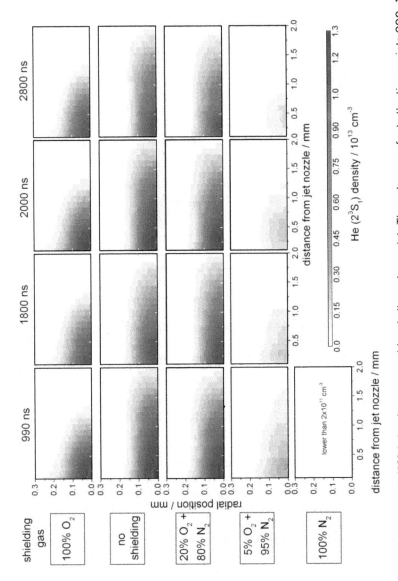

FIGURE 7.24 Spatio-temporal He(2^3S_1) density measured in a helium plasma jet. The columns refer to the time points 990, 1800, 2000, and 2800 ns. (From Winter, J. et al., *Plasma Sour. Sci. Technol.*, 24, 025015, 2015. © IOP Publishing. Reproduced with permission. All rights reserved.)

N$_2$ shield gas
max: 10^{13} cm^{-3} (3-dec)

O$_2$ shield gas
max: 10^{13} cm^{-3} (3-dec)

negative   positive   negative   positive

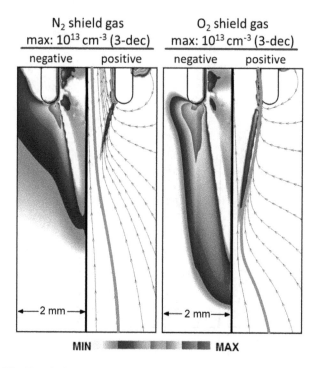

←—2 mm—→   ←—2 mm—→

MIN ▬▬ ▬ ▬ ▬ ▬ ▬ MAX

FIGURE 7.25 Electric field line distribution and negative and positive charge densities in the effluent of a helium plasma jet in N$_2$ and O$_2$ surrounding. Electric field line focusing is highlighted for a selected field line marked in grey according to [15]. (From Schmidt-Bleker, A. et al., *Plasma Sources Sci. Technol.*, 24, 035022, 2015. © IOP Publishing. Reproduced with permission. All rights reserved.)

within nanoseconds. The energy is transferred within microseconds to the surrounding species. **Figure 7.26** shows the development of argon metastables measured by tunable diode laser-absorption spectroscopy compared to simulations (dashed line). Since plasma pathways occur dominantly where argon is majority species with >99% fraction, electrons mostly collide with argon species and electron energy is thus mostly transferred to argon. A cross section analysis leads to the finding that dissociation and excitation of air species dominantly occurs through argon impact.

Figure 7.27 shows different species groups and their respective generation process comparing excited argon reactions and electron impact for three different humid air admixtures (10^{-4}, 10^{-3}, and 10^{-2}). For air impurities in the argon effluent of concentrations less than 0.1%, generation of atomic oxygen (ground state and excited), atomic hydrogen or hydroxyl radical from air humidity, atomic nitrogen and nitrogen metastable molecules N$_2$(A) are predominantly generated through argon collisions. Only singlet oxygen (a-Δ and b-Σ state) is generated through electron impact. At air impurity of 1%, metastable oxygen atoms, nitrogen molecules and evidently singlet oxygen are also

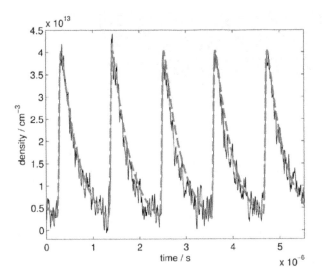

FIGURE 7.26 Measurement (continuous line) and simulation (dashed line) of argon metastable concentrations in the effluent of an argon atmospheric pressure ionization wave plasma jet. (From Schmidt-Bleker, A. et al., *Plasma Sour. Sci. Technol.*, 25, 015005, 2016. Attribution 3.0 Unported (Creative Commons Attribution 3.0 licence in Ref. [67]), http://creativecommons.org/licenses/by/3.0/, 2017.)

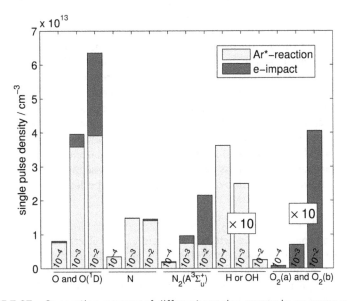

FIGURE 7.27 Generation process of different species groups in an argon plasma with air impurities of different concentrations. (From Schmidt-Bleker, A. et al., *Plasma Sour. Sci. Technol.*, 25, 015005, 2016; Attribution 3.0 Unported (Creative Commons Attribution 3.0 licence in Ref. [67]), http://creativecommons.org/licenses/by/3.0/, 2017.)

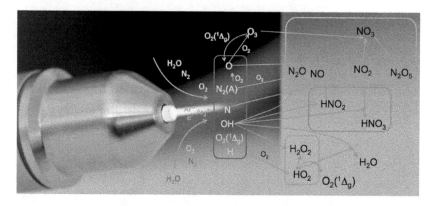

FIGURE 7.28 Reaction kinetics in the kINPen and in comparable argon ionization wave plasma jets. (From Reuter, S. et al., *J. Phys. D Appl. Phys.*, 51, 2018; Attribution 3.0 Unported (Creative Commons Attribution 3.0 licence in Ref. [67]), http://creativecommons.org/licenses/by/3.0/, 2017.)

generated significantly or entirely (for the case of singlet oxygen molecules) by electron collisions with air species.

For modelling, this allows to assume excited argon species as energy source for subsequent primary chemical species generation. To study the reaction kinetics in the effluent and far field region, first diffusion of ambient species into the effluent and subsequent dissociation generates all primary species responsible for the downstream reaction kinetics.

Figure 7.28 shows dominant reaction pathways and species interaction in the effluent region, in the near field and in the far field of the jet, where only decaying reaction chemistry occurs. The far field region is characterized by the absence of most direct plasma species. It is, however, strongly influenced by the plasma processes and from gas phase diagnostics a multitude of plasma processes can be identified. In the far field region, long living reactive species are present. These long living reactive species can be studied by a variety of methods. Advantage of long living species is that they can be collected for measurements. Multi-pass cell absorption spectroscopy is one of the best choice methods for detection of long living reactive species.

7.6.1 Multi-Pass Absorption Spectroscopy

Long-living species can be detected remotely, e.g., by electrochemical probes, by sample surfaces as reactivity probes [272], by gas chromatography, by mass spectrometry, by chemical and electrochemical sensors [273] or by multi-pass absorption spectroscopy. The latter has the advantage to be highly sensitive and absolutely calibrated. With broad band absorption spectroscopy, multiple species can be detected simultaneously and chemical reaction pathways can be identified.

Since many reactions occur downstream the plasma effluent, it is helpful to apply chemical kinetics modelling to analyze measurement data. A recent

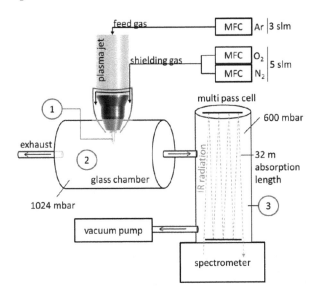

FIGURE 7.29 Schematics of the FTIR measurement setup. The encircled labels 1–3 correspond to the simulation steps of a kinetic model used for the data evaluation. (From Schmidt-Bleker, A. et al., *Plasma Sour. Sci. Technol.*, 25, 015005, 2016; Attribution 3.0 Unported (Creative Commons Attribution 3.0 licence in Ref. [67]), http://creativecommons.org/licenses/by/3.0/, 2017.)

example is Fourier transform infrared multi-pass absorption spectroscopy performed on an argon atmospheric pressure plasma jet. For this, the plasma jet is guided into a sampling gas chamber. The plasma jet is equipped with a curtain device ensuring constant operation conditions. From the sampling chamber, a fraction is probed into a multi-pass absorption cell connected to an FTIR absorption spectrometer (see **Figure 7.29**).

The infrared absorption spectrometer can detect multiple species relevant in atmospheric pressure plasma jets (see **Figure 7.30**). Obviously not all chemical pathways can be identified because multiple reactions of different educts can lead to the same product.

Taking a closer look at the nitrogen species reaction pathways demonstrates this clearly [147]. Various reactive species pathways lead to formation of long living NO_2. Typical generation of nitric oxide occurs through recombination of N and O. An alternative pathway is reaction of excited N_2 with atomic oxygen. Fast oxidation of NO by atomic oxygen or OH leading to NO_2 can be substituted by oxidation of NO with ozone, or through oxidation of HNO_2 by atomic oxygen.

Knowing the reaction network allows to control the reactive species output of plasma jets. Tuning the chemistry from an ozone dominated composition (mostly achieved when cold plasma jets are operated in ambient conditions) to a nitric oxide dominated chemistry can be achieved by generating atomic nitrogen in the core plasma [274] with pure nitrogen admixture (of 1%)

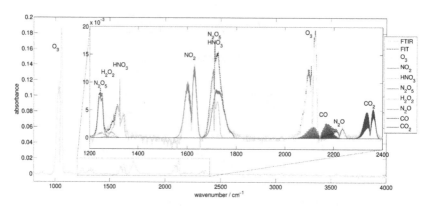

FIGURE 7.30 Absorbance as measured by FTIR spectroscopy and as obtained from the fitting procedure for individual species densities for a shielding gas composition of $O_2/N_2 = 0.2$. (From Schmidt-Bleker, A. et al., *Plasma Sour. Sci. Technol.*, 25, 015005, 2016; Attribution 3.0 Unported (Creative Commons Attribution 3.0 licence in Ref. [67]), http://creativecommons.org/licenses/by/3.0/, 2017.)

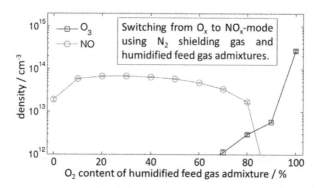

FIGURE 7.31 Switching plasma chemistry from O_x generation mode to NO_x mode by combination of feed gas and surrounding composition adjustment. (From Schmidt-Bleker, A. et al., *Plasma Process. Polym.*, 13, 1120–1127, 2016. Copyright © 2016 by John Wiley Sons. Reprinted by permission of John Wiley & Sons.)

in humidified argon and a pure nitrogen shielding gas (see **Figure 7.31**). The plasma generates more NO than O_3 with oxygen admixture fractions of up to 80% (of the 1% molecular feed gas admixture). At higher oxygen fractions, ozone generation dominates the processes. The water admixture to the feed gas leads to formation of hydroxyl radicals and HO_2 [184]. Both species are sinks for atomic oxygen leading to the generation of molecular oxygen and more OH, or H respectively [274].

Adding water to the plasma jet's chemistry can influence the reaction chemistry drastically. It is, however, important where and how the water is added.

The admixture of water to the core plasma (through feed gas admixture) leads to a multifold higher generation of hydroxyl radicals [275] than adding the same amount of water through ambient humidity. This is true for those cases, where the dominant energy dissipation occurs in the plasma core region.

It is important to note that plasma jets operated over a liquid surface in ambient air are strongly influenced by the evaporated humidity above the liquid interface. It was found that humidity above a liquid surface results in a stronger generation of NO_2 in the far field [277]. The reason for this is that evaporated humidity enters the zone of noble gas in the jet effluent. This leads to a different relative ratio of humidity to indiffusing air, which leads to a different downstream reaction kinetics favoring NO_x species. This phenomenon has to be taken into account, when fundamental *in-vitro* plasma studies in plasma medicine are performed.

8

From Plasma to Liquid

ESPECIALLY IN NEW APPLICATIONS such as plasmas for medicine, liquid diagnostics and surface diagnostics are vital. This chapter gives a brief introduction to possible diagnostics for plasma induced reactive species generation in liquid and reports on important traps and disadvantages of diagnosing plasma-liquid interaction. Section 8.1 describes selected possible diagnostic methods of the liquid phase and Section 8.2 summarizes effects of plasma jets interacting with liquids on the example of the kINPen. The latter section was published in a recent review on the plasma source [15].

8.1 Diagnostics of Liquids

8.1.1 Dye-Based Techniques

For liquids, a simple diagnostic technique is the use of dye-based assays. A dye molecule changes its absorption spectrum or its fluorescent emission upon excitation depending on the reactions it had with the molecules to be detected. This technique works well in known conditions. Typically, the colorimetric response is valid for a certain concentration regime of the probed species as well as for a certain pH-value range. Several colorimetric assays have successfully been applied to plasma treated liquid. For example, measurement of hydrogen peroxide was performed by an AmplexRed® assay which converts the non-fluorescent AmplexRed® compound (10-acetyl-3,7-dihydroxyphenoxazine) in the presence of hydrogen peroxide and horseradish peroxidase to the fluorescent resorufin, which can be detected by two methods. The first is through absorption measurements at 570 nm, the second by fluorescence detection using excitation at 570 nm and emission at 585 nm. In plasma treatment of RPMI (Roswell Park Memorial Institute) cell

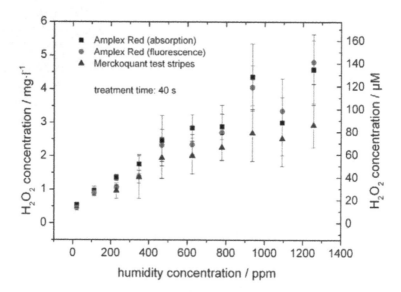

FIGURE 8.1 H_2O_2 concentration in a plasma-treated liquid increases with feed gas humidity. Merckoquant stripes and AmplexRed® assay were used for H_2O_2 detection. (From Winter, J. et al., *J. Phys. D Appl. Phys.*, 46, 295401, 2013. © IOP Publishing. Reproduced with permission. All rights reserved.)

culture medium by an argon plasma jet with varying humidity concentrations in the feed gas, hydrogen peroxide concentrations were measured by the absorption method as well as by the fluorescence method. Additionally, test stripes were used (see **Figure 8.1**). Test stripe assays are also based on colorimetric techniques, due to inherent inaccuracy, however, it might be advisable to compare results with other techniques.

A further technique for detection of hydrogen peroxide uses reactions of H_2O_2 with a titanium sulfonate reagent [278]. This has been applied to plasma liquid studies (see e.g., [279]). Titanium(IV) oxysulfate ($TiOSO_4$) dissolves in sulphuric environment and forms ions according to **Reaction 8.1**:

$$TiOSO_4 + H_2O \leftrightarrow \left[Ti(OH)_3(H_2O)_3 \right]^+ + HSO_4^- \qquad (8.1)$$

The product reacts with hydrogen peroxide according to **Reaction 8.2**:

$$H_2O\left[Ti(OH)_3(H_2O)_3 \right]^+ + H_2O_2 \leftrightarrow \left[Ti(O)_2(OH)(H_2O)_3 \right]^+ + 2H_2O \quad (8.2)$$

$\left[Ti(O)_2(OH)(H_2O)_3 \right]^+$ is a yellow orange compound that can be detected colorimetrically.

Especially plasma jets operated in an air environment generate nitrite and nitrate in plasma treated liquids. These can be determined by the so-called Griess assay. Only nitrite can be determined directly by the Griess assay.

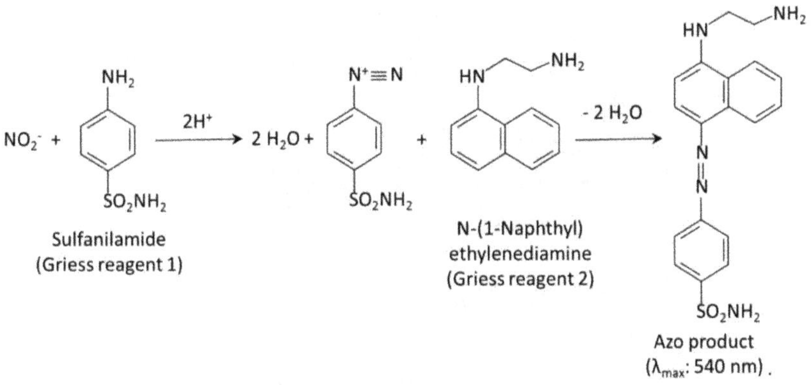

FIGURE 8.2 Determination of nitrite concentration with Griess reagent 1 and 2. (From Jablonowski, H. et al., *Biointerphases*, 10, 029506, 2015. With permission. Copyright 2015, American Vacuum Society.)

For this, the nitrite is brought into contact with sulfanilamide (Griess reagent 1) and afterwards with N-(1-Naphthyl)ethylendiamine (Griess reagent 2) forming a deeply purple azo compound that can be detected colorimetrically at 540 nm (see **Figure 8.2**). Nitrate concentration is determined by transforming nitrite to nitrate using a nitrate reductase enzyme and the related cofactor. From the measurements of nitrite concentration and nitrite concentration plus nitrite formed from nitrate, nitrate concentration can be determined.

However, dye-based assays lack in selectivity. Structural changes in most dye molecules that lead to the desired change in absorbance or fluorescence of certain wavelengths can be achieved not only through a single compound. Although the manufacturers claim selectivity for their chemical probes, a plasma environment challenges this claim. This can be seen in spurious ozone concentration measurements by an ozone assay using iodine bleaching. Measurements were reported that showed an iodine response to plasma treatment, even, when no ozone was present. Since this bleaching process can also be triggered by atomic oxygen and other oxidizing species generated by plasma jets, it must be carefully evaluated if the assay can be used.

Also, the response of H_2DCFDA, a probe for measurement of OH concentrations, is sensitive to a broad spectrum of reactive species. H_2DCFDA has been used to measure peroxynitrite in plasma treated liquid [280]. **Figure 8.3** shows the response (normalized to hydroxyl radical signal) of H_2DCFDA to various reactive oxygen and nitrogen species (RONS) [15]. It is important to note that for the RONS shown in **Figure 8.3** -depending on their respective concentration that can differ by orders of magnitude – similar fluorescent signals can originate from entirely different species compositions (see [15]).

Due to these difficulties with colorimetric and fluorescent chemical probes, it is important to identify reactive species selectively. One method is the detection of degradation products with a finger print specific to certain educts. This can be, e.g., degradation of phenol by reactive plasma species and detection of the products by high performance liquid chromatography (HPLC) with mass

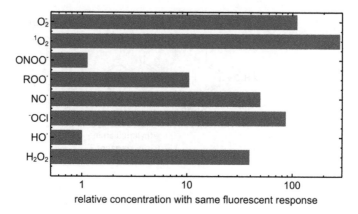

FIGURE 8.3 Cross reactivity for H_2DCFDA relative concentration yielding the same fluorescent signal. (Data from [281]; Reuter, S. et al., *J. Phys. D Appl. Phys.*, 51, 2018. Attribution 3.0 Unported (Creative Commons Attribution 3.0 licence in Ref. [67]), http://creativecommons.org/licenses/by/3.0/, 2017.)

spectrometric detection, as performed for ozone in [282]. But also this method is susceptible to cross reactivities, e.g., by OH [283]. Apart from absorption spectroscopy which, however, has broad absorption features that are hard to distinguish [284], a successful method is electron paramagnetic resonance spectroscopy, described in the following.

8.1.2 Radicals in Gas and Liquid

One of the most reactive classes of species generated from plasmas are radicals. These atoms or molecules have an unpaired electron which makes them highly reactive. Radicals highly relevant for plasma medicine are the ·OH radical or the NO· radical, where the dot marks the position of the unpaired electron. The method of choice for detection of radicals is electron paramagnetic resonance spectroscopy (EPR), also called electron spin resonance spectroscopy (ESR). EPR determines the splitting of degenerate energy levels into several sublevels within a magnetic field due to the Zeeman effect. The resulting spectra are characteristic for each radical. In a magnetic field, the electron's magnetic moment m_s can align itself either parallel or antiparallel to the field. The magnetic moment is then $m_s = 1/2$ for parallel and $m_s = -1/2$ for antiparallel alignment each having different energy calculated by:

$$E = m_s g_e \mu_B B_0 \tag{8.3}$$

g_e is the Landé factor linking the magnetic moment of the electron to the spin angular momentum quantum number (2.0023 for a free electron), μ_B is the Bohr magneton $e\hbar/2m_e$ and B_0 is the magnetic field. For the two states of parallel and antiparallel alignment of the electron's magnetic moment, an energy difference of the two resulting energy states is consequentially proportional to

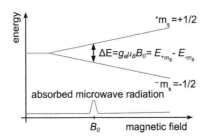

FIGURE 8.4 Principle of EPR measurements (notations see text).

the applied magnetic field according to $\Delta E = g_e \mu_B B_0$. In EPR spectroscopy, the magnetic field matches the criterion for resonance energy, an absorption line is observed (see **Figure 8.4**).

Due to the readily availability of components, X-B and EPR spectrometers operating at a frequency of 10 GHz are most common. The components required for X-Band EPR were initially developed for radar applications. EPR spectroscopy can be used in gas, liquid, and solid environments and is a well known surface characterization technique (see e.g., [285–289]). In the gas phase, oxygen and nitrogen atomic radical concentrations were determined by EPR spectroscopy in the afterglow of a microwave discharge first in [290] and later in [291], where the reaction kinetics were compared to model calculations. In plasma medicine, one of the first studies to perform EPR spectroscopy is [292,293]. The known approach to perform EPR spectroscopy in the afterglow of a microwave plasma was combined with inactivation of fungal spores of *Penicillium digitatum* [294]: The decay of the EPR signal attributed to intracellular semiquinone radicals was observed as a function of oxygen plasma exposure.

Detecting transient species in liquids requires a more careful approach, as radicals such as ˙OH have a low lifetime and diagnostics can rarely be performed in situ and requires liquid handling steps until measurements can be started. In a specifically developed technique called spin-trapping (see e.g., [295]), nitroso or nitrone compounds form a relative stable spin adduct, which is paramagnetic and thus EPR active. This technique stabilizes radicals with a typical lifetime of µs to lifetimes of the spin adduct of minutes to hours, enough to perform EPR studies ex situ. The advantage of spin trap EPR over, e.g., colorimetric or fluorometric methods, where chemical compounds of the probed molecule with the detected species exhibit optically different properties, is its specificity of detection. Due to the detected splitting of the energy levels, the EPR spectra can be attributed to an adduct with the specific radical. In recent years, this method, which has been used successfully in biomedical research [296], has gained focus interest in plasma medicine [297].

The experimental procedure is to add the spin trap to the plasma treated medium, and then place the medium in a glass cuvette, which is inserted into the resonator of the EPR spectrometer (see e.g., [298]). The choice of spin trap is essential for the quality of the measurements [299] as it can drastically reduce background noise and spurious signals (see **Figure 8.5**).

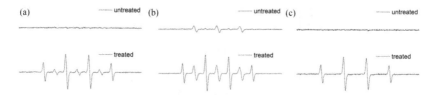

FIGURE 8.5 EPR spectrum of DMPO/OH before and after plasma treatment for spin traps from three different producers. (From Tresp, H. et al., *Plasma Med.*, 3, 45–55, 2013. © IOP Publishing. Reproduced with permission. All rights reserved.)

Not only the spin trap but also the used liquid strongly influences the measurement results [300]. The use of enzymatic activities can help to verify the results [301]. As a remote technique, EPR can determine the effect of plasma jet generated VUV radiation on radical generation in a small enclosed volume of liquid [302].

8.1.3 Plasma Liquid Interface Diagnostics

Plasma liquid interaction is to a large part determined by processes immediately at the plasma liquid interface. Interaction between plasma and liquid occurs through photons and electric fields, through reactive species flux, and through further processes described in Section 8.2. Remote plasma treatment in combination with UV radiation utilizes the high reactivity of plasma generated reactive species and the photochemistry at the water surface by the UV-radiation [303]. It was shown in a corona discharge over a water surface that ionic wind generates a force of 100 μN [304]. The flux of plasma generated reactive species alone influences the structure at the plasma liquid interface [305]. While many important processes take place at the gas/plasma to liquid transition, few diagnostics are sensitive to this region. A diagnostic technique intrinsically sensing only the liquid surface is vibrational sum frequency generation occurring in symmetry breaking environments as provided by the gas–liquid surface [305] used this technique to study water structures. The spectra from sum frequency generation can be analyzed to provide the polar orientation, molecular conformation, and average tilt angle of the surface molecules to the surface normal [306].

With the setup used in [305] (equivalent to the sketch shown in **Figure 8.6**) the influence of neutral reactive plasma generated species on the surface structure of water was investigated.

In plasma liquid interaction, it is relevant whether the plasma has direct contact to the liquid surface, or not. Especially ionic species are important in the first case. In [307], a liquid surface streamer is studied for its OH generation and liquid motion induced by the ionic wind of the streamer. When the plasma has contact to the liquid (touching plasma), charged species play a major role. Especially $H_3O^+_{aq}$, $O_3^-_{aq}$, and $O_2^-_{aq}$ contribute to reactive species generation [308]. Touching plasma increases especially HO_2 and OH generation in the liquid for the case of a helium plasma in air over water.

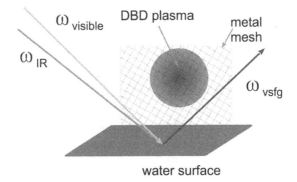

water surface

FIGURE 8.6 Vibrational sum frequency spectroscopy on a water surface as described in [305]. A dielectric barrier discharge is operated perpendicular to the surface. Ions are shielded from the water surface by a grounded metal mesh. Neutral reactive species are allowed to diffuse towards the water surface.

8.2 Interaction with Liquids

Plasma liquid interaction has gained the focus of attention not least due to groundbreaking research in the field of plasma medicine. Several recent reviews as well as roadmaps on plasmas and plasma liquid interaction give an overview [283,309,310]. Plasma liquid interaction combines disciplines from physics, chemistry, and biology. In the following, studies performed on the kINPen interacting with liquids are presented. Section 8.2 describes in three parts the liquid interaction processes of the kINPen that have been studied so far, namely reactive species generation by photons, mass flow processes, and chemical interaction processes.

Figure 8.7 shows a sketch of the kINPen and four separate relevant zones in plasma jet liquid interaction. Section (A) represents the plasma core region, which is the source of all subsequent reaction pathways and Section (B) denotes the plasma gas phase. Section (C) shows the plasma/gas/liquid interface. Section (D) is the liquid phase.

Sections (B), (C), and (D) are closely connected through various processes [311]. At the interface, seven dominant processes can occur (see **Figure 8.7**): Mass transfer (1) [312] depending on the Henry constants of the respective species as well as on gas and liquid flow (2), photolysis (3) [302], positive ions and clusters (4) that can lead to sputtering processes releasing water, gases, or even electrons from the liquid, negative ions, clusters, and cluster transport [313] (5), evaporation (6) [49,314], and electron impact or transport (7) [315,316]. In treatment of liquids by plasma jets, a distinction has to be made between touching and non-touching plasma jets [308]. In plasma liquid interaction, some parallels can be drawn to so-called advanced oxidation techniques that use, e.g., photon reactions to generate oxidative compounds [317].

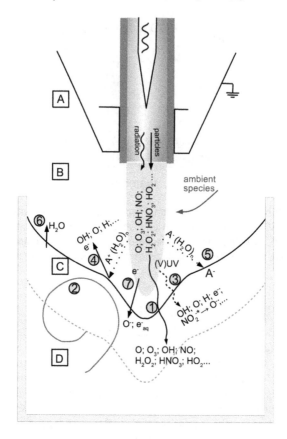

FIGURE 8.7 Interaction processes of treatment of liquids by the kINPen. Four regions are marked: (A) core plasma region (B) effluent and plasma gas phase interaction zone (C) plasma/gas-liquid interface region (D) bulk liquid region. (From Reuter, S. et al., *J. Phys. D Appl. Phys.*, 233001, 51, 2018. Attribution 3.0 Unported (Creative Commons Attribution 3.0 licence in Ref. [67]), http://creativecommons.org/licenses/by/3.0/, 2017)

8.2.1 (V)UV Radiation

In the interaction of plasmas with liquids, VUV radiation is responsible for generation of reactive species from photochemistry as shown by, e.g., model calculations [318]. VUV induced liquid chemistry has to occur within the first 10 μm of the liquid layer, as here 90% of the initial VUV radiation is absorbed according to absorption cross sections of water for VUV [319,320]. Bond dissociation energies of the majority of simple molecules lie in the UV spectral region [321]. The absorption coefficient of water rises over six orders of magnitude from wavelengths of 200 nm down to 155 nm [322]. The quantum yield for water homolysis is unity at 124 nm and is reduced down to 0.3 at 185 nm [323]. VUV-photolysis of water dominantly yields OH and H radicals, hydrated electrons are generated on a smaller scale [323] (see **Table 8.1**). The subsequent reactions of these primary

Table 8.1   Quantum Yield Φ of Photolytic Reactions (pH Value Listed Where Known)

Reaction	Φ	Reference
$H_2O + h\nu_{172nm} \rightarrow OH + H^+ + e^-$	0.05	[326]
$H_2O + h\nu_{185nm} \rightarrow OH + H^+ + e^-$	0.045	[326]
$H_2O + h\nu_{172nm} \rightarrow OH + H$	0.45	[322,326]
$H_2O + h\nu_{185nm} \rightarrow OH + H$	0.33	[326]
$H_2O_2 + h\nu_{172nm} \rightarrow OH + OH$	0.5	[322,327]
$H_2O_2 + h\nu_{185nm} \rightarrow OH + OH$	0.5	[328]
$H_2O_2 + h\nu_{254nm} \rightarrow OH + OH$	$0.49_{pH\,2.7}$	[327]
$NO_2^- + h\nu_{254nm} \rightarrow ... \rightarrow NO + OH$	$0.046_{pH\,1.4}$	[324] and refs. therein
$NO_2^- + h\nu_{346.5nm} \rightarrow ... \rightarrow NO + OH$	$0.347_{pH\,2.0}$	[324] and refs. therein
$NO_3^- + h\nu_{254nm} \rightarrow ... \rightarrow NO_2^- + \frac{1}{2}O_2$	$0.17_{pH\,11.5}$	[324] and refs. therein
$NO_3^- + h\nu_{313nm} \rightarrow ... \rightarrow NO_2^- + \frac{1}{2}O_2$	$0.021_{pH\,11.7}$	[324] and refs. therein
$O_3 + h\nu_{254nm} \rightarrow O(^1D) + O_2$	0.64	[329]

Source: Reuter, S. et al., *J. Phys. D Appl. Phys.*, 233001, 51, 2018. Attribution 3.0 Unported. Creative Commons Attribution 3.0 licence in Ref. [67], http://creativecommons.org/licenses/by/3.0/, 2017.

radicals yield hydroperoxyl radicals, superoxide and oxide radical anions, hydroxide and hydroperoxide anions, and protons, leading finally to molecular oxygen, hydrogen peroxide, molecular hydrogen, and water ([323] and references therein).

Additional components other than water represent further precursors for photolytically generated reactive species. Photolysis of nitrite, for example, will result in the formation of nitric oxide radicals and atomic oxygen ion radicals, which in water will lead to the formation of OH [324]. Especially at wavelengths higher than 200 nm, depending on the water composition, this pathway can dominate OH generation. Photo-dissociation can result in excited products. For nitrite photochemistry, this results in a branching of possible reaction products [324]. Also for ozone, the relatively weak O_2-O bond of ~1 eV results in excess photon-energy from UV-photons which can lead to excitation of the fragment species [325].

In [302], studies on the kINPen interacting with water and more complex biological relevant liquids showed the effect of plasma generated VUV radiation on reactive species generation. For this, the kINPen was compared in two plasma liquid treatment modes: plasma jet in direct contact with liquid on the one hand and on the other hand plasma jet interaction with a glass-, quartz-, or MgF_2-plate

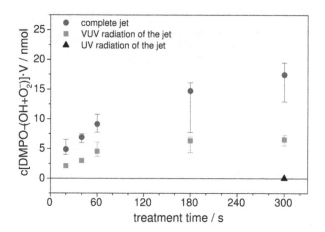

FIGURE 8.8 Oxygen radicals generated by all reactive components of the plasma jet (complete jet) and by VUV/UV-radiation only (5 mL ultrapure water for complete jet measurements, 80 μL ultrapure water in a micro-chamber covered with quartz or MgF$_2$ window for UV and VUV radiation studies). (From Jablonowski, H. et al., *Phys. Plasmas*, 22, 122008, 2015; Attribution 3.0 Unported (Creative Commons Attribution 3.0 licence in Ref. [67]), http://creativecommons.org/licenses/by/3.0/, 2017.)

covered liquid to study solely the effect of (V)UV radiation. It was shown that plasma generated VUV radiation can generate oxygen radicals (see **Figure 8.8**) as well as H$_2$O$_2$. Oxygen radicals were measured by electron paramagnetic resonance spectroscopy using spin trap compounds (see Section 8.1.2). H$_2$O$_2$ generation by VUV radiation was in the order of mmol, while OH and O$_2^-$-spin trap adduct concentrations were in the order of nmol [302]. It is expected that photolytically generated H$_2$O$_2$ originates from OH-radicals, which means that twice as much OH than H$_2$O$_2$ should be observed. Why this was, however, not reflected in the measurements, could be due to the missing convection in the MgF$_2$-covered liquid for the VUV treatment leading to a depletion of spin trap used for the radical detection. It can be assumed that locally much more radicals are generated, but cannot be detected by the chosen method. The study performed in [302] revealed that in buffered medium as well as in cell culture medium (V) UV-induced photochemistry generates the same amount of spin trap radical adduct as in pure water, while direct plasma jet treatment generates increasingly more spin-trap oxygen radical adducts in the more complex media.

To summarize: VUV radiation contributes to reactive species generation dominantly through dissociation of water, even in more complex media, and has to be taken into account when plasma liquid interaction is studied.

8.2.2 Flow Effects

In [330] the plasma jet kINPen interacting with a liquid surface was compared to a pure argon flow on the liquid surface. While the liquid recirculation induced was independent of plasma-on or -off, the dimple in the water surface induced by the gas flow was about 30% less pronounced with the plasma switched on compared to the -off case. This was attributed to either a net electrical field of the plasma sheath or a charge transfer leading to an upwards-directed drag on the water surface.

A planar laser sheet was used to illuminate the liquid. The kINPen was operated at a lower gas flux of 1.9 slm. The highest velocity derived from the laser scattering on micro bubbles was 12 mm/s. A methyl orange test, which indicates acidification by a colour change from red to orange, was performed [330]. It revealed that acidification occurs at diffusion time constants. This means that for high gas flow as used in the kINPen, transport in liquids needs to be considered. In addition to the stochastic position of the argon plasma streamer on the surface, liquid convection results in a continuously renewed liquid volume. This has a significant effect on the reaction kinetics, as high density gradients are levelled out. In [331] a model of streamer induced liquid chemistry generated by a stationary streamer on the one hand and by a moving streamer on the other were compared and strong differences in the end products have been observed. Additionally, stirring and high gas flow result in a degasing of the liquid as was shown in [332], where the oxygen content of cell culture medium was reduced by half through 60 s plasma treatment. 25% oxygen loss already occurs by pure argon flow treatment and can be attributed to flow induced degassing. The additional 25% oxygen loss either originates from different flow conditions with the plasma switched on, or it is caused by chemical reactions. Recently, it was found that treatment of water with the kINPen leads to an enrichment with heavy isotopes, namely ^{2}H and ^{18}O [333]. Heavy isotope enrichment by plasma was significantly higher than enrichment with heavy isotopes by a comparable gas flow without plasma. Dominant reactive species and reaction processes interlinking gas phase and liquid phase chemistry are described in the following.

8.2.3 Reactive Species

Plasma treatment of liquids generates reactivity through the processes mentioned in **Figure 8.7**. In [149] it was demonstrated that the kINPen generates hydrogen peroxide in plasma treated liquids. It could be shown that the net production rate of H_2O_2 in the gas phase is the same as in the liquid phase (see **Figure 8.9**). The production rate in the gas phase was calculated from the measured H_2O_2 concentration and the gas flow velocity. The production rate in the liquid phase was calculated from the measured H_2O_2 concentration and the treatment time. It was found that hydrogen peroxide production, as well as OH generation in the gas phase, increases linearly with humidity admixture to the argon feed gas in the studied regime of up to 1600 ppm water concentration. Humidity was measured with a dew-point hygrometer. The same results were found in a study on OH in [210]. Here, for higher feed gas admixtures, the OH concentration saturates. In [149], it was furthermore shown that hydrogen peroxide can be assumed to be dominantly responsible for a reduction in cell viability of human skin cells as studied by a redox based cell viability assay [334]. A study of water admixture to the feed gas versus admixture to the surrounding shielding gas showed that feed gas water admixture led to an about

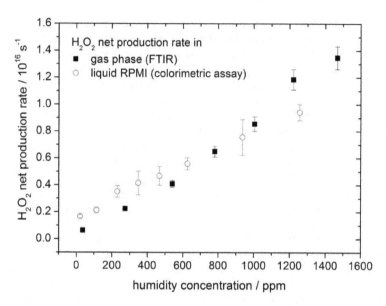

FIGURE 8.9 Net production rate of H_2O_2 in the gas phase and in the liquid phase (5 mL complete cell growth medium (Roswell Park Memorial Institute) RPMI 1640 + 8% fetal calf serum + 1% penicillin/streptomycin solution). (From Winter, J. et al., *J. Phys. D Appl. Phys.*, 47, 285401, 2014. © IOP Publishing. Reproduced with permission. All rights reserved.)

20 times higher hydrogen peroxide production than ambient humidity [275]. From an application point of view, this means that the plasma jet can be operated with no relevant impact by changes in the atmospheric humidity conditions. There is a low impact of ambient humidity since the major part of the energy dissipation occurs in the core plasma region inside the capillary which is protected from changes in the ambient. Feed gas humidity, however, can have a significant effect, e.g., in plasma medicine, due to the resulting OH formation: Plasma generated OH radicals can cause lipid peroxidation processes. In a biophysical phospholipid liposome cell model, it was shown that plasma treatment initiated a macroscopic structural change of the liposomes [335]. While structural changes of the phospholipid layer typically originate from lipid tail group oxidation, a recent study comparing plasma oxidized lipids with molecular dynamics simulations showed that also head group oxidation leads to structural changes of the lipid layer [336].

Hydrogen peroxide is expected to be produced either in the gas phase through a recombination of OH, or in the interphase and liquid phase region. However, while H_2O_2 concentration increases linearly with increasing feed gas humidity in the liquid and in the gas-phase (see **Figure 8.9**), OH radical concentration increases linearly in the gas phase, but remains constant in the liquid phase independent of feed gas humidity. The linearly increasing H_2O_2 concentration does thus not originate dominantly from aqueous OH. These findings accord with studies of H_2O_2 formation in sonochemistry [337] stating that hydrogen peroxide if formed from OH is not originating from the liquid but the cavitation phase. H_2O_2 has a high Henry's constant, which is almost 7 orders of magnitude higher than that of ozone [338]. Therefore, gas phase H_2O_2 is very likely to be transferred to the treated liquid. In [337], three additional generation mechanisms for hydrogen peroxide are proposed that are similarly possible in plasma liquid interaction as in sonochemistry. Hydrogen peroxide can be produced by the dismutation of superoxide anion, a process that is also observed in biological systems [339]. A second generation pathway is through atomic oxygen, where recombination with water forms hydrogen peroxide. The study in [337] found that this reaction occurs fast and can compete in its reaction rate with the formation of ozone through the reaction of atomic oxygen with molecular oxygen. A third pathway is through the recombination of water molecules with excited water molecules as earlier studies in radiology found [340]. In all pathways, the generation of hydrogen peroxide from water molecules is independent of the pH values in the range from 6 to 9 [340].

On several reactive species other than hydrogen peroxide, the pH value has a larger influence. The pH value of unbuffered aqueous solutions changes through plasma treatment. In many cases, the liquid becomes more acidic. Several studies have tried to identify the cause for the acidification, with peroxinitrous acid being discussed as one of the candidate species for this so-called plasma acid. To study the effect of nitrogen and oxygen reactive species of liquid treatment with the kINPen, curtain gas variations were performed. A variation of the curtain gas composition from pure oxygen to pure nitrogen resulted in a decrease of pH. The lowest pH value was achieved at an oxygen to nitrogen ratio that resembles the composition of air [341]. In carbonate buffered medium, kINPen treatment has led to an increase of pH value, which was attributed to a degassing of the carbonate buffer [341].

The lowest pH value correlates with the highest amount of nitrite and nitrate at 25% oxygen and 75% nitrogen, close to air composition conditions (see **Figure 8.10**). The study was performed in non-buffered saline solution. For formation of nitrite and nitrate, both nitrogen and oxygen are required, which is why at 0% nitrogen, neither nitrite nor nitrate was detected. In case of a pure nitrogen curtain, however, nitrite and nitrate were produced. A possible explanation might be the influence of nitrogen metastables species [213,342] or VUV radiation resulting in oxygen species generation from water [302]. Gas phase and liquid phase chemistry are linked and gas phase ROS and RNS dynamics for varying gas curtain correlate with liquid phase ROS and RNS dynamics. Measured nitrite and nitrate concentrations in sodium chloride solution [279] shown in **Figure 8.10** exhibit a comparable dynamic as HNO_2 and HNO_3 and NO_2 in the gas phase [147]. In [147], gas phase species were measured in an FTIR multi-pass cell (setup shown in **Figure 7.29**) as a function of varying shielding gas composition and simulated with a model. In the simulation, the concentration maximum of HNO_3 densities is slightly shifted towards higher nitrogen concentrations in the shielding gas in comparison to the maximum of HNO_2. The same trend is observed for nitrite and nitrate in the liquid phase, respectively. While the measurements are performed at

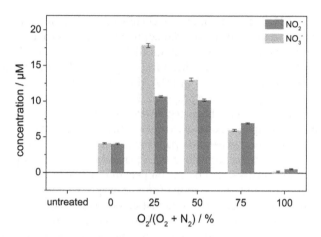

FIGURE 8.10 Nitrite and nitrate values in liquid (5 mL sodium chloride solution at treatment times of 10 minutes) measured for the kINPen. The experiments were performed with pure argon feed gas. (From Jablonowski, H. et al., *Biointerphases*, 10, 029506, 2015. With permission. Copyright 2015, American Vacuum Society.)

different conditions, and the simulations could not yet be validated, it is still noteworthy that similar trends can be observed, allowing conclusions about generation mechanisms: Generation mechanisms of nitric and nitrous acid in the gas phase are dominated by the following reactions:

$$NO + OH + M \rightarrow HNO_2 + M \tag{8.4}$$

$$NO_2 + OH + M \rightarrow HNO_3 + M \tag{8.5}$$

HNO_2 can be transformed to HNO_3 by **Reaction 8.6** followed by **Reaction 8.5** [274].

$$HNO_2 + O \rightarrow OH + NO_2 \tag{8.6}$$

The reaction of hydrogen peroxide with nitrite leads to the generation of peroxynitrous acid according to **Reaction 8.7** [282].

$$NO_2^- + H_2O_2 + H^+ \rightarrow ONOOH + H_2O \tag{8.7}$$

which can lead to the formation of nitrate. In a buffered system, H^+ can be considered constant and **Reaction 8.7** can be described with a pseudo second order rate constant [282]. From the time development of hydrogen peroxide, nitrite, and nitrate, the formation of peroxynitrite can thus be determined [282]. Peroxinitrous acid decomposes into biologically active OH and NO_2 radicals:

$$ONOOH \rightarrow OH + NO_2 \tag{8.8}$$

Studies with the kINPen on bacterial inactivation mechanisms [343] show the often-observed increasing bacterial inactivation efficacy of plasma treated

Escherichia Coli bacterial physiological solution with increasing treatment times. Measurements of reactive species in the liquid media show increasing nitrite, nitrate, and hydrogen peroxide. The work suggests that bacterial inactivation mechanisms are at least partially due to OH and NO_2 radicals from **Reaction 8.8** from post discharge chemistry.

Control over nitrogen and oxygen species allows controlling the biologic impact of plasma treatment: In [279], it was found that the condition of 25% oxygen to 75% nitrogen in the curtain gas shown in **Figure 8.10** led to a maximal inactivation of bacteria, while at the same time having the least impact on human skin cells. This study has the potential to form the basis for a tailored plasma wound treatment: Switching the plasma composition to achieve bacterial inactivation, or human cell inactivation (e.g., for cancer treatment), or stimulation of cell proliferation (wound healing) may in future be a unique feature of plasma based therapy. Using feedback signals, tailored plasma chemistry will allow treating diseased tissue with certain parameters and healthy tissue with different parameters within the same treatment.

Open questions still remain on the role of short living active plasma species, not least due to the difficulties in their detection. Short living plasma species can be excited species, charged species, atomic or molecular radicals, and other highly reactive molecules. Ions can be found in the effluent of the kINPen. Their concentration, if derived from quasi-neutrality, lies around a density of $10^{12}\,cm^{-3}$ with quick recombination in the effluent. Ions can be stabilized by cluster formation and their role in plasma liquid interaction still needs to be investigated. While plasma generated radicals and atoms can reach concentrations that are three orders of magnitude higher, ions can still play a vital role. Negative ions, for example, reduce the size of nano bubbles [344]—bubbles with a diameter smaller than a micrometer [345]—and thus reduce stable pockets of gas reservoirs that can contribute to plasma generated liquid chemistry.

Recently, it was found that plasma generated neutrals change the surface structure at the air water interface [305]. It was found in tropospheric chemistry that both hydrophobic and hydrophilic species saturate at the air liquid interface [346], which shows its significance for chemical reactions.

A special role must be attributed to the class of metastable neutrals generated by plasma. The kINPen is, depending on the operating gases and parameters, a source of ample metastable species including helium [172], argon [147], oxygen, and nitrogen metastables [213]. These species carry an energy in the range of molecular bonds or ionization energy levels and have a long lifetime that can reach minutes. These properties make metastable species key candidates for an energy transfer to superficial layers of liquids or biological tissue. Metastables may be involved in generation of reactive species in plasma treated liquids, as discussed in a study of kINPen liquid treatment [342]. In medicine, plasma generated singlet oxygen is stated to play a dominant role in tumour reduction [347] based on an inactivation of membrane bound catalase [348].

Especially for an identification of the role of short living species in plasma liquid interaction, important information can be gained by comparison of plasma sources that differ only in few aspects. In [349], the kINPen was compared to an RF argon plasma jet of similar build. Main differences are the needle electrode position, the operating frequency of 13.56 MHz compared to 1.1 MHz for the kINPen, the gas flow, and the grounded electrode size. In [349], the findings of [277] were confirmed that cell viability in in-vitro plasma treatment experiments depends purely on the H_2O_2 generation in the treated medium: Both jets exhibited a distinct difference in hydrogen peroxide production as a function of distance. The RF jet exhibits a step in hydrogen peroxide concentration and the kINPen generates a more or less constant value. The stepwise concentration development indicates that short living species are involved in hydrogen peroxide generation. At distances where the short living species are not present anymore, the kINPen generates more hydrogen peroxide compared to the RF jet. This may indicate a gas born generation of hydrogen peroxide. Although the authors did not identify the cause for the step in the hydrogen peroxide concentration, several hypotheses can be made:

The most distinctive difference is the excitation frequency of the two jets. The excitation frequency in atmospheric pressure discharges influences the plasma sheath thickness. The higher the frequency, the closer electrons are to the electrode surface [350]. Another (possible) difference between the two jets is that one jet might come into contact with the liquid surface (depending on the distance), while the other one might not. A model calculation showed that touching versus non-touching case can lead to differences in hydrogen peroxide production by orders of magnitude [308]. An influx of electrons from the plasma may lead to the formation of solvated electrons [351]. Electron beam irradiation—although at completely different energy regimes—is known to generate hydrogen peroxide [352]. The different excitation frequencies of the plasma jets can furthermore result in a higher atomic oxygen concentration for the RF jet [349]. Atomic oxygen has been identified to play a role in plasma initiated liquid chemistry [353,354] and, as previously stated, can lead to the formation of hydrogen peroxide [337]. From a parameter study of the RF jet, it was followed that dominant reaction pathways include interaction of atomic oxygen and chlorine forming Cl_2^- and ClO^-.

A further, highly reactive short living species is the hydroperoxy radical (HO_2), which has been measured in the kINPen effluent, recently. HO_2 concentrations were determined by cavity enhanced absorption spectroscopy [184].

In liquid media, hydroperoxy radicals are in equilibrium with superoxide anions [355] according to the pH value via

$$O_2^- + H^+ \leftrightarrow HO_2 \qquad pKa = 4.8. \qquad (8.9)$$

Formation of hydroperoxy radicals from superoxide anions at pH values lower than 4.8 lead to a pH dependence of bactericidal activity of plasma activated liquid [355]. Both species are highly active and have a low lifetime in liquids [356]. O_2^- is generated by plasma in the gas phase and its concentration in the liquid can be influenced through a variation of the ambient around the plasma effluent. O_2^- can also be generated in liquid through reactions of peroxinitric acid with hydrogen peroxide [302,357]. For the kINPen,

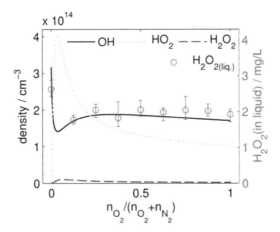

FIGURE 8.11 Comparison of the gas phase model and measurements in the liquid phase [358] (liquid measurements performed in 5 mL sodium chloride solution treated at 9 mm distance to the kINPen nozzle according to [300]). (Reproduced with kind permission of A. Schmidt-Bleker.)

it could be shown in a model-based evaluation of liquid measurements of hydrogen peroxide that HO_2 plays a role in the gas phase reaction pathways that lead to the generation of H_2O_2 via OH. The comparison of gas phase chemistry with liquid analysis connected the reactions of OH, H_2O_2, and HO_2. The study [358] showed that the development of hydrogen peroxide in plasma treated buffered medium for varying oxygen to nitrogen ratio in a gas curtain can be explained by the reactions involving these three species. The OH dynamics calculated by the reaction model [147] follows the hydrogen peroxide concentration in the liquid (see **Figure 8.11**), suggesting that gas phase OH and liquid phase hydrogen peroxide are strongly linked. Interestingly, a sharp increase by a factor of 1.5 towards 0% of oxygen in the hydrogen peroxide concentration was reproduced by the model calculations. The explanation for this increase was given by the model to origin from a large increase in HO_2 concentration for small oxygen concentrations in the curtain gas [358]. HO_2 concentrations in the proposed quantities were confirmed by cavity enhanced absorption spectroscopy [184]. The results shown in **Figure 8.11** support the assumption that an enriched concentration of gas phase OH at the gas liquid interface is a potential origin for liquid hydrogen peroxide.

Summarizing, it can be stated that the role of the gas liquid interface in plasma liquid interaction is still a topic of present studies [283,311]. A lot can be gained from including perspectives from adjacent research fields such as electron beam studies [352], sonochemistry [359], or photocatalytic chemistry, where plasma generated electrons replace photocatalytically generated electrons and similar processes occur e.g., in hydrogen peroxide formation [360]. The interdisciplinary exchange at the boundary of physics and chemistry will help revealing further fundamental mechanisms of plasma liquid interaction.

References for Part III

1. F. Iza, G.J. Kim, S.M. Lee, J.K. Lee, J.L. Walsh, Y.T. Zhang and M.G. Kong, Microplasmas: Sources, particle kinetics, and biomedical applications, *Plasma Process. Polym.* 5 (2008) 322–344. doi:10.1002/ppap.200700162

2. V. Léveillé and S. Coulombe, Design and preliminary characterization of a miniature pulsed RF APGD torch with downstream injection of the source of reactive species, *Plasma Sour. Sci. Technol.* 14 (2005) 467–476. doi:10.1088/0963-0252/14/3/008

3. A. Schutze, J.Y. Jeong, S.E. Babayan, P. Jaeyoung, G.S. Selwyn and R.F. Hicks, The atmospheric-pressure plasma jet: A review and comparison to other plasma sources, *IEEE Trans. Plasma Sci.* 26 (1998) 1685–1694. doi:10.1109/27.747887

4. E. Stoffels, A.J. Flikweert, W.W. Stoffels and G.M.W. Kroesen, Plasma needle: A non-destructive atmospheric plasma source for fine surface treatment of (bio)materials, *Plasma Sour. Sci. Technol.* 11 (2002) 383–388. doi:10.1088/0963-0252/11/4/304

5. K. Niemi, S. Reuter, L. Schaper, N. Knake, V. Schulz-von der Gathen and T. Gans, Diagnostics on an atmospheric pressure plasma jet, *J. Phys. Conf. Ser.* 71 (2007) 012012. doi:10.1088/1742-6596/71/1/012012

6. S. Reuter, K. Niemi, V. Schulz-von der Gathen and H.F. Döbele, Generation of atomic oxygen in the effluent of an atmospheric pressure plasma jet, *Plasma Sour. Sci. Technol.* 18 (2009) 015006. doi:10.1088/0963-0252/18/1/015006

7. X. Lu, M. Laroussi and V. Puech, On atmospheric-pressure non-equilibrium plasma jets and plasma bullets, *Plasma Sour. Sci. Technol.* 21 (2012) 034005. doi:10.1088/0963-0252/21/3/034005

8. Z. Machala, L. Chládeková and M. Pelach, Plasma agents in bio-decontamination by dc discharges in atmospheric air, *J. Phys. D Appl. Phys.* 43 (2010). doi:10.1088/0022-3727/43/22/222001

9. S. Reuter, J. Winter, A. Schmidt-Bleker, D. Schroeder, H. Lange, N. Knake, V. Schulz-von der Gathen and K.D. Weltmann, Atomic oxygen in a cold argon plasma jet: TALIF spectroscopy in ambient air with modelling and measurements of ambient species diffusion, *Plasma Sour. Sci.Technol.* 21 (2012) 024005. doi:10.1088/0963-0252/21/2/024005

10. X. Lu and S. Wu, On the active species concentrations of atmospheric pressure nonequilibrium plasma jets, *IEEE Technol. Plasma Sci.* 41 (2013) 2313–2326. doi:10.1109/tps.2013.2268579

11. P. Bruggeman and R. Brandenburg, Atmospheric pressure discharge filaments and microplasmas: Physics, chemistry and diagnostics, *J. Phys. D Appl. Phys.* 46 (2013) 464001. doi:10.1088/0022-3727/46/46/464001

12. R. Ono, Optical diagnostics of reactive species in atmospheric-pressure non-thermal plasma, *J. Phys. D Appl. Phys.* 49 (2016) 083001. doi:10.1088/0022-3727/49/8/083001

13. T. von Woedtke, S. Reuter, K. Masur and K.D. Weltmann, Plasmas for medicine, *Phys. Rep.* 530 (2013) 291–320. doi:10.1016/j.physrep.2013.05.005

14. D.B. Graves, Low temperature plasma biomedicine: A tutorial review, *Phys. Plasmas* 21 (2014) 080901. doi:10.1063/1.4892534

15. S. Reuter, T. von Woedtke and K.-D. Weltmann, The kINPen—A review on physics and chemistry of the atmospheric pressure plasma jet and its applications, *J. Phys. D Appl. Phys.* 51 (2018). doi:10.1088/1361-6463/aab3ad

16. V. Schulz-von der Gathen, L. Schaper, N. Knake, S. Reuter, K. Niemi, T. Gans and J. Winter, Spatially resolved diagnostics on a microscale atmospheric pressure plasma jet, *J. Phys. D Appl. Phys.* 41 (2008) 194004. doi:10.1088/0022-3727/41/19/194004

17. G. Dilecce, Optical spectroscopy diagnostics of discharges at atmospheric pressure, *Plasma Sour. Sci. Technol.* 23 (2014). doi:10.1088/0963-0252/23/1/015011

18. C.O. Laux, T.G. Spence, C.H. Kruger and R.N. Zare, Optical diagnostics of atmospheric pressure air plasmas, *Plasma Sour. Sci. Technol.* 12 (2003) 125–138. doi:10.1088/0963-0252/12/2/301

19. N. Sadeghi, Molecular spectroscopy techniques applied for processing plasma diagnostics, *J. Plasma Fusion Res.* 80 (2004) 767–776.

20. M. Simek, Optical diagnostics of streamer discharges in atmospheric gases, *J. Phys. D Appl. Phys.* 47 (2014). doi:10.1088/0022-3727/47/46/463001

21. G.D. Stancu, F. Kaddouri, D.A. Lacoste and C.O. Laux, Atmospheric pressure plasma diagnostics by OES, CRDS and TALIF, *J. Phys. D Appl. Phys.* 43 (2010) 124002. doi:10.1088/0022-3727/43/12/124002

22. H.R. Griem, *Principles of Plasma Spectroscopy*, Cambridge University Press, Cambridge, UK, 1997.

23. S. Reuter, J.S. Sousa, G.D. Stancu and J.-P. Hubertus van Helden, Review on VUV to MIR absorption spectroscopy of atmospheric pressure plasma jets, *Plasma Sour. Sci. Technol.* 24 (2015) 054001. doi:10.1088/0963-0252/24/5/054001

24. H.R. Griem, *Principles of Plasma Spectroscopy*, Cambridge University Press, Cambridge, UK, 1964.

25. W. Demtröder, *Laser Spectroscopy*, Vol. 1, Springer, Berlin, Germany, 2008.

26. P.J. Bruggeman, N. Sadeghi, D.C. Schram and V. Linss, Gas temperature determination from rotational lines in non-equilibrium plasmas: A review, *Plasma Sour. Sci. Technol* 23 (2014). doi:10.1088/0963-0252/23/2/023001

27. A.Y. Nikiforov, C. Leys, M.A. Gonzalez and J.L. Walsh, Electron density measurement in atmospheric pressure plasma jets: Stark broadening of hydrogenated and non-hydrogenated lines, *Plasma Sour. Sci. Technol.* 24 (2015) 034001. doi:10.1088/0963-0252/24/3/034001

28. L.S. Rothman, C.P. Rinsland, A. Goldman, S.T. Massie, D.P. Edwards, J.M. Flaud, A. Perrin et al., The Hitran molecular spectroscopic database and Hawks (Hitran Atmospheric Workstation): 1996 edition, *J. Quant. Spectrosc. Radiat. Transf.* 60 (1998) 665–710. doi:10.1016/s0022-4073(98)00078-8

29. H.J. Kunze, *Introduction to Plasma Spectroscopy*, Vol. 56, Springer, Berlin, Germany, 2009.

30. A.F.H. van Gessel and P.J. Bruggeman, Thermalization of rotational states of NO A(2)Sigma(+)(v=0) in an atmospheric pressure plasma, *J. Chem. Phys.* 138 (2013) doi:10.1063/1.4802959

31. S. Hübner, N. Sadeghi, E.A.D. Carbone and J.J.A.M. van der Mullen, Density of atoms in Ar*(3p54s) states and gas temperatures in an argon surfatron plasma measured by tunable laser spectroscopy, *J. Appl. Phys.* 113 (2013). doi:10.1063/1.4799152

32. C. Penache, M. Miclea, A. Bräuning-Demian, O. Hohn, S. Schössler, T. Jahnke, K. Niemax and H. Schmidt-Böcking, Characterization of a high-pressure microdischarge using diode laser atomic absorption spectroscopy, *Plasma Sour. Sci. Technol.* 11 (2002) 476–483. doi:10.1088/0963-0252/11/4/314

33. A.V. Pipa, Y.Z. Ionikh, V.M. Chekishev, M. Dünnbier and S. Reuter, Resonance broadening of argon lines in a micro-scaled atmospheric pressure plasma jet (argon μAPPJ), *Appl. Phys. Lett.* 106 (2015) 244104. doi:10.1063/1.4922730

34. A. Kramida, Y. Ralchenko, J. Reader and 'NIST-ASD-Team', 2015 NIST Atomic Spectra Database (Version 5.2) www.nist.gov/pml/data/asd.cfm Access: 2015; Last Update: September 2014.

35. G. Herzberg, *Molecular Spectra and Molecular Structure*, Vol. 1–3, Krieger Publishing Company, Malabar, FL, 1989.

36. G.D. Stancu, J. Ropcke and P.B. Davies, Line strengths and transition dipole moment of the nu2 fundamental band of the methyl radical, *J. Chem. Phys.* 122 (2005) 14306. doi:10.1063/1.1812755

37. L.S. Rothman, The Hitran Database, 2014. http://www.cfa.harvard.edu/hitran/ Access: 2014; Last Update: October 11, 2013.

38. L.S. Rothman, I.E. Gordon, Y. Babikov, A. Barbe, D. Chris Benner, P.F. Bernath, M. Birk et al., The HITRAN2012 molecular spectroscopic database, *J. Quant. Spectrosc. Radiat. Transf.* 130 (2013) 4–50. doi:10.1016/j.jqsrt.2013.07.002

39. S. Welzel, S. Stepanov, J. Meichsner and J. Röpcke, Application of quantum cascade laser absorption spectroscopy to studies of fluorocarbon molecules, *J. Phys. Conf. Ser.* 157 (2009) 012010. doi:10.1088/1742-6596/157/1/012010

40. R.B. Miles, W.R. Lempert and J.N. Forkey, Laser Rayleigh scattering, *Meas. Sci. Technol.* 12 (2001) R33–R51. doi:10.1088/0957-0233/12/5/201

41. H. Kempkens and J. Uhlenbusch, Scattering diagnostics of low-temperature plasmas (Rayleigh scattering, Thomson scattering, CARS), *Plasma Sour. Sci. Technol.* 9 (2000) 492–506. doi:10.1088/0963-0252/9/4/305

42. S. Hübner, J.S. Sousa, J. van der Mullen and W.G. Graham, Thomson scattering on non-thermal atmospheric pressure plasma jets, *Plasma Sour. Sci. Technol,* 24 (2015) 054005. doi:10.1088/0963-0252/24/5/054005

43. E. Carbone and S. Nijdam, Thomson scattering on non-equilibrium low density plasmas: Principles, practice and challenges, *Plasma Phys. Control. Fusion* 57 (2015) 014026. doi:10.1088/0741-3335/57/1/014026

44. K. Muraoka and A. Kono, Laser Thomson scattering for low-temperature plasmas, *J. Phys. D Appl. Phys.* 44 (2011) 043001. doi:10.1088/0022-3727/44/4/043001

45. C.M. Penney, R.L. St. Peters and M. Lapp, Absolute rotational Raman cross sections for N_2, O_2, and CO_2, *J. Opt. Soc. Am.* 64 (1974). doi:10.1364/josa.64.000712

46. J.W. Daily, Laser induced fluorescence spectroscopy in flames, *Prog. Energy Combust. Sci.* 23 (1997) 133–199. doi:10.1016/s0360-1285(97)00008-7

47. G. Dilecce, L.M. Martini, P. Tosi, M. Scotoni and S. De Benedictis, Laser induced fluorescence in atmospheric pressure discharges, *Plasma Sour. Sci. Technol.* 24 (2015) 034007. doi:10.1088/0963-0252/24/3/034007

48. S. Okada, A. Tezaki, A. Miyoshi and H. Matsui, Product branching fractions in the reactions of $NH(a\,1\Delta)$ and $NH(X\,3\Sigma-)$ with NO, *J. Chem. Phys.* 101 (1994) 9582. doi:10.1063/1.467989

49. D. Riès, G. Dilecce, E. Robert, P.F. Ambrico, S. Dozias and J.M. Pouvesle, LIF and fast imaging plasma jet characterization relevant for NTP biomedical applications, *J. Phys. D Appl. Phys.* 47 (2014) 275401. doi:10.1088/0022-3727/47/27/275401

50. X. Lu, G.V. Naidis, M. Laroussi, S. Reuter, D.B. Graves and K. Ostrikov, Reactive species in non-equilibrium atmospheric-pressure plasmas: Generation, transport, and biological effects, *Phys. Rep.* 630 (2016) 1–84. doi:10.1016/j.physrep.2016.03.003

51. J.W. Daily, Laser induced spectroscopy in flames, *Prog. Energy Combust. Sci.* 23 (1997) 133–199.

52. W.P. Partridge, Jr. and N.M. Laurendeau, Formulation of a dimensionless overlap fraction to account for spectrally distributed interactions in fluorescence studies, *Appl. Opt.* 34 (1995) 2645–2647. doi:10.1364/AO.34.002645

53. M. Göppert-Mayer, *Über Elementarakte mit zwei Quantensprüngen, Annalen der Physik.* 401 (1931) 273–294. doi:10.1002/andp.19314010303

54. N. Merbahi, N. Sewraj, F. Marchal, Y. Salamero and P. Millet, Luminescence of argon in a spatially stabilized mono-filamentary dielectric barrier microdischarge: Spectroscopic and kinetic analysis, *J. Phys. D Appl. Phys.* 37 (2004) 1664–1678. doi:10.1088/0022-3727/37/12/011

55. S. Liu and M. Neiger, Excitation of dielectric barrier discharges by unipolar submicrosecond square pulses, *J. Phys. D Appl. Phys.* 34 (2001) 1632–1638. doi:10.1088/0022-3727/34/11/312

56. T.C. Manley, The electric characteristics of the ozonator discharge, *Trans. Electrochem. Soc.* 84 (1943) 83. doi:10.1149/1.3071556

57. D.E. Ashpis, M.C. Laun and E.L. Griebeler, Progress toward accurate measurement of dielectric barrier discharge plasma actuator power, *AIAA J.* 55 (2017) 2254–2268. doi:10.2514/1.J055816

58. R. Feng, G.S.P. Castle and S. Jayaram, Automated system for power measurement in the silent discharge, *IEEE Trans. Ind. Appl.* 34 (1998) 563–570. doi:10.1109/28.673727

59. I. Stefanovic, N.K. Bibinov, A.A. Deryugin, I.P. Vinogradov, A.P. Napartovich and K. Wiesemann, Kinetics of ozone and nitric oxides in dielectric barrier discharges in O2/NOx and N2/O2/NOx mixtures, *Plasma Sour. Sci. Technol.* 10 (2001) 406–416. doi:10.1088/0963-0252/10/3/303

60. M. Young Sun, V. Ravi, K. Ho-Chul and B.S. Rajanikanth, Abatement of nitrogen oxides in a catalytic reactor enhanced by nonthermal plasma discharge, *IEEE Trans. Plasma Sci.* 31 (2003) 157–165. doi:10.1109/tps.2003.808876

61. A.V. Pipa, T. Hoder, J. Koskulics, M. Schmidt and R. Brandenburg, Experimental determination of dielectric barrier discharge capacitance, *Rev. Sci. Instrum.* 83 (2012) 075111. doi:10.1063/1.4737623

62. M. Kettlitz, H. Höft, T. Hoder, S. Reuter, K.D. Weltmann and R. Brandenburg, On the spatio-temporal development of pulsed barrier discharges: Influence of duty cycle variation, *J. Phys. D Appl. Phys.* 45 (2012) 245201. doi:10.1088/0022-3727/45/24/245201

63. A.Y. Nikiforov, A. Sarani and C. Leys, The influence of water vapor content on electrical and spectral properties of an atmospheric pressure plasma jet, *Plasma Sour. Sci. Technol.* 20 (2011) 015014. doi:10.1088/0963-0252/20/1/015014

64. V.A. Godyak and R.B. Piejak, In situ simultaneous radio frequency discharge power measurements, *J. Vac. Sci. Technol. A* 8 (1990) 3833–3837. doi:10.1116/1.576457

65. M. Dünnbier, M.M. Becker, S. Iseni, R. Bansemer, D. Loffhagen, S. Reuter and K.D. Weltmann, Stability and excitation dynamics of an argon micro-scaled atmospheric pressure plasma jet, *Plasma Sour. Sci. Technol.* 24 (2015) 065018. doi:10.1088/0963-0252/24/6/065018

66. D. Marinov and N.S.J. Braithwaite, Power coupling and electrical characterization of a radio-frequency micro atmospheric pressure plasma jet, *Plasma Sour. Sci. Technol.* 23 (2014) 062005. doi:10.1088/0963-0252/23/6/062005

67. Attribution 3.0 Unported (CC BY 3.0), 2017. http://creativecommons.org/licenses/by/3.0/ Access: Last Update.

68. Y. Seepersad, M. Pekker, M.N. Shneider, A. Fridman and D. Dobrynin, Investigation of positive and negative modes of nanosecond pulsed discharge in water and electrostriction model of initiation, *J. Phys. D Appl. Phys.* 46 (2013) 355201. doi:10.1088/0022-3727/46/35/355201

69. N.B. Anikin, S.V. Pancheshnyi, S.M. Starikovskaia and A.Y. Starikovskii, Breakdown development at high overvoltage: Electric field, electronic level excitation and electron density, *J. Phys. D: Appl. Phys.* 31 (1998) 826–833. doi:10.1088/0022-3727/31/7/012

70. X. Lu and M. Laroussi, Dynamics of an atmospheric pressure plasma plume generated by submicrosecond voltage pulses, *J. Appl. Phys.* 100 (2006) 063302. doi:10.1063/1.2349475

71. S. Reuter, A. Schmidt-Bleker, S. Iseni, J. Winter and K.-D. Weltmann, On the Bullet-Streamer Dualism, *IEEE Trans. Plasma Sci.* 42 (2014) 2428–2429. doi:10.1109/tps.2014.2332539

72. V.A. Godyak, R.B. Piejak and B.M. Alexandrovich, Electron energy distribution function measurements and plasma parameters in inductively coupled argon plasma, *Plasma Sour. Sci. Technol.* 11 (2002) 525–543. doi:10.1088/0963-0252/11/4/320

73. M.N. Shneider and R.B. Miles, Microwave diagnostics of small plasma objects, *J. Appl. Phys.* 98 (2005) 033301. doi:10.1063/1.1996835

74. A. Shashurin, M.N. Shneider, A. Dogariu, R.B. Miles and M. Keidar, Temporary-resolved measurement of electron density in small atmospheric plasmas, *Appl. Phys. Lett.* 96 (2010) 171502. doi:10.1063/1.3389496

75. A. Shashurin, M.N. Shneider, A. Dogariu, R.B. Miles and M. Keidar, Temporal behavior of cold atmospheric plasma jet, *Appl. Phys. Lett.* 94 (2009) 231504. doi:10.1063/1.3153143

76. P. Bruggeman, T. Verreycken, M.A. Gonzalez, J.L. Walsh, M.G. Kong, C. Leys and D.C. Schram, Optical emission spectroscopy as a diagnostic for plasmas in liquids: Opportunities and pitfalls, *J. Phys. D Appl. Phys.* 43 (2010). doi:10.1088/0022-3727/43/12/124005

77. C.O. Laux, R.J. Gessman, C.H. Kruger, F. Roux, F. Michaud and S.P. Davis, Rotational temperature measurements in air and nitrogen plasmas using the first negative system of N2+, *J. Quant. Spectrosc. Radiat. Transf.* 68 (2001) 473–482. doi:10.1016/s0022-4073(00)00083-2

78. N. Balcon, A. Aanesland and R. Boswell, Pulsed RF discharges, glow and filamentary mode at atmospheric pressure in argon, *Plasma Sour. Sci. Technol.* 16 (2007) 217–225. doi:10.1088/0963-0252/16/2/002

79. J.L. Walsh, F. Iza and M.G. Kong, Characterisation of a 3 nanosecond pulsed atmospheric pressure argon microplasma, *Eur. Phys. J.* 60 (2010) 523–530. doi:10.1140/epjd/e2010-00238-9:

80. M.A. Gigosos, M.Á. González and V.n. Cardeñoso, Computer simulated Balmer-alpha, -beta and -gamma Stark line profiles for non-equilibrium plasmas diagnostics, *Spectrochim. Acta B* 58 (2003) 1489–1504. doi:10.1016/s0584-8547(03)00097-1

81. N. Konjević, M. Ivković and N. Sakan, Hydrogen Balmer lines for low electron number density plasma diagnostics, *Spectrochim. Acta B* 76 (2012) 16–26. doi:10.1016/j.sab.2012.06.026:

82. Q. Xiong, A.Y. Nikiforov, M.Á. González, C. Leys and X.P. Lu, Characterization of an atmospheric helium plasma jet by relative and absolute optical emission spectroscopy, *Plasma Sour. Sci. Technol.* 22 (2013) 015011. doi:10.1088/0963-0252/22/1/015011

References for Part III

83. M.S. Dimitrijević, M. Christova and S. Sahal-Bréchot, Stark broadening of visible Ar I spectral lines, *Physica Scripta* 75 (2007) 809–819. doi:10.1088/0031-8949/75/6/011

84. S. Djurović and N. Konjević, On the use of non-hydrogenic spectral lines for low electron density and high pressure plasma diagnostics, *Plasma Sour. Sci. Technol.* 18 (2009) 035011. doi:10.1088/0963-0252/18/3/035011

85. L. Dong, J. Ran and Z. Mao, Direct measurement of electron density in microdischarge at atmospheric pressure by Stark broadening, *Appl. Phys. Lett.* 86 (2005) 161501. doi:10.1063/1.1906299

86. Q. Wang, I. Koleva, V.M. Donnelly and D.J. Economou, Spatially resolved diagnostics of an atmospheric pressure direct current helium microplasma, *J. Phys. D Appl. Phys.* 38 (2005) 1690–1697. doi:10.1088/0022-3727/38/11/008

87. R.F.G. Meulenbroeks, M.F.M. Steenbakkers, Z. Qing, M.C.M. van de Sanden and D.C. Schram, Four ways to determine the electron density in low-temperature plasmas, *Phys. Rev. E* 49 (1994) 2272–2275. doi:10.1103/PhysRevE.49.2272

88. E. Iordanova, J.M. Palomares, A. Gamero, A. Sola and J.J.A.M. van der Mullen, A novel method to determine the electron temperature and density from the absolute intensity of line and continuum emission: Application to atmospheric microwave induced Ar plasmas, *J. Phys. D Appl. Phys.* 42 (2009) 155208. doi:10.1088/0022-3727/42/15/155208

89. K.T.A.L. Burm, Continuum radiation in a high pressure argon–mercury lamp, *Plasma Sour. Sci. Technol.* 13 (2004) 387–394. doi:10.1088/0963-0252/13/3/004

90. S. Park, W. Choe, S. Youn Moon and J. Park, Electron density and temperature measurement by continuum radiation emitted from weakly ionized atmospheric pressure plasmas, *Appl. Phys. Lett.* 104 (2014) 084103. doi:10.1063/1.4866804

91. X.-M. Zhu and Y.-K. Pu, A simple collisional–radiative model for low-pressure argon discharges, *J. Phys. D Appl. Phys.* 40 (2007) 2533–2538. doi:10.1088/0022-3727/40/8/018

92. K. Kano, M. Suzuki and H. Akatsuka, Spectroscopic measurement of electron temperature and density in argon plasmas based on collisional-radiative model, *Plasma Sour. Sci. Technol.* 9 (2000) 314–322. doi:10.1088/0963-0252/9/3/309

93. X.M. Zhu, Y.K. Pu, N. Balcon and R. Boswell, Measurement of the electron density in atmospheric-pressure low-temperature argon discharges by line-ratio method of optical emission spectroscopy, *J. Phys. D Appl. Phys.* 42 (2009) 142003. doi:10.1088/0022-3727/42/14/142003

94. M.M. Becker and D. Loffhagen, Enhanced reliability of drift-diffusion approximation for electrons in fluid models for nonthermal plasmas, *AIP Adv.* 3 (2013) 012108. doi:10.1063/1.4775771

95. Y.P. Raizer, M.N. Shneider and N.A. Yatsenko, *Radio Frequency Capacitive Discharges*, CRC Press, 1995.

96. J.R. Roth, *Industrial Plasma Engineering: Volume 2 - Applications to Nonthermal Plasma Processing*, Vol. 2, CRC Press, 2001.

97. M. Laroussi, Interaction of microwaves with atmospheric-pressure plasmas, *Int. J. Infrared Milli.* 16 (1995) 2069–2083. doi:10.1007/Bf02073410

98. M. Laroussi, Relationship between the number density and the phase shift in microwave interferometry for atmospheric pressure plasmas, *Int. J. Infrared Milli.* 20 (1999) 1501–1508. doi:10.1023/A:1021708720181

99. M. Baeva, M. Andrasch, J. Ehlbeck, D. Loffhagen and K.D. Weltmann, Temporally and spatially resolved characterization of microwave induced argon plasmas: Experiment and modeling, *J. Appl. Phys.* 115 (2014) 143301. doi:10.1063/1.4870858

100. N. Niemöller, V. Schulz-von der Gathen, A. Stampa and H.F. Döbele, A quasi-optical 1 mm microwave heterodyne interferometer for plasma diagnostics using a frequency-tripled Gunn oscillator, *Plasma Sour. Sci. Technol.* 6 (1997) 478–483. doi:10.1088/0963-0252/6/4/004

101. X.P. Lu and M. Laroussi, Electron density and temperature measurement of an atmospheric pressure plasma by millimeter wave interferometer, *Appl. Phys. Lett.* 92 (2008) 051501. doi:10.1063/1.2840194

102. K. Urabe, O. Sakai and K. Tachibana, Combined spectroscopic methods for electron-density diagnostics inside atmospheric-pressure glow discharge using He/N$_2$ gas mixture, *J. Phy. D Appl. Phys.* 44 (2011) 115203. doi:10.1088/0022-3727/44/11/115203

103. S.P. Jamison, J. Shen, D.R. Jones, R.C. Issac, B. Ersfeld, D. Clark and D.A. Jaroszynski, Plasma characterization with terahertz time–domain measurements, *J. Appl. Phys.* 93 (2003) 4334–4336. doi:10.1063/1.1560564

104. S.M. Meier, T.V. Tsankov, D. Luggenhölscher and U. Czarnetzki, Measurement of plasma densities by dual frequency multichannel boxcar THz time domain spectroscopy, *J. Phy. D Appl. Phys.* 50 (2017) 245202. doi:10.1088/1361-6463/aa708f

105. Y. Ito, O. Sakai and K. Tachibana, Measurement of electron density in a microdischarge-integrated device operated in nitrogen at atmospheric pressure using a millimetre-wave transmission method, *Plasma Sour. Sci. Technol* 19 (2010) 025006. doi:10.1088/0963-0252/19/2/025006

106. J.-Y. Choi, N. Takano, K. Urabe and K. Tachibana, Measurement of electron density in atmospheric pressure small-scale plasmas using CO2-laser heterodyne interferometry, *Plasma Sour. Sci. Technol.* 18 (2009) 035013. doi:10.1088/0963-0252/18/3/035013

107. F. Leipold, R.H. Stark, A. El-Habachi and K.H. Schoenbach, Electron density measurements in an atmospheric pressure air plasma by means of infrared heterodyne interferometry, *J. Phys. D Appl. Phys.* 33 (2000) 2268–2273. doi:10.1088/0022-3727/33/18/310

108. K. Urabe, T. Akiyama and K. Terashima, Application of phase-modulated dispersion interferometry to electron-density diagnostics of high-pressure plasma, *J. Phys. D Appl. Phys.* 47 (2014) 262001. doi:10.1088/0022-3727/47/26/262001

109. J. Schäfer, R. Foest, S. Reuter, T. Kewitz, J. Sperka and K.D. Weltmann, Laser schlieren deflectometry for temperature analysis of filamentary nonthermal atmospheric pressure plasma, *Rev. Sci. Instrum.* 83 (2012) 103506. doi:10.1063/1.4761924

110. Y. Inada, S. Matsuoka, A. Kumada, H. Ikeda and K. Hidaka, Highly sensitive Shack–Hartmann sensor for two-dimensional electron density imaging over extinguishing arc discharges, *Meas. Sci. Technol.* 25 (2014) 055201. doi:10.1088/0957-0233/25/5/055201

111. Y. Inada, S. Matsuoka, A. Kumada, H. Ikeda and K. Hidaka, Shack–Hartmann type laser wavefront sensor for measuring two-dimensional electron density distribution over extinguishing arc discharge, *J. Phys. D Appl. Phys.* 45 (2012) 435202. doi:10.1088/0022-3727/45/43/435202

112. Y. Inada, K. Aono, R. Ono, A. Kumada, K. Hidaka and M. Maeyama, Two-dimensional electron density measurement of pulsed positive primary streamer discharge in atmospheric-pressure air, *J. Phys. D Appl. Phys.* 50 (2017) 174005. doi:10.1088/1361-6463/aa65ee

113. R.P. Millane, J.J. Dolne, A. Schweinsberg and A.R. Valenzuela, Shack-Hartmann electronic densitometer (SHED), (2017) 4. doi:10.1117/12.2275156

References for Part III

114. Q.T. Algwari and D. O'Connell, Electron dynamics and plasma jet formation in a helium atmospheric pressure dielectric barrier discharge jet, *Appl. Phys. Lett.* 99 (2011) 121501. doi:10.1063/1.3628455

115. N. Knake, K. Niemi, S. Reuter, V. Schulz-von der Gathen and J. Winter, Absolute atomic oxygen density profiles in the discharge core of a microscale atmospheric pressure plasma jet, *Appl. Phys. Lett.* 93 (2008) 131503. doi:10.1063/1.2995983

116. J. Golda, J. Held, B. Redeker, M. Konkowski, P. Beijer, A. Sobota, G. Kroesen et al., Concepts and characteristics of the 'COST Reference Microplasma Jet', *J. Phys. D Appl. Phys.* 49 (2016) 084003. doi:10.1088/0022-3727/49/8/084003

117. R. Bussiahn, E. Kindel, H. Lange and K.D. Weltmann, Spatially and temporally resolved measurements of argon metastable atoms in the effluent of a cold atmospheric pressure plasma jet, *J. Phys. D Appl. Phys.* 43 (2010) 165201. doi:10.1088/0022-3727/43/16/165201

118. J.S. Sousa, G. Bauville and V. Puech, Arrays of microplasmas for the controlled production of tunable high fluxes of reactive oxygen species at atmospheric pressure, *Plasma Sour. Sci. Technol.* 22 (2013) doi:10.1088/0963-0252/22/3/035012

119. K. Ueno, K. Kamebuchi, J. Kakutani, L. Matsuoka, S. Namba, K. Fujii, T. Shikama and M. Hasuo, Laser absorption spectroscopy for measurement of He metastable atoms of a microhollow cathode plasma, *Jpn. J. Appl. Phys.* 57 (2018) 01AA03. doi:10.7567/jjap.57.01aa03

120. G.D. Stancu, M. Janda, F. Kaddouri, D.A. Lacoste and C.O. Laux, Time-resolved CRDS measurements of the N2(A3Sigma(u)+) density produced by nanosecond discharges in atmospheric pressure nitrogen and air, *J. Phys. Chem. A.* 114 (2010) 201–208. doi:10.1021/jp9075383

121. R. Ono and T. Oda, Spatial distribution of ozone density in pulsed corona discharges observed by two-dimensional laser absorption method, *J. Phys. D Appl. Phys.* 37 (2004) 730–735. doi:10.1088/0022-3727/37/5/013

122. M. Spaan, J. Leistikow, V. Schulz-von der Gathen and H.F. Döbele, Dielectric barrier discharges with steep voltage rise: Laser absorption spectroscopy of NO concentrations and temperatures, *Plasma Sour. Sci. Technol.* 9 (2000) 146–151. doi:10.1088/0963-0252/9/2/306

123. A. Wijaikhum, D. Schröder, S. Schröter, A.R. Gibson, K. Niemi, J. Friderich, A. Greb, V. Schulz-von der Gathen, D. O'Connell and T. Gans, Absolute ozone densities in a radio-frequency driven atmospheric pressure plasma using two-beam UV-LED absorption spectroscopy and numerical simulations, *Plasma Sour. Sci. Technol.* 26 (2017) 115004. doi:10.1088/1361-6595/aa8ebb

124. K. Niemi, D. O'Connell, N. de Oliveira, D. Joyeux, L. Nahon, J.P. Booth and T. Gans, Absolute atomic oxygen and nitrogen densities in radio-frequency driven atmospheric pressure cold plasmas: Synchrotron vacuum ultra-violet high-resolution Fourier-transform absorption measurements, *Appl. Phys. Lett.* 103 (2013) 034102. doi:10.1063/1.4813817

125. M. Foucher, D. Marinov, E. Carbone, P. Chabert and J.-P. Booth, Highly vibrationally excited O_2 molecules in low-pressure inductively-coupled plasmas detected by high sensitivity ultra-broad-band optical absorption spectroscopy, *Plasma Sour. Sci. Technol.* 24 (2015) 042001. doi:10.1088/0963-0252/24/4/042001

126. S. Schröter, M. Foucher, K. Niemi, J. Dedrick, N. de Oliveira, D. Joyeux, L. Nahon et al., Atomic oxygen and hydroxyl density measurements in an atmospheric pressure RF-plasma with water admixtures using UV and synchrotron VUV absorption spectroscopy, *22nd International Symposium on Plasma Chemistry*, Antwerp, Belgium, 2015.

127. K. Niemi, J. Waskoenig, N. Sadeghi, T. Gans and D. O'Connell, The role of helium metastable states in radio-frequency driven helium–oxygen atmospheric pressure plasma jets: Measurement and numerical simulation, *Plasma Sour. Sci. Technol.* 20 (2011) 055005. doi:10.1088/0963-0252/20/5/055005

128. C.-G. Schregel, E.A.D. Carbone, D. Luggenhölscher and U. Czarnetzki, Ignition and afterglow dynamics of a high pressure nanosecond pulsed helium micro-discharge: I. Electron, Rydberg molecules and He (23S) densities, *Plasma Sour. Sci. Technol.* 25 (2016) 054003. doi:10.1088/0963-0252/25/5/054003

129. B. Niermann, T. Hemke, N.Y. Babaeva, M. Böke, M.J. Kushner, T. Mussenbrock and J. Winter, Spatial dynamics of helium metastables in sheath or bulk dominated rf micro-plasma jets, *J Phys. D Appl. Phys.* 44 (2011) 485204. doi:10.1088/0022-3727/44/48/485204

130. S. De Benedictis and G. Dilecce, Physics and technology of atmospheric pressure discharges. In *Laser-Induced Fluorescence Methods for Transient Species Detection in High-Pressure Discharges*, Z. Zong (Ed.), Taylor and Francis, London, UK, 2013, pp. 261–284

131. Z. Yin, Z. Eckert, I.V. Adamovich and W.R. Lempert, Time-resolved radical species and temperature distributions in an Ar-O-2-H-2 mixture excited by a nanosecond pulse discharge, *Proc. Combust. Inst.* 35 (2015) 3455–3462. doi:10.1016/j.proci.2014.05.073

132. R. Ono, C. Tobaru, Y. Teramoto and T. Oda, Laser-induced fluorescence of N_2 metastable in N_2 pulsed positive corona discharge, *Plasma Sour. Sci. Technol.* 18 (2009) 025006. doi:10.1088/0963-0252/18/2/025006

133. M. Mrkvičková, J. Ráheľ, P. Dvořák, D. Trunec and T. Morávek, Fluorescence (TALIF) measurement of atomic hydrogen concentration in a coplanar surface dielectric barrier discharge, *Plasma Sour. Sci. Technol.* 25 (2016) doi:10.1088/0963-0252/25/5/055015

134. E. Es-sebbar, N. Gherardi and F. Massines, Effects of N_2O and O-2 addition to nitrogen Townsend dielectric barrier discharges at atmospheric pressure on the absolute ground-state atomic nitrogen density, *J. Phys. D Appl. Phys.* 46 (2013) 015202. doi:101088/0022-3727/46/1/015202

135. C. Lukas, M. Spaan, V. Schulz-von der Gathen, M. Thomson, R. Wegst, H.F. Dobele and M. Neiger, Dielectric barrier discharges with steep voltage rise: Mapping of atomic nitrogen in single filaments measured by laser-induced fluorescence spectroscopy, *Plasma Sour. Sci. Technol.* 10 (2001) 445–450. doi:10.1088/0963-0252/10/3/308

136. G. Dilecce, P.F. Ambrico and S. De Benedictis, OODR-LIF on N2(A3Σ u + in dielectric barrier discharges, *Czech. J. Phys.* 56 (2006) B690–B696. doi:10.1007/s10582-006-0272-6

137. G. Dilecce, P.F. Ambrico and S.D. Benedictis, Optical–optical double resonance LIF detection of in high pressure gas discharges, *Plasma Sour. Sci. Technol.* 14 (2005) 561–565. doi:10.1088/0963-0252/14/3/019

138. A. Lofthus and P.H. Krupenie, The spectrum of molecular nitrogen, *J. Phys. Chem. Ref. Data* 6 (1977) 113–307. doi:10.1063/1.555546

139. S.E. Babayan, G. Ding and R.F. Hicks, *Plasma Chem. Plasma Process.* 21 (2001) 505–521. doi:10.1023/a:1012094817122

140. X. Yang, M. Moravej, G.R. Nowling, S.E. Babayan, J. Panelon, J.P. Chang and R.F. Hicks, Comparison of an atmospheric pressure, radio-frequency discharge operating in the α and γ modes, *Plasma Sour. Sci. Technol.* 14 (2005) 314–320. doi:10.1088/0963-0252/14/2/013

References for Part III

141. M. Simek, V. Babický, M. Clupek, S. DeBenedictis, G. Dilecce and P. Sunka, Excitation of) and) states in a pulsed positive corona discharge in, - and -NO mixtures. *J. Phys. D Appl. Phys.* 31 (1998) 2591–2602. doi:10.1088/0022-3727/31/19/032

142. R. Ono and T. Oda, Nitrogen oxide γ-band emission from primary and secondary streamers in pulsed positive corona discharge, *J. Appl. Phys.* 97 (2005) 013302. doi:10.1063/1.1829371

143. K. Niemi, S. Reuter, L.M. Graham, J. Waskoenig and T. Gans, Diagnostic based modeling for determining absolute atomic oxygen densities in atmospheric pressure helium-oxygen plasmas, *Appl. Phys. Lett.* 95 (2009) 151504. doi 10.1063/1.3242382

144. K. Niemi, S. Reuter, L.M. Graham, J. Waskoenig, N. Knake, V. Schulz-von der Gathen and T. Gans, Diagnostic based modelling of radio-frequency driven atmospheric pressure plasmas, *J. Phys. D Appl. Phys.* 43 (2010) 124006. doi 10.1088/0022-3727/43/12/124006

145. J. Waskoenig, K. Niemi, N. Knake, L.M. Graham, S. Reuter, V. Schulz-von der Gathen and T. Gans, Diagnostic-based modeling on a micro-scale atmospheric-pressure plasma jet, *Pure Appl. Chem.* 82 (2010) 1209–1222. doi:10.1351/Pac-Con-09-11-05

146. R.E. Walkup, K.L. Saenger and G.S. Selwyn, Studies of atomic oxygen in O2+CF4 rf discharges by two-photon laser-induced fluorescence and optical emission spectroscopy, *J. Chem. Phys.* 84 (1986) 2668. doi:10.1063/1.450339

147. A. Schmidt-Bleker, J. Winter, A. Bösel, S. Reuter and K.-D. Weltmann, On the plasma chemistry of a cold atmospheric argon plasma jet with shielding gas device, *Plasma Sour. Sci. Technol.* 25 (2016) 015005. doi:10.1088/0963-0252/25/1/015005

148. M. Moselhy, R.H. Stark, K.H. Schoenbach and U. Kogelschatz, Resonant energy transfer from argon dimers to atomic oxygen in microhollow cathode discharges, *Appl. Phys. Lett.* 78 (2001) 880–882. doi:10.1063/1.1336547

149. J. Winter, H. Tresp, M.U. Hammer, S. Iseni, S. Kupsch, A. Schmidt-Bleker, K. Wende, M. Dünnbier, K. Masur, K.D. Weltmann and S. Reuter, Tracking plasma generated H2O2 from gas into liquid phase and revealing its dominant impact on human skin cells, *J. Phys. D Appl. Phys.* 47 (2014) 285401. doi:10.1088/0022-3727/47/28/285401

150. G.-B. Zhao, M.D. Argyle and M. Radosz, Optical emission study of nonthermal plasma confirms reaction mechanisms involving neutral rather than charged species, *J. Appl. Phys.* 101 (2007) 033303. doi:10.1063/1.2434002

151. A. Schmidt-Bleker, S. Reuter and K.-D. Weltmann, Non-dispersive path mapping approximation for the analysis of ambient species diffusion in laminar jets, *Phys. Fluids* 26 (2014) 083603. doi:10.1063/1.4893573

152. A. Schmidt-Bleker, S. Reuter and K.D. Weltmann, Quantitative schlieren diagnostics for the determination of ambient species density, gas temperature and calorimetric power of cold atmospheric plasma jets, *J. Phys. D Appl. Phys.* 48 (2015) 175202. doi:10.1088/0022-3727/48/17/175202

153. S. Iseni, A. Schmidt-Bleker, J. Winter, K.D. Weltmann and S. Reuter, Atmospheric pressure streamer follows the turbulent argon air boundary in a MHz argon plasma jet investigated by OH-tracer PLIF spectroscopy, *J. Phys D Appl. Phys.* 47 (2014) 152001. doi:10.1088/0022-3727/47/15/152001

154. R.D. Whalley and J.L. Walsh, Turbulent jet flow generated downstream of a low temperature dielectric barrier atmospheric pressure plasma device, *Sci. Rep.* 6 (2016) 31756. doi:10.1038/srep31756

155. Attribution 4.0 International (CC BY 4.0), 2017. http://creativecommons.org/licenses/by/4.0/ Access: Last Update.

156. E. Robert, V. Sarron, T. Darny, D. Riès, S. Dozias, J. Fontane, L. Joly and J.M. Pouvesle, Rare gas flow structuration in plasma jet experiments, *Plasma Sour. Sci. Technol.* 23 (2014) 012003. doi:10.1088/0963-0252/23/1/012003

157. M. Boselli, V. Colombo, E. Ghedini, M. Gherardi, R. Laurita, A. Liguori, P. Sanibondi and A. Stancampiano, Schlieren high-speed imaging of a nano-second pulsed atmospheric pressure non-equilibrium plasma jet, *Plasma Chem. Plasma Process.* 34 (2014) 853–869. doi:10.1007/s11090-014-9537-1

158. S. Zhang, A. Sobota, E.M. van Veldhuizen and P.J. Bruggeman, Gas flow characteristics of a time modulated APPJ: The effect of gas heating on flow dynamics, *J. Phys D-Appl. Phys.* 48 (2015) 015203. doi:10.1088/0022-3727/48/1/015203:

159. A.M. Lietz, E. Johnsen and M.J. Kushner, Plasma-induced flow instabilities in atmospheric pressure plasma jets, *Appl. Phys. Lett.* 111 (2017) 114101. doi:10.1063/1.4996192

160. Y. Bouremel, J.-M. Li, Z. Zhao and M. Debiasi, Effects of AC Dielectric Barrier Discharge plasma actuator location on flow separation and airfoil performance, in: *Proceedings of 2013 Asian-Pacific Conference on Aerospace Technology and Science*, Taiwan, 2013.

161. A. Dickenson, Y. Morabit, M.I. Hasan and J.L. Walsh, Directional mass transport in an atmospheric pressure surface barrier discharge, *Sci. Rep.* 7 (2017) 14003. doi:10.1038/s41598-017-14117-1

162. G. Uchida, K. Takenaka, K. Takeda, K. Ishikawa, M. Hori and Y. Setsuhara, Selective production of reactive oxygen and nitrogen species in the plasma-treated water by using a nonthermal high-frequency plasma jet. *Jpn. J. Appl. Phys.* 57 (2018) 0102B0104. doi:10.7567/jjap.57.0102b4

163. S. Reuter, J. Winter, S. Iseni, S. Peters, A. Schmidt-Bleker, M. Dünnbier, J. Schäfer, R. Foest and K.-D. Weltmann, Detection of ozone in a MHz argon plasma bullet jet, *Plasma Sour. Sci. Technol.* 21 (2012) 034015. doi:10.1088/0963-0252/21/3/034015

164. M.J. Pavlovich, H.-W. Chang, Y. Sakiyama, D.S. Clark and D.B. Graves, Ozone correlates with antibacterial effects from indirect air dielectric barrier discharge treatment of water, *J. Phys. D Appl. Phys.* 46 (2013) 145202. doi:10.1088/0022-3727/46/14/145202

165. H. Keller-Rudek, G.K. Moortgat, R. Sander and R. Sörensen, The MPI-Mainz UV/VIS spectral atlas of gaseous molecules of atmospheric interest, *Earth Syst. Sci. Data* 5 (2013) 365–373. doi:10.5194/essd-5-365-2013

166. D.B. Graves, The emerging role of reactive oxygen and nitrogen species in redox biology and some implications for plasma applications to medicine and biology, *J. Phys. D Appl. Phys.* 45 (2012) 263001. doi:10.1088/0022-3727/45/26/263001

167. P.F. Bird and G.L. Schott, Quantitative line absorption spectrophotometry: Absorbance of the OH radical near 3090 Å, *J. Quantitative Spectroscopy and Radiative Transfer* 5 (1965) 783–784. doi:10.1016/0022-4073(65)90021-x

168. N. Srivastava and C. Wang, Determination of OH radicals in an atmospheric pressure helium microwave plasma jet, *IEEE Trans. Plasma Sci.* 39 (2011) 918–924. doi:10.1109/tps.2010.2101618

169. P. Bruggeman, G. Cunge and N. Sadeghi, Absolute OH density measurements by broadband UV absorption in diffuse atmospheric-pressure He-H2O RF glow discharges, *Plasma Sour. Sci. Technol.* 21 (2012). doi:10.1088/0963-0252/21/3/035019

170. T. Moiseev, N.N. Misra, S. Patil, P.J. Cullen, P. Bourke, K.M. Keener and J.P. Mosnier, Post-discharge gas composition of a large-gap DBD in humid air by UV–Vis absorption spectroscopy, *Plasma Sour. Sci. Technol.* 23 (2014) 065033. doi:10.1088/0963-0252/23/6/065033

171. G. Dilecce, P.F. Ambrico, M. Simek and S. De Benedictis, LIF diagnostics of hydroxyl radical in atmospheric pressure He-H2O dielectric barrier discharges, *Chem. Phys.* 398 (2012) 142–147. doi:10.1016/j.chemphys.2011.03.012

172. J. Winter, J.S. Santos Sousa, N., A. Schmidt-Bleker, S. Reuter and V. Puech, The spatio-temporal distribution of He (23S1) metastable atoms in a MHz-driven helium plasma jet is influenced by the oxygen/nitrogen ratio of the surrounding atmosphere, *Plasma Sour. Sci. Technol.* 24 (2015) 025015. doi:10.1088/0963-0252/24/2/025015

173. J.U. White, Long optical paths of large aperture, *J. Opt. Soc. Am.* 32 (1942) 285. doi:10.1364/josa.32.000285

174. D.R. Herriott and H.J. Schulte, Folded optical delay lines, *Appl. Opt.* 4 (1965) 883. doi:10.1364/ao.4.000883

175. J. Winter, M. Hanel and S. Reuter, Novel focal point multipass cell for absorption spectroscopy on small sized atmospheric pressure plasmas, *Rev. Sci. Instrum.* 87 (2016) 043117. doi:10.1063/1.4947512

176. B.A. Paldus and A.A. Kachanov, An historical overview of cavity-enhanced methods, *Can. J. Phys.* 83 (2005) 975–999. doi:10.1139/p05-054

177. R. Zaplotnik, M. Bišćan, N. Krstulović, D. Popović and S. Milošević, Cavity ring-down spectroscopy for atmospheric pressure plasma jet analysis, *Plasma Sour. Sci. Technol.* 24 (2015) 054004. doi:10.1088/0963-0252/24/5/054004

178. J.M. Herbelin, J.A. McKay, M.A. Kwok, R.H. Ueunten, D.S. Urevig, D.J. Spencer and D.J. Benard, Sensitive measurement of photon lifetime and true reflectances in an optical cavity by a phase-shift method, *Appl. Opt.* 19 (1980) 144–147. doi:10.1364/AO.19.000144

179. A. O'Keefe and D.A.G. Deacon, Cavity ring-down optical spectrometer for absorption measurements using pulsed laser sources, *Rev. Sci. Instrum.* 59 (1988) 2544–2551. doi:10.1063/1.1139895

180. S.E. Fiedler, A. Hese and A.A. Ruth, Incoherent broad-band cavity-enhanced absorption spectroscopy, *Chem. Phys. Lett.* 371 (2003) 284–294. doi:10.1016/s0009-2614(03)00263-x

181. G. Berden and R. Engeln, *Cavity Ring-Down Spectroscopy: Techniques and Applications*, Wiley, Hoboken, NJ, 2009.

182. G. Berden, R. Peeters and G. Meijer, Cavity ring-down spectroscopy: Experimental schemes and applications, *Int. Rev. Phys. Chem.* 19 (2010) 565–607. doi:10.1080/014423500750040627

183. J.H. van Helden, N. Lang, U. Macherius, H. Zimmermann and J. Röpcke, Sensitive trace gas detection with cavity enhanced absorption spectroscopy using a continuous wave external-cavity quantum cascade laser, *Appl. Phys. Lett.* 103 (2013) 131114. doi:10.1063/1.4823545

184. M. Gianella, S. Reuter, A.L. Aguila, G.A.D. Ritchie and J.-P.H. van Helden, Detection of HO2 in an atmospheric pressure plasma jet using optical feedback cavity-enhanced absorption spectroscopy, *New J. Phys.* 18 (2016) 113027. doi:10.1088/1367-2630/18/11/113027

185. G. Gagliardi and H.-P. Loock, *Cavity-Enhanced Spectroscopy and Sensing*, Vol. 179, Springer, Berlin, Germany, 2014.

186. J.D. Allen, J.D. Durham, G.K. Schweitzer and W.E. Deeds, A new electron spectrometer design: II, *J. Electron Spectros. Relat. Phenomena* 8 (1976) 395–410. doi:10.1016/0368-2048(76)80026-6

187. J.C. Biordi, Molecular beam mass spectrometry for studying the fundamental chemistry of flames, *Prog. Energy Combust. Sci.* 3 (1977) 151–173. doi:10.1016/0360-1285(77)90002-8

188. S. Große-Kreul, S. Hübner, S. Schneider, D. Ellerweg, A. von Keudell, S. Matejčík and J. Benedikt, Mass spectrometry of atmospheric pressure plasmas, *Plasma Sour. Sci. Technol.* 24 (2015) 044008. doi:10.1088/0963-0252/24/4/044008

189. B.T.J. van Ham, S. Hofmann, R. Brandenburg and P.J. Bruggeman, In situ absolute air, O-3 and NO densities in the effluent of a cold RF argon atmospheric pressure plasma jet obtained by molecular beam mass spectrometry, *J. Phys. D Appl. Phys.* 47 (2014). doi:10.1088/0022-3727/47/22/224013

190. H. Singh, J.W. Coburn and D.B. Graves, Appearance potential mass spectrometry: Discrimination of dissociative ionization products, *J. Vac. Sci. Technol. A* 18 (2000) 299–305. doi:10.1116/1.582183

191. S. Schneider, M. Dünnbier, S. Hübner, S. Reuter and J. Benedikt, Atomic nitrogen: A parameter study of a micro-scale atmospheric pressure plasma jet by means of molecular beam mass spectrometry, *J. Phys. D Appl. Phys.* 47 (2014) 505203. doi:10.1088/0022-3727/47/50/505203

192. G. Willems, J. Benedikt and A. von Keudell, Absolutely calibrated mass spectrometry measurement of reactive and stable plasma chemistry products in the effluent of a He/H2O atmospheric plasma, *J. Phys. D Appl. Phys.* 50 (2017) 335204. doi:10.1088/1361-6463/aa77ca:

193. E. Stoffels, Y.A. Gonzalvo, T.D. Whitmore, D.L. Seymour and J.A. Rees, Mass spectrometric detection of short-living radicals produced by a plasma needle, *Plasma Sour. Sci. Technol.* 16 (2007) 549–556. doi:10.1088/0963-0252/16/3/014

194. D. Maletić, N. Puač, S. Lazović, G. Malović, T. Gans, V. Schulz-von der Gathen and Z.L. Petrović, Detection of atomic oxygen and nitrogen created in a radio-frequency-driven micro-scale atmospheric pressure plasma jet using mass spectrometry, *Plasma Phys. Control. Fusion* 54 (2012) 124046. doi:10.1088/0741-3335/54/12/124046

195. G. Malović, N. Puač, S. Lazović and Z. Petrović, Mass analysis of an atmospheric pressure plasma needle discharge, *Plasma Sour. Sci. Technol.* 19 (2010) 034014. doi:10.1088/0963-0252/19/3/034014

196. J.-S. Oh, H. Furuta, A. Hatta and J.W. Bradley, Investigating the effect of additional gases in an atmospheric-pressure helium plasma jet using ambient mass spectrometry, *Jpn. J. Appl. Phys.* 54 (2015) 01AA03. doi:10.7567/jjap.54.01aa03

197. Y. Aranda Gonzalvo, T.D. Whitmore, J.A. Rees, D.L. Seymour and E. Stoffels, Atmospheric pressure plasma analysis by modulated molecular beam mass spectrometry, *J. Vac. Sci. Technol. A* 24 (2006) 550. doi:10.1116/1.2194938

198. J. Benedikt, A. Hecimovic, D. Ellerweg and A. von Keudell, Quadrupole mass spectrometry of reactive plasmas, *J. Phys. D Appl. Phys.* 45 (2012) 403001. doi:10.1088/0022-3727/45/40/403001

199. M. Dünnbier, A. Schmidt-Bleker, J. Winter, M. Wolfram, R. Hippler, K.D. Weltmann and S. Reuter, Ambient air particle transport into the effluent of a cold atmospheric-pressure argon plasma jet investigated by molecular beam mass spectrometry, *J. Phys. D Appl. Phys.* 46 (2013) 435203. doi:10.1088/0022-3727/46/43/435203

200. J.T. Salmon and N.M. Laurendeau, Calibration of laser-saturated fluorescence measurements using Rayleigh scattering, *Appl. Opt.* 24 (1985) 65. doi:10.1364/ao.24.000065

201. S. Iseni, S. Zhang, A.F.H. van Gessel, S. Hofmann, B.T.J. van Ham, S. Reuter, K.D. Weltmann and P.J. Bruggeman, Nitric oxide density distributions in the effluent of an RF argon APPJ: Effect of gas flow rate and substrate, *New J. Phys.* 16 (2014) 123011. doi:10.1088/1367-2630/16/12/123011

202. J. Vorac, P. Dvorak, V. Prochazka, J. Ehlbeck and S. Reuter, Measurement of hydroxyl radical (OH) concentration in an argon RF plasma jet by laser-induced fluorescence, *Plasma Sour. Sci. Technol.* 22 (2013) 025016. doi:10.1088/0963-0252/22/2/025016

203. S. Yonemori, Y. Nakagawa, R. Ono and T. Oda, Measurement of OH density and air–helium mixture ratio in an atmospheric-pressure helium plasma jet, *J. Phys. D Appl. Phys.* 45 (2012) 225202. doi:10.1088/0022-3727/45/22/225202

204. O. Tochikubo and K. Nishijima, Sodium intake and cardiac sympatho-vagal balance in young men with high blood pressure, *Hypertens Res.* 27 (2004) 393–398. doi:10.1143/JJAP.43.315

205. T. Verreycken, R.M. van der Horst, N. Sadeghi and P.J. Bruggeman, Absolute calibration of OH density in a nanosecond pulsed plasma filament in atmospheric pressure He-H2O: Comparison of independent calibration methods, *J. Phys. D Appl. Phys.* 46 (2013). doi:10.1088/0022-3727/46/46/464004

206. S.I. Chu, M. Yoshimine and B. Liu, Ab initio study of the X2Π and A2Σ+ states of OH. I. Potential curves and properties, *J. Chem. Phys.* 61 (1974) 5389–5395. doi:10.1063/1.1681891

207. M.J. Dunn and A.R. Masri, A comprehensive model for the quantification of linear and nonlinear regime laser-induced fluorescence of OH under A2Σ+←X2Π(1,0) excitation, *Appl. Phys. B* 101 (2010) 445–463. doi:10.1007/s00340-010-4129-0

208. Q. Xiong, A.Y. Nikiforov, L. Li, P. Vanraes, N. Britun, R. Snyders, X.P. Lu and C. Leys, Absolute OH density determination by laser induced fluorescence spectroscopy in an atmospheric pressure RF plasma jet, *Eur. Phys. J. D* 66 (2012). doi:10.1140/epjd/e2012-30474-8

209. G. Dilecce and S. De Benedictis, Laser diagnostics of high-pressure discharges: Laser induced fluorescence detection of OH in He/Ar-H2O dielectric barrier discharges, *Plasma Phys. Control. Fusion* 53 (2011). doi:10.1088/0741-3335/53/12/124006

210. T. Verreycken, R. Mensink, R. van der Horst, N. Sadeghi and P.J. Bruggeman, Absolute OH density measurements in the effluent of a cold atmospheric-pressure Ar-H2O RF plasma jet in air, *Plasma Sour. Sci. Technol.* 22 (2013). doi:10.1088/0963-0252/22/5/055014

211. R. Ono and T. Oda, Dynamics and density estimation of hydroxyl radicals in a pulsed corona discharge, *J. Phys. D Appl. Phys.* 35 (2002) 2133–2138. doi:10.1088/0022-3727/35/17/309

212. A. Broc, S. De Benedictis and G. Dilecce, LIF investigations on NO, O and N in a supersonic N-2/O-2/NO RF plasma jet, *Plasma Sour. Sci. Technol.* 13 (2004) 504–514. doi:10.1088/0963-0252/13/3/017

213. S. Iseni, P.J. Bruggeman, K.-D. Weltmann and S. Reuter, Nitrogen metastable (N$_2$(A^{3}Σ+$_u$)) in a cold argon atmospheric pressure plasma jet: Shielding and gas composition, *Appl. Phys. Lett.* 108 (2016) 184101. doi:10.1063/1.4948535

214. A. Ershov and J. Borysow, Dynamics of OH (X2Pi, v=0) in high-energy atmospheric pressure electrical pulsed discharge, *J. Phys. D Appl. Phys.* 28 (1995) 68–74. doi:10.1088/0022-3727/28/1/012

215. T. Verreycken, N. Sadeghi and P.J. Bruggeman, Time-resolved absolute OH density of a nanosecond pulsed discharge in atmospheric pressure He-H2O: Absolute calibration, collisional quenching and the importance of charged species in OH production, *Plasma Sour. Sci. Technol.* 23 (2014). doi:10.1088/0963-0252/23/4/045005

216. J. Voráč, A. Obrusník, V. Procházka, P. Dvořák and M. Talába, Spatially resolved measurement of hydroxyl radical (OH) concentration in an argon RF plasma jet by planar laser-induced fluorescence, *Plasma Sour. Sci. Technol.* 23 (2014) 025011. doi:10.1088/0963-0252/23/2/025011

217. K. Urabe, Y. Ito, K. Tachibana and B.N. Ganguly, Behavior of N2+ions in He microplasma jet at atmospheric pressure measured by laser induced fluorescence spectroscopy, *Appl. Phys. Exp.* 1 (2008). doi:10.1143/apex.1.066004

218. J. Luque, R.J.H. Klein-Douwel, J.B. Jeffries, G.P. Smith and D.R. Crosley, Quantitative laser-induced fluorescence of CH in atmospheric pressure flames, *Appl. Phys. B* 76 (2003) 715–715. doi:10.1007/s00340-002-1038-x:

219. K. Niemi, V. Schulz-von der Gathen and H.F. Döbele, Absolute atomic oxygen density measurements by two-photon absorption laser-induced fluorescence spectroscopy in an RF-excited atmospheric pressure plasma jet, *Plasma Sour. Sci. Technol.* 14 (2005) 375–386. doi:10.1088/0963-0252/14/2/021

220. G. Dilecce, M. Vigliotti and S. De Benedictis, A TALIF calibration method for quantitative oxygen atom density measurement in plasma jets, *J. Phys. D Appl. Phys.* 33 (2000) L53–L56. doi:10.1088/0022-3727/33/6/101

221. A. Goehlich, T. Kawetzki and H.F. Döbele, On absolute calibration with xenon of laser diagnostic methods based on two-photon absorption, *J. Chem. Phys.* 108 (1998) 9362. doi:10.1063/1.476388

222. K. Niemi, V. Schulz-von der Gathen and H.F. Döbele, Absolute calibration of atomic density measurements by laser-induced fluorescence spectroscopy with two-photon excitation, *J. Phys. D Appl. Phys.* 34 (2001) 2330–2335. doi:10.1088/0022-3727/34/15/312

223. S. Zhang, A.F.H. van Gessel, S.C. van Grootel and P.J. Bruggeman, The effect of collisional quenching of the O 3p P-3(J) state on the determination of the spatial distribution of the atomic oxygen density in an APPJ operating in ambient air by TALIF, *Plasma Sour. Sci. Technol.* 23 (2014). doi:10.1088/0963-0252/23/2/025012

224. Q. Xiong, H. Liu, N. Britun, A.Y. Nikiforov, L. Li, Q. Chen and C. Leys, Time-selective TALIF spectroscopy of atomic oxygen applied to an atmospheric pressure argon plasma jet, *IEEE Trans. Plasma Sci.* 44 (2016) 2745–2753. doi:10.1109/tps.2016.2545862

225. A.V. Klochko, J. Lemainque, J.P. Booth and S.M. Starikovskaia, TALIF measurements of oxygen atom density in the afterglow of a capillary nanosecond discharge, *Plasma Sour. Sci. Technol.* 24 (2015). doi:10.1088/0963-0252/24/2/025010

226. T. Oda, Y. Yamashita, K. Takezawa and R. Ono, Oxygen atom behaviour in the nonthermal plasma, *Thin Solid Films* 506 (2006) 669–673. doi:10.1016/j.tsf.2005.08.266

227. V. Schulz-von der Gathen, V. Buck, T. Gans, N. Knake, K. Niemi, S. Reuter, L. Schaper and J. Winter, Optical diagnostics of micro discharge jets, *Contrib. Plasma Phys.* 47 (2007) 510–519. doi:10.1002/ctpp.200710066

228. E. Wagenaars, T. Gans, D. O'Connell and K. Niemi, Two-photon absorption laser-induced fluorescence measurements of atomic nitrogen in a radio-frequency atmospheric-pressure plasma jet, *Plasma Sour. Sci. Technol.* 21 (2012) 042002. doi:10.1088/0963-0252/21/4/042002

229. S. Mazouffre, C. Foissac, P. Supiot, P. Vankan, R. Engeln, D.C. Schram and N. Sadeghi, Density and temperature of N atoms in the afterglow of a microwave discharge measured by a two-photon laser-induced fluorescence technique, *Plasma Sour. Sci. Technol.* 10 (2001) 168–175. doi:10.1088/0963-0252/10/2/306

230. P. Dvořák, M. Talába, A. Obrusník, J. Kratzer and J. Dědina, Concentration of atomic hydrogen in a dielectric barrier discharge measured by two-photon absorption fluorescence, *Plasma Sour. Sci. Technol.* 26 (2017). doi:10.1088/1361-6595/aa76f7

231. F. Gaboriau, U. Cvelbar, M. Mozetic, A. Erradi and B. Rouffet, Comparison of TALIF and catalytic probes for the determination of nitrogen atom density in a nitrogen plasma afterglow, *J. Phys. D Appl. Phys.* 42 (2009). doi:10.1088/0022-3727/42/5/055204

232. Y. Inoue and R. Ono, Measurement of singlet delta oxygen in an atmospheric-pressure helium–oxygen plasma jet, *J. Phys. D Appl. Phys.* 50 (2017). doi:10.1088/1361-6463/aa6c53

233. M. Laroussi and X. Lu, Room-temperature atmospheric pressure plasma plume for biomedical applications, *Appl. Phys. Lett.* 87 (2005) 113902. doi:10.1063/1.2045549

234. X. Lu, S. Wu, P.K. Chu, D. Liu and Y. Pan, An atmospheric-pressure plasma brush driven by sub-microsecond voltage pulses, *Plasma Sour. Sci. Technol.* 20 (2011) 065009. doi:10.1088/0963-0252/20/6/065009

235. M. Sneep and W. Ubachs, Direct measurement of the Rayleigh scattering cross section in various gases, *J. Quant. Spectrosc. Radiat. Transf.* 92 (2005) 293–310. doi:10.1016/j.jqsrt.2004.07.025

236. H.M. Katsch, T. Sturm, E. Quandt and H.F. Döbele, Negative ions and the role of metastable molecules in a capacitively coupled radiofrequency excited discharge in oxygen, *Plasma Sour. Sci. Technol.* 9 (2000) 323–330. doi:10.1088/0963-0252/9/3/310

237. J.A. Wagner and H.M. Katsch, Negative oxygen ions in a pulsed RF-discharge with inductive coupling in mixtures of noble gases and oxygen, *Plasma Sour. Sci. Technol.* 15 (2006) 156–169. doi:10.1088/0963-0252/15/1/022

238. S. Nemschokmichal, R. Tschiersch and J. Meichsner, The influence of negative ions in helium–oxygen barrier discharges: III. Simulation of laser photodetachment and comparison with experiment, *Plasma Sour. Sci. Technol.* 26 (2017). doi:10.1088/1361-6595/aa8e1b:

239. T. Yamamoto and M. Okubo. Nonthermal plasma technology. In: *Advanced Physicochemical Treatment Technologies: Handbook of Environmental Engineering*, Vol. 5. L.K. Wang, Y.T. Hung, and N.J. Shammas (Eds.), Humana Press. 2007. doi:https://doi.org/10.1007/978-1-59745-173-4_4

240. A. Schmidt-Bleker, S.A. Norberg, J. Winter, E. Johnsen, S. Reuter, K.D. Weltmann and M.J. Kushner, Propagation mechanisms of guided streamers in plasma jets: The influence of electronegativity of the surrounding gas, *Plasma Sour. Sci. Technol.* 24 (2015) 035022. doi:10.1088/0963-0252/24/3/035022

241. K. McKay, J.L. Walsh and J.W. Bradley, Observations of ionic species produced in an atmospheric pressure pulse-modulated RF plasma needle, *Plasma Sour. Sci. Technol.* 22 (2013) 035005. doi:10.1088/0963-0252/22/3/035005

242. M. Pavlik and J.D. Skalny, Generation of [H3O]+.(H2O)n clusters by positive corona discharge in air, *Rapid Commun. Mass Spectrom.* 11 (1997) 1757–1766. doi:10.1002/(sici)1097-0231(19971030)11:16<1757::aid-rcm16>3.0.co;2-8

243. P. Bruggeman, F. Iza, D. Lauwers and Y.A. Gonzalvo, Mass spectrometry study of positive and negative ions in a capacitively coupled atmospheric pressure RF excited glow discharge in He-water mixtures, *J. Phys. D Appl. Phys.* 43 (2010). doi:10.1088/0022-3727/43/1/012003

244. M. Dünnbier, Plasma jets for life science applications: Characterisation and tuning of the reactive species composition, Institute of Physics, Ernst-Moritz-Arndt-Universität Greifswald, Greifswald, Germany, 2015.

245. T. Kawasaki, Y. Arai and T. Takada, Two-dimensional measurement of electrical surface charge distribution on insulating material by electrooptic pockels effect, *Jpn. J. Appl. Phys.* 30 (1991) 1262–1265. doi:10.1143/jjap.30.1262

246. M. Bogaczyk, R. Wild, L. Stollenwerk and H.E. Wagner, Surface charge accumulation and discharge development in diffuse and filamentary barrier discharges operating in He, N_2 and mixtures, *J. Phys. D Appl. Phys.* 45 (2012) 465202. doi:10.1088/0022-3727/45/46/465202

247. R. Wild, T. Gerling, R. Bussiahn, K.D. Weltmann and L. Stollenwerk, Phase-resolved measurement of electric charge deposited by an atmospheric pressure plasma jet on a dielectric surface, *J. Phys. D Appl. Phys.* 47 (2014) 042001. doi:10.1088/0022-3727/47/4/042001

248. R. Tschiersch, S. Nemschokmichal, M. Bogaczyk and J. Meichsner, Surface charge measurements on different dielectrics in diffuse and filamentary barrier discharges, *J. Phys. D Appl. Phys.* 50 (2017) 105207. doi:10.1088/1361-6463/aa5605

249. A. Sobota, O. Guaitella and E. Garcia-Caurel, Experimentally obtained values of electric field of an atmospheric pressure plasma jet impinging on a dielectric surface, *J. Phys. D Appl. Phys.* 46 (2013) 372001. doi:10.1088/0022-3727/46/37/372001

250. E. Slikboer, E. Garcia-Caurel, O. Guaitella and A. Sobota, Charge transfer to a dielectric target by guided ionization waves using electric field measurements, *Plasma Sour. Sci. Technol.* 26 (2017) 035002. doi:10.1088/1361-6595/aa53fe

251. O. Guaitella and A. Sobota, The impingement of a kHz helium atmospheric pressure plasma jet on a dielectric surface, *J. Phys. D Appl. Phys.* 48 (2015) 255202. doi:10.1088/0022-3727/48/25/255202

252. G. Gaborit, J. Dahdah, F. Lecoche, P. Jarrige, Y. Gaeremynck, E. Duraz and L. Duvillaret, A nonperturbative electrooptic sensor for in situ electric discharge characterization, *IEEE Trans. Plasma Sci.* 41 (2013) 2851–2857. doi:10.1109/tps.2013.2257874

253. T. Darny, J.M. Pouvesle, V. Puech, C. Douat, S. Dozias and E. Robert, Analysis of conductive target influence in plasma jet experiments through helium metastable and electric field measurements, *Plasma Sour. Sci. Technol.* 26 (2017) 045008. doi:10.1088/1361-6595/aa5b15

254. A. Begum, M. Laroussi and M.R. Pervez, Atmospheric pressure He-air plasma jet: Breakdown process and propagation phenomenon, *AIP Adv.* 3 (2013). doi:10.1063/1.4811464

255. B.R. Weatherford, E.V. Barnat and J.E. Foster, Two-dimensional laser collision-induced fluorescence measurements of plasma properties near an RF plasma cathode extraction aperture, *Plasma Sour. Sci. Technol.* 21 (2012). doi:10.1088/0963-0252/21/5/055030

256. T. Kampschulte, J. Schulze, D. Luggenhölscher, M.D. Bowden and U. Czarnetzki, Laser spectroscopic electric field measurement in krypton, *New J. Phys.* 9 (2007) 18. doi:10.1088/1367-2630/9/1/018

257. B.M. Goldberg, P.S. Böhm, U. Czarnetzki, I.V. Adamovich and W.R. Lempert, Electric field vector measurements in a surface ionization wave discharge, *Plasma Sour. Sci. Technol.* 24 (2015) 055017. doi:10.1088/0963-0252/24/5/055017

258. T. Ito, K. Kobayashi, U. Czarnetzki and S. Hamaguchi, Rapid formation of electric field profiles in repetitively pulsed high-voltage high-pressure nanosecond discharges, *J. Phys. D Appl. Phys.* 43 (2010) 062001. doi:10.1088/0022-3727/43/6/062001

259. V.P. Gavrilenko, E.B. Kupriyanova, D.P. Okolokulak, V.N. Ochkin, S.Y. Savinov, S.N. Tskhai and A.N. Yarashev, Generation of coherent IR light on a dipole-forbidden molecular transition with biharmonic pumping in a static electric field, *JETP Lett.* 56 (1992).

260. M. van der Schans, P. Böhm, J. Teunissen, S. Nijdam, W. Ijzerman and U. Czarnetzki, Electric field measurements on plasma bullets in N2 using four-wave mixing, *Plasma Sour. Sci. Technol.* 26 (2017) 115006. doi:10.1088/1361-6595/aa9146

261. A. Dogariu, B.M. Goldberg, S. O'Byrne and R.B. Miles, Species-independent femtosecond localized electric field measurement, *Phys. Rev. Appl.* 7 (2017). doi:10.1103/PhysRevApplied.7.024024

262. J.F. Ward and I.J. Bigio, Molecular second- and third-order polarizabilities from measurements of second-harmonic generation in gases, *Phys. Rev. A* 11 (1975) 60–66. doi:10.1103/PhysRevA.11.60

263. B.M. Goldberg, A. Dogariu and R.B. Miles, Electric field measurements in nanosecond pulsed discharges using a femtosecond laser, *AIAA Aerospace Sciences Meeting*, Kissimmee, FL, 2018. doi:10.2514/6.2018-1432

264. R.S. Finn and J.F. Ward, DC-induced optical second-harmonic generation in the inert gases, *Phys. Rev. Lett.* 26 (1971) 285–289. doi:10.1103/PhysRevLett.26.285

265. J.S. Foster, Application of quantum mechanics to the stark effect in helium, *Proc. Royal Soc. A: Math. Phys. Eng. Sci.* 117 (1927) 137–163. doi:10.1098/rspa.1927.0171

266. M.M. Kuraica and N. Konjević, Electric field measurement in the cathode fall region of a glow discharge in helium, *Appl. Phys. Lett.* 70 (1997) 1521–1523. doi:10.1063/1.118606

267. B.M. Obradović, S.S. Ivković and M.M. Kuraica, Spectroscopic measurement of electric field in dielectric barrier discharge in helium, *Appl. Phys. Lett.* 92 (2008) 191501. doi:10.1063/1.2927477

268. S.S. Ivković, B.M. Obradović and M.M. Kuraica, Electric field measurement in a DBD in helium and helium–hydrogen mixture, *J. Phys. D Appl. Phys.* 45 (2012) 275204. doi:10.1088/0022-3727/45/27/275204

269. A. Sobota, O. Guaitella, G.B. Sretenović, I.B. Krstić, V.V. Kovačević, A. Obrusník, Y.N. Nguyen, L. Zajíčková, B.M. Obradović and M.M. Kuraica, Electric field measurements in a kHz-driven He jet—The influence of the gas flow speed, *Plasma Sour. Sci. Technol.* 25 (2016) 065026. doi:10.1088/0963-0252/25/6/065026

270. G.B. Sretenovic, I.B. Krstic, V.V. Kovacevic, B.M. Obradovic and M.M. Kuraica, Spectroscopic study of low-frequency helium DBD plasma jet, *IEEE Trans. Plasma Sci.* 40 (2012) 2870–2878. doi:10.1109/tps.2012.2219077

271. T. Gerling, A.V. Nastuta, R. Bussiahn, E. Kindel and K.D. Weltmann, Back and forth directed plasma bullets in a helium atmospheric pressure needle-to-plane discharge with oxygen admixtures, *Plasma Sour. Sci. Technol.* 21 (2012) 034012. doi:10.1088/0963-0252/21/3/034012

272. O. Ozgen, E.A. Aksoy, V. Hasirci and N. Hasirci, Surface characterization and radical decay studies of oxygen plasma-treated PMMA films, *Surf. Interface Anal.* 45 (2013) 844–853. doi:10.1002/sia.5181

273. U. Cvelbar, K. Ostrikov, A. Drenik and M. Mozetic, Nanowire sensor response to reactive gas environment, *Appl. Phys. Lett.* 92 (2008). doi:10.1063/1.2905265

274. A. Schmidt-Bleker, R. Bansemer, S. Reuter and K.-D. Weltmann, How to produce an NOx- instead of Ox-based chemistry with a cold atmospheric plasma jet, *Plasma Process. Polym.* 13 (2016) 1120–1127. doi:10.1002/ppap.201600062

275. S. Reuter, J. Winter, S. Iseni, A. Schmidt-Bleker, M. Dunnbier, K. Masur, K. Wende and K.-D. Weltmann, The influence of feed gas humidity versus ambient humidity on atmospheric pressure plasma jet-effluent chemistry and skin cell viability, *IEEE Trans. Plasma Sci.* 43 (2015) 3185–3192. doi:10.1109/tps.2014.2361921

276. L. Hansen, A. Schmidt-Bleker, R. Bansemer, H. Kersten, K.-D. Weltmann and S. Reuter, Influence of a liquid surface on the NOx production of a cold atmospheric pressure plasma jet, *J. Phys D.* submitted (2018).

277. J. Winter, K. Wende, K. Masur, S. Iseni, M. Dunnbier, M.U. Hammer, H. Tresp, K.D. Weltmann and S. Reuter, Feed gas humidity: A vital parameter affecting a cold atmospheric-pressure plasma jet and plasma-treated human skin cells, *J. Phys. D Appl. Phys.* 46 (2013) 295401. doi:10.1088/0022-3727/46/29/295401

278. G. Eisenberg, Colorimetric determination of hydrogen peroxide, *Ind. Eng. Chem. Anal. Ed.* 15 (1943) 327–328. doi:10.1021/i560117a011

279. H. Jablonowski, M.A. Hansch, M. Dunnbier, K. Wende, M.U. Hammer, K.D. Weltmann, S. Reuter and T. Woedtke, Plasma jet's shielding gas impact on bacterial inactivation, *Biointerphases* 10 (2015) 029506. doi:10.1116/1.4916533

280. B. Tarabová, P. Lukeš, M.U. Hammer, H. Jablonowski, T. von Woedtke, S. Reuter and Z. Machala, Fluorescent measurements of peroxynitrites in cold air plasma treated aqueous solution. in preparation (2018).

281. G.Y. Wiederschain, *Invitrogen Molecular Probes Handbook: A Guide to Fluorescent Probes and Labeling Technologies*, 11th edition. 2010.

282. P. Lukes, E. Dolezalova, I. Sisrova and M. Clupek, Aqueous-phase chemistry and bactericidal effects from an air discharge plasma in contact with water: Evidence for the formation of peroxynitrite through a pseudo-second-order post-discharge reaction of H_2O_2 and HNO_2, *Plasma Sour. Sci. Technol.* 23 (2014) 015019. doi:10.1088/0963-0252/23/1/015019

283. P.J. Bruggeman, M.J. Kushner, B.R. Locke, J.G.E. Gardeniers, W.G. Graham, D.B. Graves, R.C.H.M. Hofman-Caris et al., Plasma–liquid interactions: A review and roadmap, *Plasma Sour. Sci. Technol.* 25 (2016) 053002. doi:10.1088/0963-0252/25/5/053002

284. J.-S. Oh, E.J. Szili, K. Ogawa, R.D. Short, M. Ito, H. Furuta and A. Hatta, UV–vis spectroscopy study of plasma-activated water: Dependence of the chemical composition on plasma exposure time and treatment distance, *Jpn. J. Appl. Phys.* 57 (2018). doi:10.7567/jjap.57.0102b9

285. J. Janca, P. Stahel, F. Krcma and L. Lapcik, Plasma surface treatment of textile fibres for improvement of car tires, *Czech J. Phys.* 50 (2000) 449–452. doi:10.1007/BF03165927

286. C. Elsner, M. Lenk, L. Prager and R. Mehnert, Windowless argon excimer source for surface modification, *Appl. Surf. Sci.* 252 (2006) 3616–3624. doi:10.1016/j.apsusc.2005.05.071

287. Y. Kamiura, M. Ogasawara, K. Fukutani, T. Ishiyama, Y. Yamashita, T. Mitani and T. Mukai, Enhancement of blue emission from GaN films and diodes by water vapor remote plasma treatment, *Phys. B* 401 (2007) 331–334. doi:10.1016/j.physb.2007.08.180

288. A. Kondyurin, I. Kondyurina and M. Bilek, Radiation damage of polyethylene exposed in the stratosphere at an altitude of 40 km, *Polym. Degrad. Stabil.* 98 (2013) 1526–1536. doi:10.1016/j.polymdegradstab.2013.04.008

289. I. Novak, A. Popelka, A.S. Luyt, M.M. Chehimi, M. Spirkova, I. Janigova, A. Kleinova et al., Adhesive properties of polyester treated by cold plasma in oxygen and nitrogen atmospheres, *Surf. Coat. Tech.* 235 (2013) 407–416. doi:10.1016/j.surfcoat.2013.07.057

290. A.A. Westenberg and N. de Haas, Quantitative measurements of gas phase O and N atom concentrations by ESR, *J. Chem. Phys.* 40 (1964) 3087. doi:10.1063/1.1724954

291. M. Mrázková, P. Vašina, V. Kudrle, A. Tálský, C.D. Pintassilgo and V. Guerra, On the oxygen addition into nitrogen post-discharges, *J. Phys. D Appl. Phys.* 42 (2009) 075202. doi:10.1088/0022-3727/42/7/075202

292. J. Pan, P. Sun, Y. Tian, H.X. Zhou, H.Y. Wu, N. Bai, F.X. Liu, W.D. Zhu, J. Zhang, K.H. Becker and J. Fang, A novel method of tooth whitening using cold plasma microjet driven by direct current in atmospheric-pressure air, *IEEE Trans. Plasma Sci.* 38 (2010) 3143–3151. doi:10.1109/Tps.2010.2066291

293. N. Bai, P. Sun, H.X. Zhou, H.Y. Wu, R.X. Wang, F.X. Liu, W.D. Zhu, J.L. Lopez, J. Zhang and J. Fang, Inactivation of staphylococcus aureus in water by a cold, $He/O-2$ atmospheric pressure plasma microjet, *Plasma Process. Polym.* 8 (2011) 424–431. doi:10.1002/ppap.201000078

294. K. Ishikawa, H. Mizuno, H. Tanaka, K. Tamiya, H. Hashizume, T. Ohta, M. Ito et al., Real-time in situ electron spin resonance measurements on fungal spores of Penicillium digitatum during exposure of oxygen plasmas, *Appl. Phys. Lett.* 101 (2012). doi:10.1063/1.4733387

295. E.G. Janzen and B.J. Blackburn, Detection and identification of short-lived free radicals by an electron spin resonance trapping technique, *J Am. Chem. Soc.* 90 (1968) 5909–5910. doi:10.1021/ja01023a051

296. A.J. Hoff, *Advanced EPR: Applications in Biology and Biochemistry*, Elsevier Science, St. Louis, MI, 2012.

297. Y. Sun, S. Yu, P. Sun, H.Y. Wu, W.D. Zhu, W. Liu, J. Zhang, J. Fang and R.Y. Li, Inactivation of Candida biofilms by non-thermal plasma and its enhancement for fungistatic effect of antifungal drugs, *Plos One* 7 (2012). doi:10.1371/journal.pone.0040629

298. H.Y. Wu, P. Sun, H.Q. Feng, H.X. Zhou, R.X. Wang, Y.D. Liang, J.F. Lu, W.D. Zhu, J. Zhang and J. Fang, Reactive oxygen species in a non-thermal plasma microjet and water system: Generation, conversion, and contributions to bacteria inactivation-an analysis by electron spin resonance spectroscopy, *Plasma Process. Polym.* 9 (2012) 417–424. doi:10.1002/ppap.201100065

299. H. Tresp, M.U. Hammer, J. Winter, K.D. Weltmann and S. Reuter, Quantitative detection of plasma-generated radicals in liquids by electron paramagnetic resonance spectroscopy, *J. Phys. D Appl. Phys.* 46 (2013) 435401. doi:10.1088/0022-3727/46/43/435401

300. H. Tresp, M.U. Hammer, K.-D. Weltmann and S. Reuter, Effects of atmosphere composition and liquid type on plasma-generated reactive species in biologically relevant solutions, *Plasma Med.* 3 (2013) 45–55. doi:10.1615/PlasmaMed.2014009711

301. A. Tani, S. Fukui, S. Ikawa and K. Kitano, Diagnosis of superoxide anion radical induced in liquids by atmospheric-pressure plasma using superoxide dismutase, *Jpn. J. Appl. Phys.* 54 (2015). doi:10.7567/Jjap.54.01af01

302. H. Jablonowski, R. Bussiahn, M.U. Hammer, K.-D. Weltmann, T. von Woedtke and S. Reuter, Impact of plasma jet VUV-radiation on reactive oxygen species generation in bio-relevant liquids, *Phys. Plasmas* 22 (2015) 122008. doi:10.1063/1.4934989

303. T. Sakakura, S. Uemura, M. Hino, S. Kiyomatsu, Y. Takatsuji, R. Yamasaki, M. Morimoto and T. Haruyama, Excitation of H2O at the plasma/water interface by UV irradiation for the elevation of ammonia production, *Green Chem.* 20 (2018) 627–633. doi:10.1039/C7GC03007J

304. H. Kawamoto and S. Umezu, Electrohydrodynamic deformation of water surface in a metal pin to water plate corona discharge system, *J. Phys. D Appl. Phys.* 38 (2005) 887–894. doi:10.1088/0022-3727/38/6/017

305. T. Kondo, M. Tsumaki, W.A. Diño and T. Ito, Influence of reactive gas-phase species on the structure of an air/water interface, *J. Phys. D Appl. Phys.* 50 (2017) 244002. doi:10.1088/1361-6463/aa7159

306. A.G. Lambert, P.B. Davies and D.J. Neivandt, Implementing the theory of sum frequency generation vibrational spectroscopy: A tutorial review, *Appl. Spectrosc. Rev.* 40 (2005) 103–145. doi:10.1081/asr-200038326

307. S. Kanazawa, H. Kawano, S. Watanabe, T. Furuki, S. Akamine, R. Ichiki, T. Ohkubo, M. Kocik and J. Mizeraczyk, Observation of OH radicals produced by pulsed discharges on the surface of a liquid, *Plasma Sour. Sci. Technol.* 20 (2011) 034010. doi:10.1088/0963-0252/20/3/034010

308. S.A. Norberg, W. Tian, E. Johnsen and M.J. Kushner, Atmospheric pressure plasma jets interacting with liquid covered tissue: Touching and not-touching the liquid, *J. Phys. D Appl. Phys.* 47 (2014) 475203. doi:10.1088/0022-3727/47/47/475203

309. J.-L. Brisset and J. Pawlat, Chemical effects of air plasma species on aqueous solutes in direct and delayed exposure modes: Discharge, post-discharge and plasma activated water, *Plasma Chem. Plasma Process.* 36 (2015) 355–381. doi:10.1007/s11090-015-9653-6

310. I. Adamovich, S.D. Baalrud, A. Bogaerts, P.J. Bruggeman, M. Cappelli, V. Colombo, U. Czarnetzki et al., The 2017 plasma roadmap: Low temperature plasma science and technology, *J. Phys. D Appl. Phys.* 50 (2017) 323001. doi:10.1088/1361-6463/aa76f5

311. A.D. Lindsay, D.B. Graves and S.C. Shannon, Fully coupled simulation of the plasma liquid interface and interfacial coefficient effects, *J. Phys. D Appl. Phys.* 49 (2016) 235204. doi:10.1088/0022-3727/49/23/235204

312. A. Lindsay, C. Anderson, E. Slikboer, S. Shannon and D. Graves, Momentum, heat, and neutral mass transport in convective atmospheric pressure plasma-liquid systems and implications for aqueous targets, *J. Phys. D Appl. Phys.* 48 (2015) 424007. doi:10.1088/0022-3727/48/42/424007

313. C.A.J. van Gils, S. Hofmann, B.K.H.L. Boekema, R. Brandenburg and P.J. Bruggeman, Mechanisms of bacterial inactivation in the liquid phase induced by a remote RF cold atmospheric pressure plasma jet, *J. Phys. D Appl. Phys.* 46 (2013) 175203. doi:10.1088/0022-3727/46/17/175203

314. I. Yagi, R. Ono, T. Oda and K. Takaki, Two-dimensional LIF measurements of humidity and OH density resulting from evaporated water from a wet surface in plasma for medical use, *Plasma Sour. Sci. Technol.* 24 (2014) 015002. doi:10.1088/0963-0252/24/1/015002

315. P. Rumbach, D.M. Bartels, R.M. Sankaran and D.B. Go, The effect of air on solvated electron chemistry at a plasma/liquid interface, *J. Phys. D Appl. Phys.* 48 (2015) 424001. doi:10.1088/0022-3727/48/42/424001

316. P. Rumbach, M. Witzke, R.M. Sankaran and D.B. Go, Decoupling interfacial reactions between plasmas and liquids: Charge transfer vs plasma neutral reactions, *J. Am. Chem. Soc.* 135 (2013) 16264–16267. doi:10.1021/ja407149y:

317. J.R. Bolton and S.R. Cater, *Surface and Aquatic Photochemistry*, G.R. Helz (Ed.), et al., Lewis Publishers, Boca Raton, FL, 1994, pp. 467–490.

References for Part III

318. W. Tian and M.J. Kushner, Atmospheric pressure dielectric barrier discharges interacting with liquid covered tissue, *J. Phys. D Appl. Phys.* 47 (2014) 165201. doi:10.1088/0022-3727/47/16/165201

319. K. Watanabe and M. Zelikoff, Absorption coefficients of water vapor in the vacuum ultraviolet, *J. Opt. Soc. Am.* 43 (1953) 753. doi:10.1364/josa.43.000753

320. C.A. Cantrell, A. Zimmer and G.S. Tyndall, Absorption cross sections for water vapor from 183 to 193 nm, *Geophys. Res. Lett.* 24 (1997) 2195–2198. doi:10.1029/97gl02100

321. B. Darwent, *Bond Dissociation Energies in Simple Molecules*, National Standards Reference Data Sheets – U.S. National Bureau of Standards 31 (1970).

322. F. Crapulli, D. Santoro, M.R. Sasges and A.K. Ray, Mechanistic modeling of vacuum UV advanced oxidation process in an annular photoreactor, *Water Res.* 64C (2014) 209–225. doi:10.1016/j.watres.2014.06.048

323. G. Heit, A. Neuner, P.-Y. Saugy and A.M. Braun, Vacuum-UV (172 nm) actinometry. The quantum yield of the photolysis of water, *J. Phys. Chem. A* 102 (1998) 5551–5561. doi:10.1021/jp980130i:

324. J. Mack and J.R. Bolton, Photochemistry of nitrite and nitrate in aqueous solution: A review, *J. Photochem. Photobiol. A* 128 (1999) 1–13. doi:10.1016/s1010-6030(99)00155-0

325. R.P. Wayne, The photochemistry of ozone, *Atmos. Environ.* (1967) 21 (1987) 1683–1694. doi:10.1016/0004-6981(87)90107-7

326. M.C. Gonzalez and A.M. Braun, Vacuum-UV photolysis of aqueous solutions of nitrate: Effect of organic matter I. Phenol, *J. Photochem. Photobiol. A* 93 (1996) 7–19. doi:10.1016/1010-6030(95)04127-3

327. G. Wittmann, I. Horváth and A. Dombi, UV-induced decomposition of ozone and hydrogen peroxide in the aqueous phase at pH 2-7, *Ozone: Sci. Eng.* 24 (2002) 281–291. doi:10.1080/01919510208901619

328. G. Imoberdorf and M. Mohseni, Kinetic study and modeling of the vacuum-UV photoinduced degradation of 2,4-D, *Chem. Eng. J.* 187 (2012) 114–122. doi:10.1016/j.cej.2012.01.107

329. F.J. Beltran, G. Ovejero, J.F. Garcla-Araya and J. Rivast, Oxidation of polynuclear aromatic hydrocarbons in water. 2. UV radiation and ozonation in the presence of UV radiation, *Ind. Eng. Chem. Res.* 34 (1996) 1607–1615. doi:10.1021/ie00044a013

330. J.F.M. van Rens, J.T. Schoof, F.C. Ummelen, D.C. van Vugt, P.J. Bruggeman and E.M. van Veldhuizen, Induced liquid phase flow by RF Ar cold atmospheric pressure plasma jet, *IEEE Trans. Plasma Sci.* 42 (2014) 2622–2623. doi:10.1109/tps.2014.2328793

331. W. Tian and M.J. Kushner, Long-term effects of multiply pulsed dielectric barrier discharges in air on thin water layers over tissue: Stationary and random streamers. *J. Phys. D Appl. Phys.* 48 (2015) 494002. doi:10.1088/0022-3727/48/49/494002

332. M. Hoentsch, R. Bussiahn, H. Rebl, C. Bergemann, M. Eggert, M. Frank, T. von Woedtke and B. Nebe, Persistent effectivity of gas plasma-treated, long time-stored liquid on epithelial cell adhesion capacity and membrane morphology, *Plos One* 9 (2014) e104559. doi:10.1371/journal.pone.0104559

333. D. Mance, H. Geilmann, W.A. Brand, T. Kewitz and H. Kersten, Changes of ^{2}H and ^{18}O abundances in water treated with non-thermal atmospheric pressure plasma jet, *Plasma Process. Polym.* 14 (2017) 1600239. doi:10.1002/ppap.201600239

334. K. Wende, S. Reuter, T. von Woedtke, K.-D. Weltmann and K. Masur, Redox-based assay for assessment of biological impact of plasma treatment, *Plasma Process. Polym.* 11 (2014) 655–663. doi:10.1002/ppap.201300172

335. M.U. Hammer, E. Forbrig, S. Kupsch, K.-D. Weltmann and S. Reuter, Influence of plasma treatment on the structure and function of lipids, *Plasma Med.* 3 (2013) 97–114. doi:10.1615/PlasmaMed.2014009708

336. M. Yusupov, K. Wende, S. Kupsch, E.C. Neyts, S. Reuter and A. Bogaerts, Effect of head group and lipid tail oxidation in the cell membrane revealed through integrated simulations and experiments, *Sci. Rep.* 7 (2017) 5761. doi:10.1038/s41598-017-06412-8

337. M. Anbar and I. Pecht, On the sonochemical formation of hydrogen peroxide in water, *J. Phys. Chem.* 68 (1964) 352–355. doi:10.1021/j100784a025

338. Y.F. Yue, S. Mohades, M. Laroussi and X. Lu, Measurements of plasma-generated hydroxyl and hydrogen peroxide concentrations for plasma medicine applications, *IEEE Trans. Plasma Sci.* 44 (2016) 2754–2758. doi:10.1109/tps.2016.2550805

339. Y. Nakamura, S. Ohtaki, R. Makino, T. Tanaka and Y. Ishimura, Superoxide anion is the initial product in the hydrogen peroxide formation catalyzed by NADPH oxidase in porcine thyroid plasma membrane, *J. Biol. Chem.* 264 (1989) 4759–4761.

340. M. Anbar, S. Guttmann and G. Stein, Radiolysis of aqueous solutions highly enriched in O^{18}, *J. Chem. Phys.* 34 (1961) 703–711. doi:10.1063/1.1731664

341. L. Bundscherer, S. Bekeschus, H. Tresp, S. Hasse, S. Reuter, K.-D. Weltmann, U. Lindequist and K. Masur, Viability of human blood leukocytes compared with their respective cell lines after plasma treatment, *Plasma Med.* 3 (2013) 71–80. doi:10.1615/PlasmaMed.2013008538

342. S. Bekeschus, S. Iseni, S. Reuter, K. Masur and K.-D. Weltmann, Nitrogen shielding of an argon plasma jet and its effects on human immune cells, *IEEE Trans. Plasma Sci.* 43 (2015) 776–781. doi:0.1109/tps.2015.2393379

343. E. Dolezalova and P. Lukes, Membrane damage and active but noncultur-able state in liquid cultures of Escherichia coli treated with an atmospheric pressure plasma jet, *Bioelectrochemistry* 103 (2015) 7–14. doi:10.1016/j.bioelechem.2014.08.018

344. B. Bhushan, Y. Pan and S. Daniels, AFM characterization of nanobubble formation and slip condition in oxygenated and electrokinetically altered fluids, *J. Colloid Interface Sci.* 392 (2013) 105–116. doi:10.1016/j.jcis.2012.09.077

345. F.Y. Ushikubo, T. Furukawa, R. Nakagawa, M. Enari, Y. Makino, Y. Kawagoe, T. Shiina and S. Oshita, Evidence of the existence and the stability of nano-bubbles in water, *Colloids Surf. A* 361 (2010) 31–37. doi:10.1016/j.colsurfa.2010.03.005

346. R. Vácha, P. Slavíček, M. Mucha, B.J. Finlayson-Pitts and P. Jungwirth, Adsorption of atmospherically relevant gases at the air/water interface: Free energy profiles of aqueous solvation of N_2, O_2, O_3, OH, H_2O, HO_2, and H_2O_2, *J. Phys. Chem. A* 108 (2004) 11573–11579. doi:10.1021/jp046268k

347. G. Bauer and D.B. Graves, Mechanisms of selective antitumor action of cold atmospheric plasma-derived reactive oxygen and nitrogen species, *Plasma Process. Polym.* 13 (2016) 1157–1178. doi:10.1002/ppap.201600089

348. G. Bauer, Increasing the endogenous NO level causes catalase inactivation and reactivation of intercellular apoptosis signaling specifically in tumor cells, *Redox Biol.* 6 (2015) 353–371. doi:10.1016/j.redox.2015.07.017

349. K. Wende, P. Williams, J. Dalluge, W. Van Gaens, H. Aboubakr, J. Bischof, T. von Woedtke et al., Identification of the biologically active liquid chemistry induced by a nonthermal atmospheric pressure plasma jet, *Biointerphases* 10 (2015). doi:10.1116/1.4919710

350. D.W. Liu, F. Iza and M.G. Kong, Evolution of atmospheric-pressure RF plasmas as the excitation frequency increases, *Plasma Process. Polym.* 6 (2009) 446–450. doi:10.1002/ppap.200930009

351. P. Rumbach, D.M. Bartels, R.M. Sankaran and D.B. Go, The solvation of electrons by an atmospheric-pressure plasma, *Nat. Commun.* 6 (2015) 7248. doi:10.1038/ncomms8248

352. K. Sehested, O.L. Rasmussen and H. Fricke, Rate constants of OH with HO_2, O_2^-, and $H_2O_2^+$ from hydrogen peroxide formation in pulse-irradiated oxygenated water, *J. Phys. Chem.* 72 (1968) 626–631. doi:10.1021/j100848a040

353. M.M. Hefny, C. Pattyn, P. Lukes and J. Benedikt, Atmospheric plasma generates oxygen atoms as oxidizing species in aqueous solutions, *J. Phys. D Appl. Phys.* 49 (2016) 404002. doi:10.1088/0022-3727/49/40/404002

354. D.T. Elg, Y.-W. Yang and D.B. Graves, Production of TEMPO by O atoms in atmospheric pressure non-thermal plasma-liquid interactions, *J. Phys. D Appl. Phys.* (2017). doi:10.1088/1361-6463/aa8f8c

355. E. Takai, S. Ikawa, K. Kitano, J. Kuwabara and K. Shiraki, Molecular mechanism of plasma sterilization in solution with the reduced pH method: Importance of permeation of HOO radicals into the cell membrane, *J. Phys. D Appl. Phys.* 46 (2013) 295402. doi:10.1088/0022-3727/46/29/295402

356. S. Marklund, Spectrophotometric study of spontaneous disproportionation of superoxide anion radical and sensitive direct assay for superoxide dismutase, *J. Biol. Chem.* 251 (1976) 1504–1501.

357. S. Ikawa, A. Tani, Y. Nakashima and K. Kitano, Physicochemical properties of bactericidal plasma-treated water, *J. Phys. D Appl. Phys.* 49 (2016) 425401. doi:10.1088/0022-3727/49/42/425401

358. A. Schmidt-Bleker, *Investigations on Cold Atmospheric Pressure Plasma Jets for Medical Applications*, Institute of Physics, Ernst-Moritz-Arndt-Universität Greifswald, Greifswald, Germany, 2015.

359. P. Sathishkumar, R.V. Mangalaraja and S. Anandan, Review on the recent improvements in sonochemical and combined sonochemical oxidation processes – A powerful tool for destruction of environmental contaminants, *Renew. Sust. Energy Rev.* 55 (2016) 426–454. doi:10.1016/j.rser.2015.10.139

360. P. Pichat, *Photocatalysis and Water Purification: From Fundamentals to Recent Applications*, Wiley-VCH, 2013.

PART IV

Application of N-APPJs in Medicine – Cancer Applications

N-APPJs for Cancer Applications

9.1 Introduction

Thermal plasmas and low-pressure plasmas have been used for biomedical applications for several decades [1–3]. Low-pressure oxygen plasma work was conducted by NASA for the destruction of biological matter for space applications. In addition, in-depth investigations of low-pressure plasmas for medical sterilization applications were conducted in the early 2000s [4]. The medical applications of thermal plasma involved the use of plasma for cauterization and blood coagulation such as in the Argon Plasma Coagulator [5].

In the mid-1990s experiments were conducted that showed that nonthermal atmospheric pressure plasmas can be used to inactivate bacteria [6]. Based on these early results, the Physics and Electronics Directorate of the U.S. Air Force Office of Scientific Research (AFOSR) funded proof of principle research work that started in 1997. The main goals of the AFOSR's program were to use low-temperature plasmas (LTP) to disinfect and treat soldiers' wounds as well as the sterilization/decontamination of biotic and abiotic surfaces. Other research efforts conducted in Russia showed that plasma-generated nitric oxide (NO) plays a crucial role to enhance Phagocytosis and accelerate the proliferation of fibroblasts. Both *in vitro* and *in vivo* experiments were conducted by these investigators who called their approach "plasmadynamics therapy" of wounds [7]. Finally, and around 2002–2003 a research group from the Netherlands reported that low-temperature plasma can be used to detach mammalian cells without causing necrosis and under some conditions may lead to apoptosis [8].

These groundbreaking early research efforts showed that nonthermal plasma can interact with both prokaryotes and eukaryotes to induce various biological outcomes. Inspired by the results of the above-mentioned works, many laboratories started various investigations of the biomedical applications of plasma, thus forming a global research community that pushed the field forward. Many advances were achieved by various investigators and the field came to be known as "Plasma Medicine" [9–18]. In addition, in the late 2000s, several plasma sources were approved for use: in 2008 the FDA of the USA approved the Rhytec Portrait® (plasma jet) for use in dermatology. This was followed by other devices (for various medical applications) such as the Bovie J-Plasma®, and in Germany the medical device certification class IIa of the kINPen® (plasma jet) in 2013. **Figure 9.1** is a timeline graph showing the major milestones in the development of the field of plasma medicine.

9.1.1 Uses of N-APPJs in Plasma Medicine

Today, research topics in plasma medicine include the following applications of low-temperature plasmas in biology and medicine [9,10,16–45]:

- Sterilization, disinfection, and decontamination
- Plasma-aided wound healing
- Plasma dentistry
- Cancer applications or "Plasma Oncology"
- Plasma pharmacology
- Plasma treatment of implants for biocompatibility

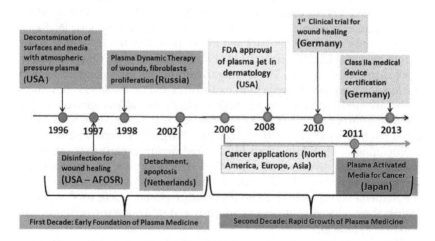

FIGURE 9.1 Timeline showing some major milestones of the field of plasma medicine. The cancer applications started around 2006 with several groups worldwide contributing various breakthroughs that pushed the research forward.

In this chapter we only discuss the cancer application of nonthermal atmospheric pressure plasma. This application has witnessed great advances during the last decade and promises to introduce a new and effective therapy to treat various cancers.

9.1.2 Cell Death: Apoptosis and Necrosis

Apoptosis is a natural programmed cell death in multicellular organisms. Apoptosis involves a cascade of biochemical events that result in changes in cell morphology followed by cell death. Apoptotic cells are removed by microphages and other adjacent cells. Cells that undergo apoptosis exhibit blebbing, cell shrinkage, changes in cell membrane, changes in attachment, and nuclear and DNA fragmentation. Apoptosis is a necessary process in the evolution of multicellular organisms. It is the absence of apoptosis that ultimately leads to accumulation of aberrant cells and cancers. Cancer cells acquire the ability to avoid natural cell death by blocking or disabling the apoptotic pathways. Under this scenario tumor suppressor gene function is inhibited as well as the expression of anti-apoptotic molecules. Armed with this ability, cancer cells often become resistant to chemotherapeutic drugs.

Necrosis, which is an unorganized and premature cell death, occurs usually after cell injury. Necrosis commonly results from external factors, such as infections, toxins, or traumas. After necrosis cellular debris and the rapid release of intracellular enzymes cause inflammation and damage to the surrounding healthy tissue.

9.2 Cancer, Cancer Progression, and Present Therapies

To maintain a regulated cell population in tissues, normal healthy cells grow, divide and die in an orderly fashion. Normal cells undergo apoptosis during development and aging as a mechanism to maintain healthy tissues and control proliferation. The apoptotic signaling cascade is initiated either by extrinsic or intrinsic pathways. The extrinsic pathway requires a ligand and cell-surface receptors to elicit caspase activity for programmed cell death, while the intrinsic pathway is cued by specific stimuli that involves mitochondrial cell signaling resulting in programmed cell death. In addition, the cell growth, migration, adhesion, signal transduction, and cell-cell interactions involve specific proteins and mediators produced by individual cells in order to achieve the coordinated process of proliferation, apoptosis, and normal cell function.

When normal cell signals become aberrant, and cell growth proceeds in an uncontrollable way, local or metastatic disease can result. Cancer cells can result from mutations in the DNA or from damaged DNA by exposure to external factors, such as ionizing radiation. Killing aberrant cells restoring balance and inhibiting uncontrolled proliferation are the goals of general or focused cancer therapies.

9.2.1 Types of Cancer

Cytotoxic T-cells, cytotoxic lymphocytes, and natural killer cells are the body's defense system against tumors. However, cancerous cells are able to overwhelm the immune system, which can result in solid or liquid tumors. The following are examples of cancers.

Carcinomas result in external or internal body surfaces such as lung, breast, gastro-intestinal tract, colon, and prostate cancer. Sarcomas are found in the supporting tissues bone, connective tissue, cartilage, fat, and muscle. Lymphomas are cancers that arise in the lymph nodes. Leukemia is the accumulation of immature blood white cells. Germ cell tumor derives from totipotent cells, which have the ability to divide and produce all the differentiated cells in an organism, including extraembryonic tissue cells. Blastic tumor (blastoma) resembles an immature or embryonic tissue.

The diversity of cancer results in very different diseases because they all begin with different cells, growing at different rates and therefore require specific treatment. The National Cancer Institute estimates that over 1.6 million men and women are diagnosed with cancer every year, resulting in the death of over 35% of this population. In addition, the increase in drug resistant cancer cells, the cost of various treatments, and the harsh side effects of some treatments are a great burden to the patients and their families. The following is a brief summary of the present cancer treatment modalities.

9.2.2 Current Treatment Methods

Current treatment methods include chemotherapy, radiation therapy, photodynamic therapy, immunotherapy, and surgery. In chemotherapy, chemical agents are used to suppress or eradicate cancer. Chemotherapy uses alkylating agents that damage DNA, antimetabolites that prevent DNA or RNA synthesis, natural products such as a plant alkaloids that interfere with cell mitotic processes to prevent cellular reproduction, cytotoxic antibiotics that interfere with DNA or RNA synthesis, hormones, or analogs such as glucocorticoids that restore the normal balance of signals, or biologic response modifiers such as interferons or monoclonal antibodies [55]. Some agents are cell-cycle specific (M phase, G1 phase, S phase or G2 phase) while others, such as alkylating agents, are cytotoxic to the cells.

Radiation treatment is typically in the form of waves or particles that transfer energy resulting in ionization of molecules. In the cell, this results in the degradation of histones and depolymerization of DNA. Radiation therapy can be delivered using an external beam of radiation or by incorporating radiation beads surgically placed at a site inside the body to treat a cancerous lesion. Sometimes radioactive fluids are directly delivered to an organ that contains the cancerous cells, and in some instances a radiation device is placed into a body cavity and removed after a treatment period. Both chemotherapy and radiation therapies can have serious side effects, but these are outside of the scope of this discussion.

Cancer cells/lesions can also be removed surgically. This typically involves heat or cold temperatures that result in cell death accompanied by an inflammatory response. Cauterization, also known as electrocautery, utilizes the conduction of heat from a metal probe heated by electric current. The process results in the destruction of tissue at the location where the probe is applied. Heat can also be delivered by a laser, which delivers intensely focused heat to inactivate cellular components. The greatest thermal damage occurs in cells with low water content resulting in interruption of cellular replication by dehydration, necrosis and inflammation. On the other hand, cryotherapy uses very cold temperatures in the form of liquid nitrogen to destroy tissues.

Photodynamic therapy (PDT) uses a substance known as a photosensitizer which binds to a target cell and can be activated by light at a specific wavelength. After absorption of light, the photosensitizer interacts with cellular molecules and generates free radicals that have a cytotoxic effect on the cells. PDT has been used to treat cancerous skin lesions and a variety of other cancers [58]. Tumor cells are destroyed by a multiple mode of action involving necrosis, apoptosis, and decrease in tumor vasculature.

Immunotherapy is designed to enhance or suppress immune responses to disease. Cancer immunotherapy works by stimulating the immune system to recognize and destroy aberrant cancerous cells. Lymphocytes, natural killer cells, macrophages, etc., are activated to target antigen that are expressed by cancer cells. Today there are several approaches involving chemokines, bacterial membrane, interferons, etc., that are undergoing clinical trials.

9.3 Destruction of Cancer Cells by N-APPJs

Starting around the mid-2000s several investigators reported on successful tests which showed that low temperature plasmas (LTP) can destroy cancerous cells *in vitro*. By 2010 some *in vivo* work was reported, which showed that plasma can reduce the size of cancer tumors in animal models. The following is a limited selection of the *in vitro* work that was reported in the last few years. Schlegel's group reported (University of Munich) on glioblastoma [39–40], Fridman's group (Drexel University) reported on melanoma [46], Keidar's group (George Washington University) reported on papilloma and carcinoma [14,42,47], Baek's group (University of Tennessee) reported on colorectal cancer [48], Hori's group (Nagoya University) reported on ovarian cancer and glioblastoma [35,36,49], Laroussi's group (Old Dominion University) reported on prostate cancer cells, squamous cell carcinoma, and leukemia [13,15,37,50], and Huang et al. (Chinese Academy of Science) reported on lung cancer [51]. In addition, *in vivo* (animal model) work was reported by Pouvesle's group (GREMI) [41], Keidar's group (George Washington University) [14,42] and Kim's group (Clemson University) [52]. **Table 9.1** is a summary of the above information.

To illustrate the efficacy of N-APPJs to kill cancer cells, an example showing results obtained on a cancer cell line are presented here. The cancer cells used are from a carcinoma cell line. This is followed by a section on indirect exposure of cancer cells via what is referred to as Plasma Activated Media (PAM). In this method, first a biological liquid medium is exposed to LTP

Table 9.1 Select List of Some of the Cancers That Were Treated by Low Temperature Plasma

Cancer	Reference	Possible Mechanisms
Glioblastoma	[39,40,49]	• Caspase activation
Melanoma	[46]	• Mitochondria dysfunction
Papilloma	[14,47]	• DNA damage
Carcinoma	[14,15,47]	• Cell cycle arrest
Prostate	[15]	• Increase in intracellular reactive oxygen species (ROS) concentration
Ovarian	[35]	
Colorectal	[48]	*Outcome:* Necrosis, apoptosis, senescence, detachment
Leukemia	[50]	
Lung	[51,52]	

and then the activated medium is applied on top of cancer cells. The N-APPJ source used in all these examples is the plasma pencil.

The first example presented here is from a study by Mohades et al. on the human bladder cancer cell line SCaBER (ATCC® HTB-3™). The cells originated from a patient with squamous cell carcinoma. These types of cells are adherent with epithelial morphology and were grown in complete growth media in a vented tissue culture flask and were incubated at 37°C in a humidified atmosphere containing 5% CO_2. The complete growth medium MEM (Minimum Essential Medium with 2 mM L-glutamine and Earle's Balanced Salt Solution) also contain 10% fetal bovine serum and 1% antibiotics (Penicillin/Streptomycin).

A Trypsinised suspension of cells with a concentration of ~10^6 cells per ml was transferred and seeded into the 24-well plate. After overnight incubation each sample in the 24-well plate was treated by the plasma pencil at different exposure times (2, 3.5, and 5 min). The distance from the tip of the nozzle of the plasma pencil and the surface of the media was 2 cm. Cells untreated by the plasma were kept as control for the cell viability assays. **Figure 9.2** shows the experiment setup.

The viability of SCaBER cells after plasma exposure was determined by Trypan blue exclusion assay. A ratio of 1:1 of the trypan blue dye (0.4% trypan blue solution) to cells in MEM solution was used. A hemocytometer under a phase-contrast bright field microscope was used to count the live versus dead cells. The dead cells retain the blue dye and appear dark under the microscope while the live cells appear clear. The numbers of live and dead cells were counted at 0, 12, 24, and 48 h post-plasma exposure. Subsequently, the total numbers of viable cells were calculated [37].

The results of cells viability are shown in **Figure 9.3** [37]. The counts immediately after plasma pencil treatment (at time 0 h) reveal no dead cells suggesting that there were no immediate effects. However, the viability of cells at

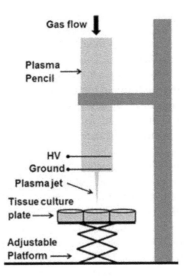

FIGURE 9.2 Schematic of the experimental set up for the exposure of cancer cells to the plasma pencil.

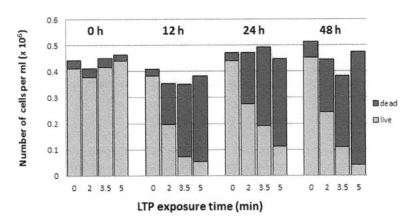

FIGURE 9.3 Cell viability of SCaBER cells reveal dead (dark bars) and live (gray bars). The viability was monitored at 0, 12, 24, and 48 h post LTP treatment. (From Mohades, S. et al., *Plasma Process. Polym.*, 11, 1150–1155, 2014.)

24 h after 2 min plasma treatment was reduced to 50% and to about 75% for 5 min treatment. This is an indication that higher doses of plasma exposure lead to higher kill. The late effects of plasma at 48 h post-plasma exposure revealed even higher kill efficacy where, 5 min plasma treatment diminished the viability of SCaBER cells to approximately 10%.

9.3.1 Plasma Activated Media

Alternative to exposing cancer cells directly to plasma, investigators have also used plasma activated media (PAM). This is done by first exposing a biological medium to LTP (such as N-APPJs plasma plumes) for various time durations, then applying the treated medium on top of cancer cells. PAM has been shown to be effective in killing cancer by several investigators. Experimental studies showed that PAM has killing and anti-proliferative effects against various cancer cell lines including ovarian cancer, gastric cancer, glioblastoma, carcinoma, and breast cancer [35–38,53,54].

Interaction of plasma with liquid media through the process of PAM preparation results in diffusion/dissolution of reactive oxygen and nitrogen species (RONS) in the treated medium. The concentrations of reactive species generated in PAM is directly related to the duration of plasma exposure, gas type and flow rate, and to the chemical composition of the medium [54]. In addition, most of the reactive species generated by plasma such as nitric oxide (NO), ozone (O_3), hydroxyl radicals (OH·), singlet oxygen ($^1O_2$), and superoxide anion (O_2^-) are highly reactive and/or are short-lived and may not diffuse deeply into the liquid [55,56]. However, some of the solvated species do react to produce reactive but more stable molecules such as hydrogen peroxide and peroxinitrite. For example, hydrogen peroxide (H_2O_2) is stable and can diffuse into liquid media relatively efficiently. Hydrogen peroxide has well-known biological implications that include the peroxidation of lipids, induction of DNA damage, and playing a role in mitogenic stimulation and cell cycle regulation. Hydrogen peroxide is generated by various reaction pathways including reactions between hydroxyl radicals:

$$OH + OH + M \rightarrow H_2O_2 + M$$

In the presence of nitrite, H_2O_2 can react to form peroxinitrite by the following reaction:

$$H_2O_2 + NO_2^- \rightarrow ONOO^- + H_2O$$

Reactions between peroxinitrite and biological cells lead to either caspase activation followed by apoptosis or to lipid peroxidation, protein nitration or oxidation, which can result in necrosis. The following section shows illustrative results of an investigation on the effects of PAM on SCaBER cells.

PAM was produced by exposing 1 mL of fresh MEM growth media (containing 10% calf bovine serum, 1% penicillin-streptomycin-glutamine) to LTP generated by the plasma pencil. Exposure times were 2, 3.5, and 5 min. Immediately after exposure PAM was transferred to wells containing the SCaBER cells and then incubated at 37°C and 5% CO_2. Cell viability at 0, 12, 24, and 48 h post PAM application was determined using trypan blue exclusion assay. The cell viability at 0 h after PAM application showed no observable effect on the cells.

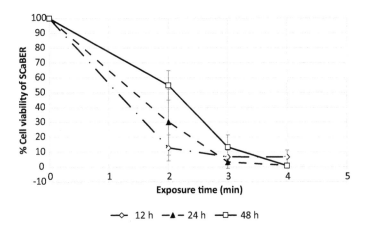

FIGURE 9.4 SCaBER cells viability evaluated at 12, 24, and 24 h post-PAM application. (From Yue, Y.F. et al., *IEEE Trans. Plasma Sci.*, 44, 2754–2758, 2016; Mohades, S. et al., *Plasma Process. Polym.*, 13, 1206, 2016.)

Figure 9.4 shows cell viability for three PAMs (created by 2, 3.5, and 5 min exposure to LTP) and at three post-PAM application times (12, 24, and 48 h) [57]. PAM created by longer exposure times to LTP exhibits higher kill rates. Cell viability evaluated at 12, 24, and 48 h post-PAM application shows similar killing trends with the kill efficiency reaching more than 90% in the case of PAM created by 5 min exposure to LTP. This study also shows that cell viability decreases for longer PAM application times, which supports the hypothesis that the killing mechanism of cancer cells by LTP is not immediate but requires biological time scales.

One of the advantages of using PAM is that it can be stored and used at later times. Mohades et al. used stored/aged PAM and evaluated its efficacy against SCaBER cells [58]. **Figure 9.5** shows the result of aged-PAM on SCaBER cell viability at 12 h post PAM treatment. PAM was stored at room temperature for 1, 8, and 12 h before the application. Comparing the cell viability outcome of SCaBER treated with aged-PAM indicates that PAM efficiency decreases with increasing storage time, depending on the duration of plasma exposure. Reduction in the efficiency of PAM is more significant at shorter exposure times such as 2 min. However, no significant reduction in the efficiency of longer exposure time PAM was observed after storage. This outcome suggests that the concentration of reactive species remains high after storage for PAM created with longer exposure times to LTP.

Mohades et al. found a correlation between the efficacy of PAM and the concentration of H_2O_2 [58]. They measured the concentration of H_2O_2 right after PAM was created and after various storage times, and they found that it decreased with storage time as did the efficacy of PAM. **Figure 9.6** shows the concentration of H_2O_2 for PAMs created with different exposure times to plasma and stored/aged for various time durations.

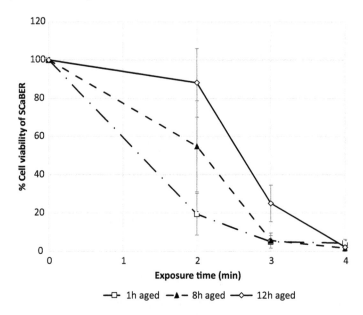

FIGURE 9.5 Effectiveness of aged-PAM to induce cell death in SCaBER cells when stored for 1, 8, and 12 h before utilizing. (From Mohades, S. et al., *Plasma Process. Polym.*, 13, 1206, 2016. Data taken at 12 h post PAM treatment.)

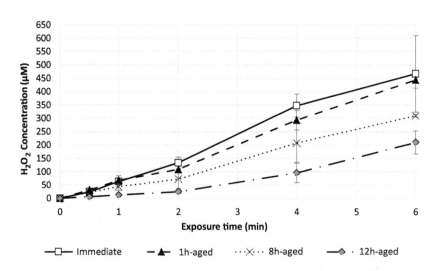

FIGURE 9.6 Concentration of H_2O_2 versus plasma exposure time measured in MEM immediately after plasma exposure and at 1, 8, and 12 h after storage. (From Mohades, S. et al., *Plasma Process. Polym.*, 13, 1206, 2016.)

9.4 Mechanisms of N-APPJs against Cancer Cells

Low-temperature plasma appears to affect cancer cells in a selective manner, mostly sparing healthy cells. In recent years several investigators reported that the same plasma exposure kills more cancer cells than their healthy counter parts. To explain this selectivity the following hypothesis was proposed: because the concentrations of RONS are much higher in cancer cells than healthy cells, LTP adds more intracellular oxidative stress, up to a level that exceeds the cancer cells defense, therefore causing their death. This was supported by experimental studies such as that of Zhao et al. [58] who used an *in vitro* model of a human hepatocellular carcinoma cell (HepG2) and found that LTP facilitated the accumulation of ROS and reactive nitrogen species (RNS) in cells and compromised their antioxidant defense system. This was evidenced by the inactivation of antioxidants such as superoxide dismutase (SOD), catalase, and glutathione (GSH). Nitrotyrosine and protein carbonyl content analysis showed that LTP caused oxidative and nitrative cell injury. This ultimately led to cellular dysfunction and to endoplasmic reticulum stress-mediated apoptosis [59]. Several other investigators reported other and/or related mechanisms whereby cancer cells show less resistance to LTP than healthy ones; these include the results of the following studies:

1. Keidar & co-workers reported that for skin cancer cells LTP causes an increase in the expression of the oxidative stress reporter γH2A.X (pSer 139) and a decrease in DNA replication in the S-phase of the cell cycle [47]. In contrast, this effect was much less pronounced in the healthy cells. So, it appears that LTP's is most effective during cellular replication, and more specifically during the DNA replication phase. Since cancer cells replicate and proliferate much more than their healthy counterpart, they show a much-elevated susceptibility to damage by LTP.

   Similar to the above study, Yan et al. [60,61] used a human hepatocellular carcinoma cell (HepG2) that was treated by a single-electrode plasma jet device. Analysis of flow cytometry data showed that plasma treatment increased the percentage of apoptotic cells being associated with cell cycle arrest at the G2/M phase. These investigators also found that the expression of the p21 Cdk inhibitor, as well as that of tumor suppressor p53, was enhanced.

2. Another mechanism is related to caspase activation. Laroussi and co-workers studied caspase activation in the case of squamous cell carcinoma and found higher level of caspase-3 activation in cancer cells treated by LTP than in control samples, indicating that LTP induces apoptosis in these cells via caspase activation [37].

3. Ishaq et al. suggested that because tumor cells are defective in several regulatory signaling pathways they exhibit metabolic imbalance which leads to a lack of cell growth regulation [62]. ROS affect the metabolism of the cancer cells by impairing redox balance which leads to slowing down or arresting the proliferation of the cells. On the other hand, healthy cells are more able to adjust/properly regulate their metabolic pathways as the ROS levels change

and therefore can cope better with plasma exposure [62]. Ishaq et al. reported that LTP up-regulates intracellular ROS levels and induce apoptosis in melanoma. They identified that LTP exposure causes a differential expression of tumor necrosis factor (TNF) family members in the cancerous cells but not in the normal cells. They found that apoptosis in the cancer cells was induced by apoptosis signal kinase 1 (ASK1) which is activated by TNF signaling [63].

The material presented above gives the reader a general idea of what is involved in the interaction of LTP with cancer cells, including some of the molecular mechanisms that are presently known to play a role in such interaction. For more in-detail coverage, the reader can consult the reviews reported in references [13,14,36,38,40,64].

9.5 Selectivity Studies

In order to achieve a safe cancer therapy, it is crucial to investigate plasma selectivity towards cancer cells. Similar plasma treatment of healthy normal cells should result in minimum damage to the cells. Using PAM, Laroussi and co-workers conducted a study where non-cancerous MDCK (Madin-Darby Canine Kidney) cells and SCaBER (from a bladder squamous cell carcinoma) epithelial cells were exposed to PAM in similar manner [58, 65]. It was found that there is an optimum dose of PAM that suppresses viability of cancer cells significantly while inducing minimum damage to normal cells. Details of this work follow:

Cell viability was quantified using The CellTiter 96® AQueous One Solution Cell Proliferation Assay (MTS) (Promega, Madison, MI, USA). The solution reagent contains a tetrazolium compound [3-(4, 5-imethylthiazol-2-yl)-5-(3-carboxymethoxyphenyl)-2-(4-sulfophenyl)-2H-tetrazolium, inner salt; MTS]. Metabolically active cells reduce the MTS tetrazolium compound into a colored formazan product that is soluble in cell culture medium. At post-treatment time, 20 µL of the MTS solution was added to 100 µL of cell culture media in each sample well of the 96-well plate and incubated for 1–2 h in an incubator. Absorbance was recorded at 490 nm using a microplate reader (AgileReader, ACTGene Inc.). All experiments were conducted in triplicate, and results are expressed as a percentage of cell viability normalized to the control. Trypan blue exclusion assay was used to count the number of live cells in a control sample in order to quantify results of MTS assay. In this dye exclusion method, nonviable cells take up the dye and turn blue in color while viable cells remain unstained since the intact membrane of live cells prevents penetration of the dye.

Figure 9.7 shows that, unlike the case of SCaBER, PAM created with up to 4 min exposure to LTP causes no or minimum killing effects on MDCK cells. However, for 6 min or longer there is substantial cell kill. It can be concluded that the MDCK normal cells exhibited a much higher tolerance to treatment by PAM than did the SCaBER cancer cells. The hypothesis is that cancer cells, which are already under high oxidative stress because of their high metabolic rate, cannot tolerate an increase in ROS. Such an increase can overwhelm the

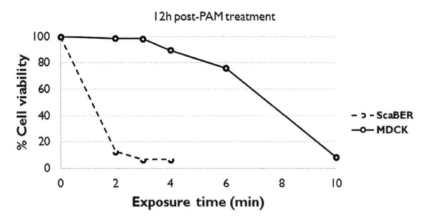

FIGURE 9.7 Cell viability of SCaBER and MDCK cells 12 h after PAM application.

cell defenses and lead to apoptosis. On the other hand, healthy cells can better tolerate the additional ROS and it takes a much more potent PAM to induce widespread cell death in these cells.

In order to assess what sub-lethal effects PAM has on normal cells Laroussi and co-workers conducted proliferation and migration investigations on MDCK cells. Time lapse imaging and immunofluorescence were the tools used to conduct these studies.

Time-lapse imaging was used to monitor changes in cells during PAM treatment [66]. MDCK cells were seeded at 80,000 cells per well in a 24-well plate overnight. Media was buffered with 10 mM HEPES solution to stabilize the pH during imaging. Since PAM with lower exposure time (3 min or below) did not induce a noticeable change in cell viability of MDCK cells while long exposure times such as 10 min PAM caused extensive cell death, 5 min plasma exposure was selected. Time-lapse microscopy of control and PAM treated samples was started right after PAM application. The 24-well plate was fixed on an automated stage at an ambient temperature of 37°C. Phase-contrast images were acquired at 10X magnification using an inverted microscope (DMi8, Leica Microsystems) equipped with a CCD camera (Andor Technology Ltd). Images were captured every 10 min for a period of 48 h of PAM treatment. ImageJ software (NIH) was used for the processing and analysis of the resulting images.

For immunofluorescence experiments, 80,000 cells per well of MDCK cells were grown overnight on a collagen coated (0.2 mg/mL col1) square coverslips which were placed in a 6-well plate. Cells were then treated, and samples were processed using the protocols detailed in [66]. After a PBS wash, cells were incubated in DAPI (4′,6-Diamidino-2-Phenylindole) diluted in PBS for 5 min. Each coverslip was mounted on a glass slide and stored overnight at 4°C prior to imaging. Fluorescence images of cells were acquired using an epi-fluorescence microscope (DMi8, Leica Microsystems) equipped with a CCD camera (Andor Technology Ltd) and 10X, 20X, and 40X objectives. The antibodies that were used are: anti-beta-catenin (BD Biosciences) to mark cell-cell adhesions, anti-paxillin (Santa Cruz Biotech) to mark cell-surface adhesions and anti-Ki-67 (Dako) to mark proliferating cell nuclei. Anti-beta-catenin and anti-paxillin antibodies were used at 1:100 dilution and anti-Ki-67 was used at 1:150 dilution. Phalloidin-488 (Molecular Probes) was used at 1:200 dilution to mark filamentous actin. The secondary antibodies used were anti-mouse and anti-rabbit IgG (Jackson ImmunoResearch), both used at 1:200 dilution.

Figure 9.8 shows images of the control and PAM treated samples at 0, 12, 24, 36, and 48 h, respectively. **Figure 9.8a** of the control sample

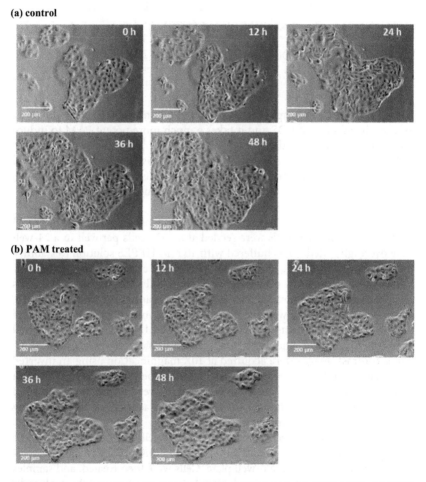

FIGURE 9.8 Phase contrast images of (a) control and (b) PAM treated MDCK cells obtained by time-lapse microscopy at different time points of 0, 12, 24, 36, and 48 h show a significant increase in cell population in the control sample compared to the PAM treated sample. Scale bar: 200 μm. (From Mohades, S. et al., *J. Phys. D Appl. Phys.*, 50, 185205, 2017.)

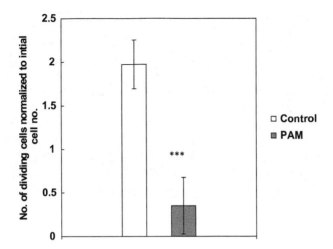

FIGURE 9.9 Number of division events normalized by initial number of cells in control and PAM as quantified from time-lapse images during 48 h of PAM treatment. Data represent averages from 6 movies from two independent experiments as means ± SD. An independent t-test indicates that cell proliferation was significantly lower in PAM treated samples (mean = 1.97, SD = 0.28) than control (mean = 0.40, SD = 0.32), $t(10) = 8.987$, *** $P < 0.001$. (From Mohades, S. et al., *J. Phys. D Appl. Phys.*, 50, 185205, 2017.)

demonstrates rapid proliferation in cells over 48 h. In contrast, cells in the PAM treated sample did not show a similar increase in population number (see **Figure 9.8b**) of the PAM sample. However, no visible cell damage or shrinkage was observed.

Figure 9.9 shows the number of division events normalized by the initial number of cells in control and PAM samples. The number of dividing cells was counted in each image sequence acquired from 48 h of time-lapse imaging. This result indicates that cell proliferation was significantly reduced in PAM treated cells ($p < 0.001$) and suggests that mitigation in cell number increase was associated with cell proliferation inhibition [66].

In order to confirm the inhibition of cell proliferation induced by PAM, the nuclear localization of Ki-67, a marker of cell proliferation, was ascertained using immunofluorescence. **Figure 9.10a** shows immunofluorescence images from control and PAM treated MDCK cells in which Ki-67 has been stained. Ki-67 staining is markedly brighter in cells that are actively involved in proliferation. The number of Ki-67 stained cells were counted in at least 8 immunofluorescence images obtained with 10x magnification from PAM and control samples.

Figure 9.10b shows the percentage of Ki-67 stained cells, on average, in immunofluorescence images: the number of Ki-67 stained cells is lower in PAM treated cells than in the control case, which indicates inhibition of proliferation [66].

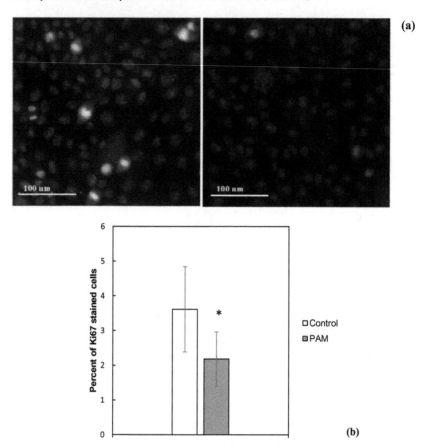

FIGURE 9.10 (a) Immunofluorescence images of a control (left) and a PAM treated (right) sample, stained for the proliferation marker Ki-67. (b) The average percentage of Ki67 stained cells in immunofluorescence images in PAM and control samples. Data represent averages from at least 8 images from two independent experiments, shown as means ± SD. * $P < 0.05$ (chi-square test). (From Mohades, S. et al., *J. Phys. D Appl. Phys.*, 50, 185205, 2017.)

To test the effects of PAM on cell migration, single cells within the MDCK islands were tracked in the control and PAM time-lapse sequences using MtrackJ (ImageJ software) and the average speed of single cells was measured. **Figure 9.11a** and **b** show the trace of a cell's migration path in control and PAM treated samples, respectively. **Figure 9.11c** shows the average speed of single cells in PAM and control samples. This result indicates that the average cell migration speed of PAM treated cells is lower than that of the control [66].

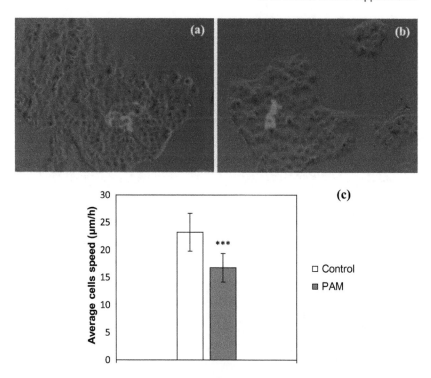

FIGURE 9.11 Trace of a cell's migration path during time-lapse imaging is shown in light grey in (a) a control sample and (b) a PAM treated sample. (c) The average speed of cells within epithelial islands of PAM and control MDCK samples. Data represent the average in μm/h $n = 20$ tracked cells of two independent experiments: means $\pm$ SD. *** $p < 0.001$ (independent t-test). (From Mohades, S. et al., *J. Phys. D Appl. Phys.*, 50, 185205, 2017.)

9.6 Attachment/Reattachment Studies

The effects of low temperature plasma on the attachment of cells to each other, to the extracellular matrix, and to a surface are important for both wound healing and cancer applications. Mohades et al. studied the effects of exposure of SCaBER cells to the plasma pencil on their morphology and their ability to attach to the surface of a culture plate [37]. **Figure 9.12** shows images of SCaBER cells at the initial time of seeding in the absence of plasma treatment and after 2 and 5 min of plasma treatment, respectively. The images were taken using phase-contrast bright-field microscope immediately after treatment, 1.5, 4.5, and 24 h after plasma exposure. The cells with round morphology at zero hour were non-attached because of the addition of trypsin. However, at later times and using their attachment machinery cells started attaching to the surface of the plate.

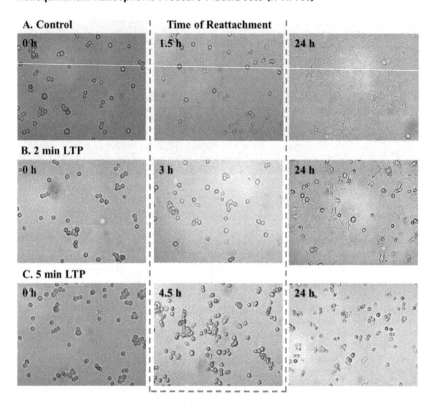

FIGURE 9.12 Images of SCaBER cells reattaching to the surface of the culture plate at different times. Images were taken at the times indicated on each image. Panel A is the control SCaBER cells that were not treated. The untreated cells adhere and are responsive at 1.5 h post reseeding. Panel B is a 2 min LTP treatment and Panel C is a 5 min LTP treatment. (From Mohades, S. et al., *Plasma Process. Polym.*, 11, 1150–1155, 2014.)

Figure 9.12a shows that after 1.5 h about 50% of the control cells started to attach to the surface and their morphology was no longer circular. However, the reattachment of plasma treated cells (2 min exposure) started only at 3 h post-treatment (**Figure 9.12b**). For the case of cells treated for 5 min, reattachment started occurring only after 4.5 h post-treatment (**Figure 9.12c**). These observations indicate that reattachment of the SCaBER cells to the surface of the plate starts later for longer plasma treatment times. In addition, the SCaBER cells, which were treated for 5 min, exhibited circular cell morphology after 24 h post-treatment indicating apoptotic or dead cells that did not reattach to the surface.

Other investigators have also shown that exposure to low temperature plasma contributes to cell detachment from the extracellular matrix as well as decreases cancer cell migration [42]. This is an important finding and indicates that plasma may have a controlling effect on cancer metastasis.

9.7 *In vivo* Studies

The first *in vivo* demonstrations of low temperature plasma anti-cancer capability were conducted by Vandamme et al. [41], Keidar et al. [42], and Kim et al. [52]. Vandamme et al. used human U87 glioblastoma xenotransplants in their study. Tumor volume measurements performed 24 h after the end of the treatment regimen revealed a significantly lower tumor volume in the treated group compared to control. These authors proposed that ROS-mediated apoptosis played a crucial role in their observed results. Other investigators also achieved encouraging results and noted that LTP treatment leads to the arrest of tumors growth on the skin of mice.

A study on a bladder tumor mouse xenograft through subcutaneous injection demonstrated the dramatic effect of plasma on tumor destruction [42] (**Figure 9.13a**): A 2 min plasma treatment led to significant decrease in tumor size to the point where the tumor could no longer be observed on the skin of mice 24 h post-treatment [42]. Similar experiments were performed on a murine melanoma model. In these experiments the tumor growth was completely inhibited over 3 weeks after plasma treatment (**Figure 9.13b**). Mice survival rates were also strongly increased as compared to the control group that did not receive plasma treatment.

Investigators have also shown that low temperature plasma treatment can activate the immune response *in vivo* to attack tumors [67–70]. Using uniform nanoseconds pulsed DBD investigators induced immunogenic cell death in tumor cells and augmented macrophage's function. In an interesting review

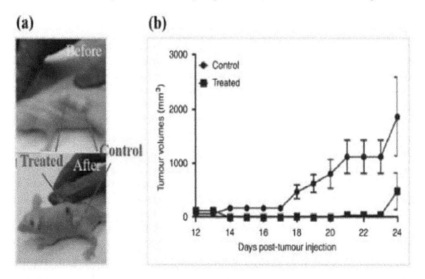

FIGURE 9.13 The anti-cancer effect of LTP in mice model. (a) Image of mouse with two tumors before and after the plasma jet treatment for 24 h. The subcutaneous tumors are grown from the seeded bladder cancer cells (SCaBER); (b) Cold plasma treatment effect on the growth of established tumor in a murine melanoma model. (From Keidar, M. et al., *Br. J. Cancer*, 105, 1295–1301, 2011.)

Miller et al. provided a comprehensive introduction to the promising cancer immunotherapy based on plasma treatment [67]. By optimizing the plasma parameters to induce immunogenic cell death in tumors locally, it is possible to trigger systemic immune responses.

9.8 RONS Penetration into Tissues

One of the main questions in plasma oncology relates to the extent of penetration of RONS into tumors. Some experimental evidence demonstrated that the effects of LTP on tumors seem to extend to the entire volume. This can be explained by the so-called "bystander effect," which involves cross cellular communication mediated by chemical signals sent by the cells in contact with plasma to cells deep within the tumor [38]. These signals initiate reactions similar to those experienced by the cells on the surface, such as the onset of apoptosis. Another possible mechanism would be that RONS generated by plasma actually do penetrate the tumor mass and react with cells much below the surface. In order to investigate this possibility investigators used various tissue models *in vitro* as well as real tissues *ex situ*.

Oh et al. used an agarose film covering a volume of deionized water contained in a quartz cuvette and measured the concentrations of RONS delivered to the deionized water using a UV-Vis spectrophotometer as well as chemical reporters for hydrogen peroxide and nitrite [71]. They found that RONS were indeed delivered to the DI water but also kept being delivered for up to 25 min after the plasma was removed. They also found that direct helium plasma treatment caused deoxygenation of the DI water while plasma treatment through the agarose film oxygenated the DI water. Based on these measurements Oh et al. concluded that exposing an open wound to plasma may lead to hypoxia (reduction in dissolved oxygen) while plasma treatment through tissue may provide improved oxygenation of biological fluids [71].

Hong et al. used a model comprising phospholipids vesicles encapsulated within a gelatin matrix and measured the delivery of reactive oxygen species (ROS) into the vesicles. This work showed that ROS can be delivered to the cells without rupturing the membranes of the vesicles [72].

To mimic biological tissue Szili et al. used gelatin gel, a derivative of collagen. These investigators reported that hydrogen peroxide (H_2O_2) in particular penetrated their tissue model for up to a 1.5 mm thickness of gelatin film [73].

To simulate more realistic conditions Duan et al. used slices of pig muscle tissue of different thicknesses placed on top of a PBS solution (see **Figure 9.14**) [74]. LTP was applied by plasma jet operated with a helium/oxygen mixture and ignited by sinusoidal high voltages at a frequency of 1 kHz. The plasma treatment times were 0, 5, 10, and 15 min.

The concentrations of H_2O_2, OH, and that of the total of ($NO_2^- + NO_3^-$) were measured for different thicknesses of the tissue slice. These investigators found that most of O_3, OH, and H_2O_2 were consumed by the tissue and could not pass through 500 µm or greater tissue thickness. On the other hand,

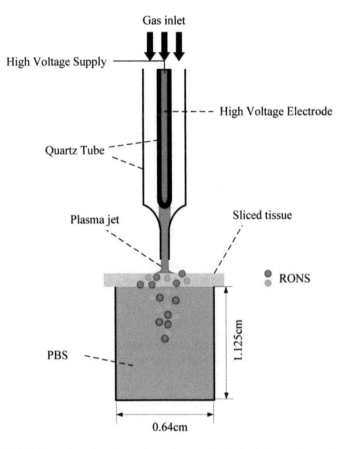

FIGURE 9.14 Experiment setup using pig muscle tissue. (From Duan, J. et al., *Phys. Plasmas*, 24, 073506, 2017.)

more than 80% of the $(NO_2^- + NO_3^-)$ penetrated a tissue slice of 500 μm thickness [28]. **Figure 9.15** shows the measured concentrations of $(NO_2^- + NO_3^-)$ as a function of tissue thickness and for three plasma treatment times (5, 10, and 15 min).

As shown in **Figure 9.9** the total concentrations of two RNS delivered to the PBS solution decrease for greater tissue thicknesses but increase for longer plasma treatment times. However, RNS were found to be able to penetrate a 500 μm thick tissue slice since the concentration of $(NO_2^- + NO_3^-)$ was comparable to the concentration when no tissue was placed on top of the PBS solution. Interestingly, ROS were found to be almost completely absorbed by the tissue.

The works presented above show that RONS do indeed penetrate tissues. However, it is important to note that some can be absorbed by the tissue within depths of only few tens of micrometers while others, such as nitrites/nitrates can penetrate much deeper. Interestingly, hydrogen peroxide was found to be

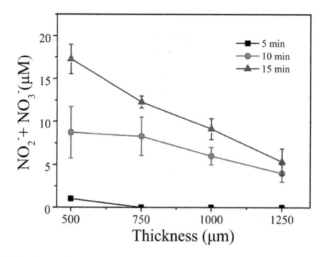

FIGURE 9.15 Total nitrite and nitrate concentration versus tissue thickness for three plasma exposure times. (From Duan, J. et al., *Phys. Plasmas*, 24, 073506, 2017.)

able to penetrate a tissue model (gelatin gel) for up to 1.5 mm in thickness but did not penetrate 500 μm thick real tissue (pig muscle).

9.9 Conclusions

The biomedical applications of low temperature plasma constitute a novel multidisciplinary field of research. Scientific investigations started in mid-1990s with experiments on the inactivation of bacteria by atmospheric pressure cold plasma and by the early 2000s the effects on mammalian cells were investigated. Various medical applications, including wound healing and cancer treatment, have led to scientific advances that may ultimately lead to innovative medical therapies. However, and to conclude, it is safe to say that a lot more work needs to be done in order to fully understand the mechanisms of action of LTP against biological cells and specifically against cancer cells. Recent works by various investigators indicate that the electric field present in the plasma and the ROS and RNS generated by the plasma play crucial roles in the observed biological effects/outcomes. In addition, there is some indication as to the selectivity of the action of LTP, but more needs to be elucidated before LTP can be declared as a technology that can form the basis of a safe and effective therapy to combat cancer.

References for Part IV

1. R. Mogul, A.A. Bolapos;shakov, S.L. Chan, R.M. Stevens, B.N. Khare, M. Meyyappan and J.D. Trent, Impact of low-temperature plasmas on *Deinococcusradiodurans* and biomolecules, *Biotechnol. Prog.* 19(3) (2003) 776–783.

2. A.A. Bol'shakov, B.A. Cruden, R. Mogul, M.V.V.S. Rao, A.P. Sharma, B.N. Khare and M. Meyyappan, Radio-frequency oxygen plasma as a sterilization source, *AIAA J.* 42(4) (2004) 823–832.

3. K.R. Stalder, Plasma characteristics of electrosurgical discharges, in: *Proceedings of the Gaseous Electronics Conference*, San Francisco, CA, p. 16 (2003).

4. M. Moisan, J. Barbeau, S. Moreau, J. Pelletier, M. Tabrizian and L'H. Yahia, Low temperature sterilization using gas plasmas: A review of the experiments, and an analysis of the inactivation mechanisms, *Int. J. Pharmaceutics* 226 (2001) 1–21.

5. D.A. Drossman, N.J. Shaheen and I.S. Grimm, Argon plasma coagulation, in: *Handbook of Gastroenterology Procedures*, 4th ed., Lippincotts, Williams & Wilkins, Philadelphia, PA, 2005, p. 235.

6. M. Laroussi, Sterilization of contaminated matter with an atmospheric pressure plasma, *IEEE Trans. Plasma Sci.* 24(3) (1996) 1188–1191.

7. A.B. Shekhter, R.K. Kabisov, A.V. Pekshev, N.P. Kozlov and Y.L. Perov, Experimental and clinical validation of plasmadynamic therapy of wounds with nitric oxide, *Bull. Exp. Biol. Med.* 126 (1998) 829–834.

8. E. Stoffels, A.J. Flikweert, W.W. Stoffels and G.M.W. Kroesen, Plasma needle: A non-destructive atmospheric plasma source for fine surface treatment of biomaterials, *Plasma Sources Sci. Technol.* 11 (2002) 383–388.

9. M. Laroussi, M. Kong, G. Morfill and W. Stolz, *Plasma Medicine: Applications of Low Temperature Gas Plasmas in Medicine and Biology*, Cambridge University Press, Cambridge, UK, 2012.

10. A. Fridman and G. Friedman, *Plasma Medicine*, Wiley, New York, 2013.

11. M. Laroussi, Low temperature plasmas for medicine? *IEEE Trans. Plasma Sci.* 37(6) (2009) 714–725.

12. G. Fridman, G. Friedman, A. Gutsol, A.B. Shekhter, V.N. Vasilets and A. Fridman, Applied plasma medicine, *Plasma Process. Polym.* 5 (2008) 503–533.

13. N. Barekzi and M. Laroussi, Effects of low temperature plasmas on cancer cells, *Plasma Proc. Polym.* 10 (2013) 1039.

14. M. Keidar, A. Shashurin, O. Volotskova, M.A. Stepp, P. Srinivasan, A. Sandler and B. Trink, Cold atmospheric plasma in cancer therapy, *Phys. Plasmas* 20 (2013) 057101.

15. M. Laroussi, S. Mohades and N. Barekzi, Killing of adherent and non-adherent cancer cells by the plasma pencil, *Biointerphases* 10 (2015) 029410.

16. M. Laroussi, Non-thermal decontamination of biological media by atmospheric pressure plasmas: Review, analysis, and prospects, *IEEE Trans. Plasma Sci.* 30(4) (2002) 1409–1415.

17. M. Laroussi, Low temperature plasma-based sterilization: Overview and state-of-the-art, *Plasma Proc. Polym.* 2(5) (2005) 391–400.

18. T. von Woedtke, S. Reuter, K. Masur and K.-D. Weltmann, Plasma for medicine, *Phys. Repts.* 530 (2013) 291.

19. M. Laroussi, Interaction of low temperature plasma with prokaryotic and eukaryotic cells, in: *Proceedings 61st Gaseous Electronics Conference*, Dallas, TX, p. 20 (2008).

20. G. Isbary, G. Morfill, H.U. Schmidt, M. Georgi, K. Ramrath, J. Heinlin et al. A first prospective randomized controlled trial to decrease bacterial load using cold atmospheric argon plasma on chronic wounds in patients, *Br. J. Dermatol.* 163 (2010) 78.

21. A.D. Morris, G.B. McCombs, T. Akan, W. Hynes, M. Laroussi and S.L. Tolle, Cold plasma technology: Bactericidal effects on *Geobacillus Stearothermophilus* and *Bacillus Cereus* microorganisms, *J. Dent. Hyg.* 83(2) (2009) 55–61.

22. D. Claiborne, G. McCombs, M. Lemaster, M.A. Akman and M. Laroussi, Low temperature atmospheric pressure plasma enhanced tooth whitening: The next generation technology, *Int. J. Dent. Hyg.* (2013). doi:10.1111/idh.12031.

23. M. Laroussi and N. Barekzi, Effects of low temperature plasma on two Eukaryotic cell lines: Epithelial cells and prostate cancer cells, in: *Proceedings of the International Conference on Phenomena in Ionized Gases*, Granada, Spain, July 2013.

24. N. Barekzi and M. Laroussi, Fibropblasts cell morphology altered by low temperature atmospheric pressure plasma, *IEEE Trans. Plasma Sci.* 42 (2014) 2738.

25. M. Laroussi, E. Karakas and W. Hynes, Influence of cell type, initial concentration, and medium on the inactivation efficiency of low temperature plasma, *IEEE Trans. Plasma Sci.* 39(11) (2011) 2960–2961.

26. G. Fridman, A. Brooks, M. Galasubramanian, A. Fridman, A. Gutsol, V. Vasilets, H. Ayan and G. Friedman, Comparison of direct and indirect effects of non-thermal atmospheric-pressure plasma on bacteria, *Plasma Process. Polym.* 4 (2007) 370–375.

27. A. Shashurin, M. Keidar, S. Bronnikov, R.A. Jurjus and M.A. Stepp, Living tissue under treatment of cold plasma atmospheric jet, *Appl. Phys. Lett.* 93 (2008) 181501.

28. X. Yan, F. Zou, S. Zhao, X. Lu, G. He, Z. Xiong, Q. Xiong et al., On the mechanism of plasma inducing cell apoptosis, *IEEE Trans. Plasma Sci.* 38 (2010) 9.

29. Z. Xiong, Y. Cao, X. Lu and T. Du, Plasmas in tooth root canal, *IEEE Trans. Plasma Sci.* 39 (2011) 968.

30. J.L. Zimmermann, T. Shimizu, V. Boxhammer and G.E. Morfill, Disinfection through different textiles using low-temperature atmospheric pressure plasma, *Plasma Process. Polym.* 9 (2012) 792–798.

31. N.Y. Babaeva and M.J. Kushner, Reactive fluxes delivered by dielectric barrier discharge filaments to slightly wounded skin, *J. Phys. D Appl. Phys.* 46 (2013) 025401.

32. K.-D. Weltmann, E. Kindel, R. Brandenburg, C. Meyer, Bussiahn, C. Wilke and T. von Woedtke, Atmospheric pressure plasma jet for medical therapy: Plasma parameters and risk estimation, *Contrib. Plasma Phys.* 49(9) (2009) 631–640.

33. J. Ehlbeck, U. Schnabel, M. Polak, J. Winter, T. von Woedtke, R. Brandenburg, T. von dem Hagen and K.-D. Weltmann, Low temperature atmospheric pressure plasma sources for microbial decontamination, *J. Phys. D Appl. Phys.* 44 (2011) 013002.

34. G.B. McCombs, M. Darby and M. Laroussi, Dental applications, in: *Plasma Medicine: Applications of Low temperature Gas Plasmas in Medicine and Biology*, M. Laroussi, M. Kong, G. Morfill, and W. Stolz (Eds.), Cambridge University Press, Cambridge, UK, 2012.

35. F. Utsumi, H. Kjiyama, K. Nakamura, H. Tanaka, M. Mizuno, K. Ishikawa, H. Kondo, H. Kano, M. Hori and F. Kikkawa, Effect of indirect nonequilibrium atmospheric pressure plasma on anti-proliferative activity against chronic chemo-resistant ovarian cancer cells *in vitro* and *in vivo*, *PLoS One* 8 (2013) e81576.

36. H. Tanaka, M. Mizuno, K. Ishikawa, K. Takeda, K. Nakamura, F. Utsumi, H. Kajiyama et al., Plasma medical science for cancer therapy: Toward cancer therapy using nonthermal atmospheric pressure plasma, *IEEE Trans. Plasma Sci.* 42 (2014) 3760.

37. S. Mohades, N. Barekzi and M. Laroussi, Efficacy of low temperature plasma against SCaBER cancer cells, *Plasma Process. Polym.* 11(12) (2014) 1150–1155.

38. M. Laroussi, From killing bacteria to destroying cancer cells: Twenty years of plasma medicine, *Plasma Process. Polym.* 11(12) (2014) 1138–1141.

39. J. Köritzer, V. Boxhammer, A. Schäfer, T. Shimizu, T.G. Klämpfl, Y.-F. Li, C. Welz, S. Schwenk-Zieger, G.E. Morfill, J.L. Zimmermann and J. Schlegel, Restoration of sensitivity in chemo-resistant glioma cells by cold atmospheric plasma, *Plos One* 8(5) (2013) e64498.

40. J. Schlegel, J. Koritzer and V. Boxhammer, Plasma in cancer treatment, *Clin. Plasma Med.* 1 (2013) 2.

41. M. Vandamme, E. Robert, S. Pesnele, E. Barbosa, S. Dozias, J. Sobilo, S. Lerondel, A. Le Pape and J.-M. Pouvesle, Antitumor effects of plasma treatment on U87 glioma xenografts: Preliminary results, *Plasma Process. Polym.* 7(3–4) (2010) 264–273.

42. M. Keidar, R. Walk, A. Shashurin, P. Srinivasan, A. Sandler, S. Dasgupta, R. Ravi, R. Guerrero-Preston and B. Trink, Cold plasma selectivity and the possibility of a paradigm shift in cancer therapy, *Br. J. Cancer* 105(9) (2011) 1295–1301.

43. M. Laroussi and M. Keidar, Plasma & cancer, *Plasma Process. Polym.* 11(12) (2014) 1118–1119.

44. Y.Z. Tang, X.P. Lu, M. Laroussi and F.C. Dobbs, Sublethal and killing effects of atmospheric pressure, non-thermal plasma on eukaryotic microalgae in aqueous media, *Plasma Process. Polym.* 5(6), (2008) 552–558.

45. K. Oehmigen, M. Hähnel, R. Brandenburg, C. Wilke, K.-D. Weltmann and T. von Woedtke, The role of acidification for antimicrobial activity of atmospheric pressure plasma in liquids, *Plasma Process. Polym.* 7 (2010) 250.

46. G. Fridman, A. Shereshevsky, M.M. Jost, A. Brooks, A. Fridman, A. Gutsol, V. Vasilets and G. Friedman, Floating electrode dielectric barrier discharge plasma in air promoting apoptotic behavior in melanoma skin cancer cell lines, *Plasma Chem. Plasma Process.* 27(2) (2007) 163.

47. O. Volotskova, T.S. Hawley, M.A. Stepp and M. Keidar, Targeting the cancer cell cycle by cold atmospheric plasma, *Sci. Rep.* 2 (2012) 636.

48. C.-H. Kim, J.H. Bahn, S.-H. Lee, G.-Y. Kim, S.-I. Jun, K. Lee and S.J. Baek, Induction of cell growth arrest by atmospheric non-thermal plasma in colorectal cancer cells, *J. Biotechnol.* 150 (2010) 530.

49. H. Tanaka, M. Mizuno, K. Ishikawa, K. Nakamura, F. Utsumi, H. Kajiyama, H. Kano, S. Maruyama, F. Kikkawa and M. Hori, Cell survival and proliferation signaling pathways are downregulated by plasma activated medium in glioblastoma brain tumor cells, *Plasma Med.* 2 (2012) 201.

50. N. Barekzi and M. Laroussi, Dose-dependent killing of leukemia cells by low-temperature plasma, *J. Phys. D Appl. Phys.* 45 (2012) 422002.

51. J. Huang, H. Li, W. Chen, G.-H. Lv, X.-Q. Wang, G.-P. Zhang, K. Ostrikov, P.-Y. Wang and S.-Z. Yang, Dielectric barrier discharge plasma in Ar/O_2 promoting apoptosis behavior in A549 cancer cells, *Appl. Phys. Lett.* 99 (2011) 253701.

References for Part IV

52. J.Y. Kim, J. Ballato, P. Foy, T. Hawkins, Y. Wei, J. Li and S.O. Kim, Apoptosis of lung carcinoma cells induced by a flexible optical fiber-based cold microplasma, *Biosens. Bioelectron.* 28 (2011) 333.

53. T. Sato, M. Yokoyama and K. Johkura, A key inactivation factor of HeLa cell viability by a plasma flow, *J. Phys. D, Appl. Phys.* 44 (2011) 372001.

54. S. Mohades, M. Laroussi, J. Sears, N. Barekzi and H. Razavi, Evaluation of the effects of a plasma activated medium on cancer cells, *Phys. Plasmas* 22 (2015) 122001.

55. F.A. Villamena, *Chemistry of Reactive Species*, John Wiley & Sons, Hoboken, NJ, 2013.

56. D.B. Graves, Oxy-nitroso shielding burst model of cold atmospheric plasma therapeutics, *Clin. Plasma Med.* 2 (2014) 38.

57. Y.F. Yue, S. Mohades, M. Laroussi and X. Lu, Measurements of plasma-generated hydroxyl and hydrogen peroxide concentrations for plasma medicine applications, *IEEE Trans. Plasma Sci.* 44(11) (2016) 2754–2758.

58. S. Mohades, N. Barekzi, H. Razavi, V. Maramuthu and M. Laroussi, Temporal evaluation of antitumor efficiency of plasma activated media, *Plasma Process. Polym.* 13 (2016) 1206.

59. S. Zhao, Z. Xiong, X. Mao, D. Meng, Q. Lei, Y. Li, P. Deng et al., Atmospheric pressure room temperature plasma jets facilitate oxidative and nitrative stress and lead to endoplasmic reticulum stress dependent apoptosis in HepG2 cells, *PLoS One* 8 (2013) e73665.

60. X. Yan, F. Zou, S. Zhao, X. Lu, G. He, Z. Xiong, Q. Xiong et al., On the mechanism of plasma inducing cell apoptosis, *IEEE Trans. Plasma Sci.* 38 (2010) 2451.

61. X. Yan, Z. Xiong, F. Zou, S. Zhao, X. Lu, G. Yang, G. He and K. Ostrikov, Plasma-induced death of HepG2 cancer cells: Intracellular effects of reactive species, *Plasma Process. Polym.* 9 (2012) 59.

62. M. Ishaq, M. Evans and K. Ostrikov, Effects of atmospheric gas plasmas on cancer cell signaling, *Int. J. Cancer* 134 (2014) 1517.

63. M. Ishaq, S. Kumar, H. Varinli, Z.J. Han, A.E. Rider, M. Evans, A.B. Murphy and K. Ostrokov, Atmospheric gas plasma-induced ROS production activates TNS-ASK1 pathway for the induction of melanoma cancer cell apoptosis, *Mol. Biol. Cells* 25 (2014) 1523.

64. C. Hoffmann, C. Berganza and J. Zhang, Cold atmospheric plasma: Methods of production and application in dentistry and oncology, *Med. Gas Res.* 3 (2013) 21.

65. M. Laroussi and V. Maruthamuthu, On the exposure of epithelial cells to PAM, in: *Proceedings of the Int. Conf. Plasma Medicine*, Philadelphia, PA (June 2018).

66. S. Mohades, M. Laroussi and V. Maruthamuthu, Moderate plasma activated media supresses proliferation and migration of MDCK epithelial cells, *J. Phys. D Appl. Phys.* 50 (2017) 185205.

67. V. Miller, A. Lin and A. Fridman, Why target immune cells for plasma treatment of cancer, *Plasma Chem. Plasma Process.* 36 (2016) 259.

68. A. Lin, B. Truong, A. Pappas, K. Kirifides, A. Oubarri, S. Chen, D. Dobrynin et al., Uniform nanosecond pulsed dielectric barrier discharge plasma enhances anti-tumor effects by induction of immunogenic cell death in tumors and stimulation of macrophages, *Plasma Process. Polym.* 12 (2015) 1392.

69. V. Miller, A. Lin, G. Fridman, D. Dobrynin and A. Fridman, Plasma stimulation of migration of macrophages, *Plasma Process. Polym.* 11 (2014) 1193.

70. M. Laroussi, X. Lu and M. Keidar, Perspective: The physics, diagnostics, and applications of atmospheric pressure low temperature plasma sources used in plasma medicine, *J. Appl. Phys.* 122 (2017) 020901.

71. J.-S. Oh, E.J. Szili, N. Gaur, S.-H. Hong, H. Futura, H. Kurita, A. Mizuno, A. Hatta and R.D. Short, How to assess the plasma delivery of RONS into tissue fluid and tissue, *J. Phys. D Appl. Phys.* 49 (2016) 304005.

72. S.-H. Hong, E.J. Szili, A. Toby, A. Jenkins and R.D. Short, Ionized gas (plasma) delivery of reactive oxygen species (ROS) into artificial cells, *J. Phys. D Appl. Phys.* 47 (2014) 362001.

73. E.J. Szili, J.W. Bradley and R.D. Short, A "tissue model" to study the plasma delivery of reactive oxygen species, *J. Phys. D Appl. Phys.* 47 (2014) 152002.

74. J. Duan, X. Lu and G. He, On the penetration depth of reactive oxygen and nitrogen species generated by a plasma jet through real biological tissue, *Phys. Plasmas* 24 (2017) 073506.

Index

NOTE: Page numbers in italic and bold refer to figures and tables, respectively.

Index

Index